DESIGN RULES FOR A CIM SYSTEM

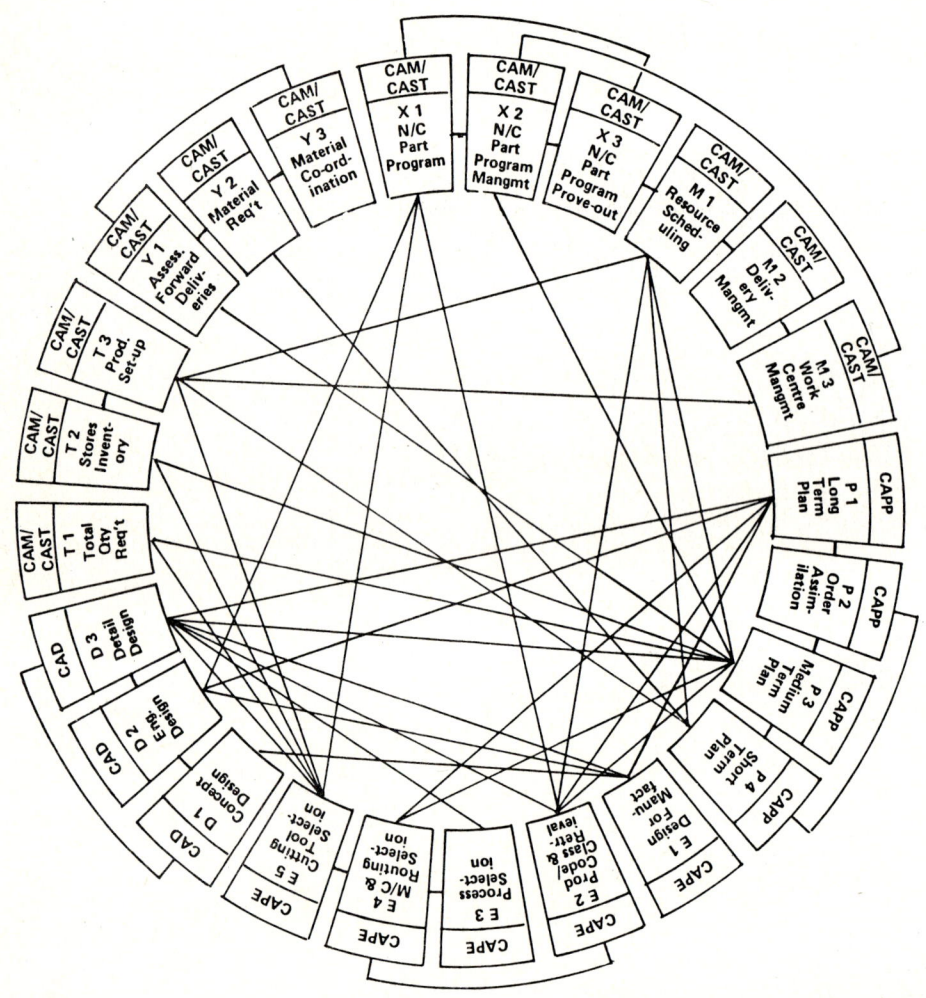

CIM Sub-Systems and Interconnections

DESIGN RULES FOR A CIM SYSTEM

Edited by

R. W. YEOMANS
*Istel Ltd.
Redditch
United Kingdom*

A. CHOUDRY
*Computer Science Department
University of Amsterdam
The Netherlands*

and

P. J. W. TEN HAGEN
*Center for Mathematics and Computer Science
Amsterdam
The Netherlands*

1985

NORTH-HOLLAND
AMSTERDAM • NEW YORK • OXFORD

© R. W. Yeomans, A. Choudry, and P. J. W. ten Hagen, 1985

All rights reserved. No part of this publication may be reproduced, stored in a retrieval system, or transmitted, in any form or by any means, electronic, mechanical, photocopying, recording or otherwise, without the prior permission of the copyright owner.

ISBN: 0 444 87812 2

Published by:
ELSEVIER SCIENCE PUBLISHERS B.V.
P.O. Box 1991
1000 BZ Amsterdam
The Netherlands

Sole distributors for the U.S.A. and Canada:
ELSEVIER SCIENCE PUBLISHING COMPANY, INC.
52 Vanderbilt Avenue
New York, N.Y. 10017
U.S.A.

This research work and its publication has been supported by the ESPRIT program of the Commission of the European Communities under contract no. ESPRIT-CIM 5.1/34.

PRINTED IN THE NETHERLANDS

Acknowledgement

The Commission of the European Communities (CEC) has exercised admirable foresight in launching the ESPRIT (European Strategic Planning for Research in Information Technology) program which made it possible to undertake such multi-disciplinary research in a joint Industrial-Academic environment on a Community wide basis. The authors very much appreciate the opportunity to have been a part of this worthy venture.
It is a pleasure to thank Ms. Patricia MacConaill, the coordinator in ESPRIT-CIM, for her sustained interest and collaboration in this work. Also the support offered by her, in transforming the Project Report into this book, is gratefully acknowledged.

Contributors

The work reported here is based on ESPRIT (European Strategic Planning for Research in Information Technology) Pilot Project 5.1/34, titled:

Design Rules for CIM (Computer Integrated Manufacturing)

awarded to Istel Ltd., Redditch, Worcestershire, England, as prime contractor (R.W. Yeomans, project leader).
The Centre for Mathematics and Computer Science (CWI) was a partner (P.J.W. ten Hagen, team manager), and FVI (Computer Science Department), University of Amsterdam was a sub-contractor (L.O. Hertzberger, team manager).
The following members of these institutes contributed to this book:

Istel Ltd	CWI	FVI
J. Barr	C.L. Blom	A. Choudry
J.J. Brett	H.J. Bos	L.O. Hertzberger
K.W. Cockbill	M. Cornelissen	F. Tuynman
R. Crees	P.J.W. ten Hagen	
D.P. Edwards	A. Janssen	
J. Gough	A.A.M. Kuijk	
W.E. Henry	W.E. van Waning	
E. Levy		
K.J. Morby		
G.S. Myatt		
D.S.T. Nunney		
H.J. Reaper		
R.W. Yeomans		

Contents

	Acknowledgement	v
	Contributors	vii
1	Introduction	1
2	Development of CIM Design Rules	9
3	Computer Aided Design (CAD)	12
4	Computer Aided Production Engineering (CAPE)	50
5	Computer Aided Production Planning (CAPP)	105
6	Computer Aided Manufacture (CAM) / Computer Aided Storage and Transportation (CAST)	147
7	Computer Aided Manufacture (CAM) / Computer Aided Storage and Transportation (CAST) Sub-topics	155
8	General Interface Rules	291
9	Development of strategies	293
10	Data strategy	297
11	Processing: state of the art	313
12	Processing strategy	322
13	Communication: state of the art	351
14	Communication strategy	365
15	Sensor Systems and Computer Integrated Manufacturing	384
16	Graphics Systems and Computer Integrated Manufacturing	403
Appendix 1	Flowcharting conventions	412
Appendix 2	Selection processes	415
Appendix 3	Computer Aided Design of Solid Objects	420
Appendix 4	Sensor Applications	431
Appendix 5	CIM in the small firm	452
	Index	454

Chapter 1

Introduction

The subject of production mechanisation is not a new one. The inventor of the first potters wheel was motivated by objectives which were not dissimilar from those of a late twentieth century Production Engineer considering the introduction of a Flexible Manufacturing System (FMS) cell - namely, how to perform more useful work with less human effort. What has changed and changed quite dramatically in recent years, is the degree or level of automation which the available technology can support. The very advanced level and sophistication of contemporary technology has led many lay people to conclude that the 'unmanned' factory is already an entirely realisable possibility - even though professional practitioners recognise that this age-old vision is, at least for many kinds of manufacturing, still a very long way off indeed. Contemporary technology does however support very advanced levels of automation for certain kinds of manufacturing.

People working in the field of advanced manufacturing technology are increasingly coming to recognise that higher levels of automation require broader levels of approach. Where the mechanisation of a given process can be realised by means of a single simple machine, the project manager can quite often restrict his principal area of interest to, say, the design and operation of just this one machine. In such cases, the mechanisation of the process is done in almost total isolation from all of the other activities which take place within the company - often even in isolation from other processes which are to be performed on the same product.

Even in the case of a single machine however, higher levels of automation demand that consideration be given to 'related' as well as to 'direct' processes and activities. For example; a company contemplating the introduction of even a single numerically controlled machine tool will be obliged to give consideration to the method for creating numerical control part programmes. This in turn may oblige the company to study the much larger subject of Computer Aided Design.

If a company introduces only a very limited number of new or modified products each week, it may be quite sufficient for that company to establish a parts programming operation which works directly from paper drawings. An operation of this kind may not however be appropriate to say a jobbing shop which needs to create many tens or even hundreds of new parts programmes each period, and which has insufficient machine capacity to devote very large amounts of machine time to the validation of each parts programme. In either event, the project manager charged with introducing the new machine will need to consider other company activities - besides the machining process itself.

Where a mechanisation project involves a number of associated machines and processes, the company will find it necessary to consider a great many related activities in order to effect a high level of automation. In the case of say a FMS cell, the company may need to introduce a large number of new systems to address such diverse activities as parts classification, capacity planning, production planning and scheduling, inventory management and control, and many other production related administrative activities, as well as many 'engineering' and 'production' activities such as product design and development, process planning, etc.

The highest levels of factory automation (which include but is by no means confined to unmanned operation) may, and almost invariably do, require a re-examination of almost every activity which takes place within the organisation concerned. The identification of this requirement led directly to the creation of the term Computer Integrated Manufacturing (CIM) as embracing almost every department and function of a manufacturing organisation. The ESPRIT (European Strategic Program for Research in Information Technology) project of the European Economic Community established a CIM Group to study this problem. The diagram opposite this page admirably illustrates the scope of CIM, as perceived by the ESPRIT-CIM group.

The comparatively low levels of automation which were introduced in the past could be, and often were, managed entirely by members of a single discipline - frequently the Production Engineering discipline. As higher levels of automation require broader levels of approach they also require the involvement of a wider range of different professional disciplines. The more advanced levels of automation which the current technology now makes possible, need to be addressed by multi-disciplinary teams - which may include not only Production Engineers but also Computer Systems Engineers; Product Design Engineers; Robotics Engineers; Production Control experts; Telecommunications experts etc.

This presents many different problems of both technical and organisational nature. Many particularly difficult problems arise as a direct result of the fact that no single vendor (no single industry even), could possibly supply all of the many different products and services which would be required for even a modestly high level automation project.

In short, a number, possibly a quite large number, of different individuals and suppliers are normally involved in the introduction of high level automation, and a wide variety of dissimilar subjects and activities need to be addressed and provided for. The administrative and organisational difficulties which these two problems occasion, are very seriously compounded by the fact that no agreed structures and definitions for CIM currently exist.

The situation in many ways resembles that of the Tower of Babel. CIM; MRP; CAD; AMT; CAM; JIT; FMS - everyone knows the terms, no-one understands what the terms mean. More precisely, many if not most attach different and often conflicting meanings to such terms. As a result, a large number of individuals, from widely different cultural backgrounds and disciplines, who are mutually dependent upon each other for very advanced and complex undertakings, are greatly handicapped by the total absence of any formal structure or even an agreed vocabulary for the undertakings they are charged with. The problem is made even worse by the lack of agreement, the lack of willingness to co-operate even, within the IT industry itself.

Whilst everyone involved in CIM recognises that CIM comprises many separate modules or sub-systems, there is no generally agreed sub-system structure, not even a generally accepted list of sub-systems. There is not indeed any accepted understanding of the range of company activities which any one sub-system should address. Do Computer Aided Draughting and Finite Element Model Generation constitute separate sub-systems ?, or are they simply two functional elements of a single sub-system ?. Different vendors group or 'package' support for different company activities in dissimilar ways. Some will provide support for a very large number of activities within a single product, whilst others will sub-divide the same range of activities to produce a range of different products. Furthermore, the groupings adopted by any-one vendor will normally differ from the way in which other vendors package these to produce competitive products.

Introduction

REFERENCE MODEL OF
MANUFACTURING CONTROL SYSTEM

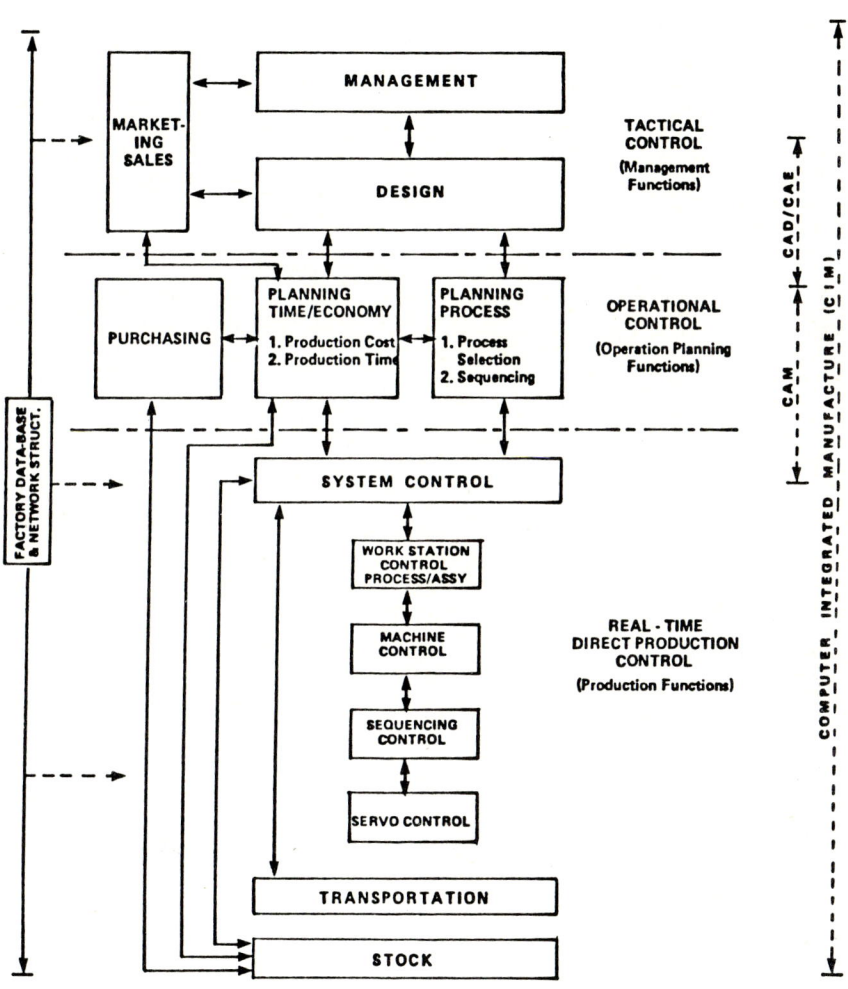

Manufacturing companies therefore invariably experience great difficulty in incorporating products from different vendors into a single composite integrated manufacturing operation. In many instances, product incompatibility is not due to accident but to design - shortsighted and damaging though this may be. Vendors who deliberately make it artificially difficult for clients to interface products from other vendors, to those which they themselves supply, damage both themselves and European manufacturing industry.

As previously stated, no single vendor, no single industry even, could possibly provide all of the different products and sub-systems which make up a total CIM operation. Such an operation may comprise computers, machine tools, robots, telecommunications equipment, automated warehousing and transportation facilities etc., along with a truly tremendous range of different software products, including facility management, control systems and production administration systems of many different kinds. Where a particular vendor designs a product in such a manner as to make it costly and difficult for a client company to interface other products from different vendors to it, it has the corollary effect of making it equally difficult for the client to add the same product to an existing composite network of systems.

Furthermore, by making it unnecessarily costly or technically difficult for client companies to implement higher levels of factory automation, vendors discourage large numbers of otherwise highly motivated organisations from doing so. This not only depresses the overall market upon which the vendors depend, but also greatly hinders the European manufacturing industry in charting a course away from labour intensive manufacturing, towards automated manufacturing, This is an essential pre-requisite to the survival of the European manufacturing industry in a competition with countries where labour costs are substantially lower.

The principal objective of this study, as an ESPRIT Pilot Project (No. 5.1/34) titled 'Design Rules for Computer Integrated Manufacturing Systems', was to propose a European CIM Systems Structure. In particular, it was proposed to address three separate but related goals:

1   To modularise the total CIM into functionally discrete sub-systems.

2   To describe the minimum functional specifications of each sub-system.

3   To identify the interrelationships that exist between any one CIM sub-system and all the other sub-systems.

The CIM-Structure presented here, is not, and could not be, either definitive or final for two reasons. Firstly, no structure for CIM can be made definitive except by general consensus; secondly, technology is dynamic and innovative and no attempt should be made to halt progress and invention at a particular point in time. Even structures which are proposed in reference to scientific subjects (such as that proposed by Carolus Linnaeus for biological classification) have to be updated in line with advances in scientific knowledge. The subject of manufacturing, even at this time, includes almost as much art as it does science - and new methods, new disciplines even, are being evolved at an ever increasing rate. As a consequence, no manufacturing structure may be considered to be scientifically 'correct' - it can only be considered to be pragmatically useful at a given point in time. It is nevertheless hoped that the proposed structure will constitute an initial European framework, which can be regularly refined and updated - in accordance with consensus wishes, and in line with evolving needs. The proposed structure is therefore intended to provide both a base and a focus for the European industry.

An undertaking to provide CIM structures which would comprehensively address all of the needs, of every branch, of every form of manufacturing industry, would be too ambitious for an initial study. It was, for this reason, necessary to limit the scope of the undertaking in three particular respects. These were:

1   the sector of industry
2   the range of company activities
3   the levels of mechanisation supported.

Each of these three constraints therefore need to be considered.

Sector of industry

In this study we decided to address only the 'machining' sector of the mechanical engineering industry. There were a number of reasons for taking this decision, as discussed below;

- There are more manufacturing organisations in Europe involved in machining than in any other single type of manufacturing operation. Machining therefore represents the single largest market for CIM system products, and would accordingly be of more immediate interest to a greater number of potential CIM system vendors, than would any other sector of manufacturing industry.

- Machining presently offers greater and more immediate scope for automation than most other types of manufacture. Machine tools, mechanised conveyancing systems, robots for material handling, automated stores for the control of tools, work-pieces, etc., of the kinds needed by the machining industry, are all readily available in the form required to support high level automation.

- Machining organisations afford greater and more immediate scope for vertical integration than most other kinds of organisation. Linking of the sub-systems used to design and evaluate products (CAD) to the sub-system used to generate numerical control programmes (CAM programming) for machining operations (eg. millings; drilling; wire spark erosion; flame cutting; turning; grinding; punching; nibbling, etc.) can be achieved more directly and easily than in, say, component assembly operations, sheet metal forming operations, etc. The linking of other sub-systems (process planning; CAM-scheduling; CAM-machining; CAM-inspection, etc.) is similarly more direct than it is in connection with other types of operation.

- Many of the sub-systems required by the machining industry will, with relatively minor modification, be appropriate to other branches of manufacturing industry. The group of sub-systems needed to support the machining industry, as presented here should therefore be viewed as a set of 'primitive generic' modules from which more complex models can be constructed.

Range of company activities

This study was designed to provide support for all of the activities that are **directly** related to the design and manufacture of machined products This project was therefore not designed to address such 'indirect' activities as Market Research; Financial and Management Accounting; Sales and Distribution; Purchasing, etc. - but was instead focussed towards those functions and activities which are normally described as the 'Engineering' activities of a company - including both Design and Production Engineering.

These activities were grouped under five principal headings or 'topics', namely;
- Computer Aided Design (CAD)
- Computer Aided Production Engineering (CAPE)
- Computer Aided Manufacturing (CAM)
- Computer Aided Storage and Transportation (CAST)
- Computer Aided Production Planning (CAPP)

Although certain of the above topic titles are already commonly used within the manufacturing industry, there is no generally accepted definition of the scope or range of company activities which are addressed within each title. Also, certain of the above titles have been expressly created for this project. A definition or description of each of these five topics is included as the introduction to each topic.

Levels of mechanisation supported

Levels of mechanisation supported in the machining sector span the entire range from an 'unlinked NC machine' to a 'Flexible Manufacturing System which can be defined as follows;

Unlinked NC machine operation relates to the use of numerically controlled machine tools, which are not served by mechanised work-piece conveyancing or work-piece loading/unloading facilities. This type of operation is normally managed and supervised by human operators, supported by computer generated outputs. These outputs typically include punched NC paper tapes, printed job set-up instructions, printed production schedules, printed job cards, etc. Production Schedules for this level of operation would normally be determined in accordance with 'planned operational sequences' as these are established by a Process Planning Engineer - alternative NC programmes are not normally provided to permit operations on any one part to be carried out in any sequence other than the one devised by the process planning engineer.

Flexible Manufacturing System (FMS) refers to an environment in which DNC (Direct Numerical Control) Machine Tools are used to carry out the machining processes - and where these Machine Tools are served by automated work-piece conveyancing, and work-piece loading/unloading facilities. Notionally, this type of operation is managed and supervised directly by intelligent and semi-intelligent devices. Production in this environment is 'sequenced' rather than scheduled and decision making related to 'the next task to be performed' is carried out only when an already initiated machining task is nearing completion. This type of operation may involve the use of complex automated support facilities such as automated warehouses, automated tool management and delivery facilities, etc.

Many of the sub-systems which are needed to support a specified level of operation could be applicable to even lower levels of operation. For example, a Computer Aided Design sub-system would be applicable to an organisation which employs lower level machines than NC - as would the production scheduling sub-system, and the systems used to create printed job cards, printed job set-up instructions, etc. Although this might be an important consideration for gradual automation of existing facilities, it was not intended that this pilot project should make **comprehensive** provision for levels of operation lower than an unlinked NC machine. The objective of the pilot project was to consider all levels of operations between an unlinked NC machine and FMS.

The ordering and structuring of CIM requires the creation of two quite different types of rules -rules that apply to particular sub-systems and interfaces, and rules that are generally applicable to all the sub-systems of CIM. To differentiate these two separate issues, the term 'Design Rules' is chosen to imply rules which relate to a particular sub-system and interface. Rules that posses universal applicability across all sub-systems and interfaces, are termed 'Maxims'. The project therefore requires the creation of both design rules and maxims.

Design rules are principally concerned with defining the functional scope of each sub-system. This document accordingly identifies every significant CIM activity, and describes which of these are to be provided for in a specific sub-system. Rules relating to particular interfaces will be principally concerned with the nature, scope and form of the data which each specified sub-system is to make available to each related sub-system. Although the authors recognised the need for rules relating to interfacing data, these could only be provided for at a high level within the scope of the project.

Vendors wishing to claim that their CIM products confirm to the EEC Design Rules would be obliged to combine functions in the groupings defined within the established design rules. In order to avoid confusion which is frequently caused by vendors questionably claiming that their particular products comply with recognised standards (as for example happened in respect to the CODASYL data base proposal), the project team recommended the creation of EEC Certification Centres. These would evaluate CIM products and authorise the use of EEC approved CIM symbol on certified products and within sales literature used to advertise these products. It is however not certain at this time if this proposal will be adopted.

To develop the maxims, which are valid across all sub-systems, it is noted that Processing, Data and Communication are universal concepts applicable to all CIM Sub-systems and accordingly maxims for these have been developed. Maxims for one of these are collectively termed as a Strategy. Three strategies are therefore necessary in reference to any CIM undertaking:

- A Processing Strategy
- A Data Strategy
- A Communications Strategy

The method chosen to formulate a strategy was to develop maxims which could help a designer of CIM sub-systems to more easily find a lasting solution. Maxims have a number of characteristics: they are generally applicable throughout CIM (hence they represent strategic maxims), they address important aspects of CIM (such as complexity of local processing), they recognise the state of the art (capacity trade-offs) and they emphasise tendencies and ways towards further interpretation in CIM (distribution of processing).

The Processing Strategy concerns the manner in which processing is to be distributed between a large number of different processing devices. These will almost certainly include one or more 'centralised' mainframe computers, a large number of mini and micro-computers of greatly varying size and type, programmable logic control units, and a very large number of 'intelligent' and non-intelligent devices -from robots to relays. The Processing Strategy must be designed to take account of differing, and sometimes conflicting objectives encountered in such systems.

The Data Strategy concerns the design and distribution of the total data such that all processors and procedures may have access to consistent and authoritative data values - particularly in reference to items of data that are of common interest to many different manufacturing functions and activities. Due to substantial differences in the way in which many sub-systems need to process basic data, certain data will have to be replicated in several different files, of many different kinds. The maxims therefore should ensure that all copies of each data item are consistently maintained to correspond with the 'master' occurrence - eg. 'latest' modification level, 'current' stock level and price, etc. It should however be noted that it is a Data strategy which is being proposed, not a Data Base Strategy. The authors believe that it would be entirely impractical to attempt to propose physical arrangements for the storage of all the data needed to support every CIM activity. It is considered to be entirely infeasible that any single cohesive data base management system, of either a centralised or a distributed nature could be developed within the foreseeable future. This is not to be taken to imply that certain activities may not share certain common files (eg. Computer Aided Design and Parts Programming using a common Product Data Base) - the authors would indeed entirely subscribe to this practice.

The communications network for a computerised manufacturing system needs to be built in such a way that all transmissions can take place within the required time frame. Between each two communicating processes using the network a protocol must be chosen. For an integrated manufacturing system, all processes communicate, at least conceptually over the same net. A communications protocol has two aspects, the general format which may exist on each level, and the particular format that two processes have agreed upon. Integrated manufacturing needs both. In addition, tools are needed to map a particular format onto a general format, so that designers only need to be concerned with application formats. Standards in this area which can be mapped onto general formats such as OSI layer protocols, are vital parts of any manufacturing system. Many of the communication strategy maxims have to do with the fast evolving state of the art in communication techniques and consequently the lack of standardised methods. It is therefore important that the maxims emphasise the special requirements of communication in CIM, so as to be able to decide where forthcoming general solutions can be adopted and where more expensive but specific solutions are necessary.

Chapter 2

Development of CIM Design Rules

Everyone involved in Computer Integrated Manufacturing recognises that CIM comprises many separate modules or sub-systems, however, the problem is that no agreed structures and definitions for CIM currently exist. There is, as stated, no generally agreed sub-system structure - not even any generally accepted list of sub-systems. The scope of any system can only be defined in terms of function - or more precisely, in terms of the particular company activities which are expressly catered for by the system. Terms such as Product Design or Process Planning, whilst useful as generic titles for general disciplines, are not sufficiently precise for the purpose of defining the scope of particular systems. The reason for this is that such terms are portmanteau expressions for large numbers of discrete and quite different activities. It would invariably be inappropriate, and almost always impractical, to attempt to provide support for say each and every different Product Design activity, within a single computer system - even if Europe-wide agreement could be reached on the total list of detailed activities which together make up the generic term Product Design.

Attempts to label particular systems by use of such terms (eg. a Computer Aided Design System, a Production Control system, etc.) have contributed greatly to the confusion and scepticism which presently exists in reference to IT(Information Technology) products developed for manufacturing industry. The scope of any CIM sub-system needs to be defined in terms of detailed company activities, which, unlike generic 'roles' are essentially non-contentious. This is not to be taken to imply that every company will or should carry out every identified activity - nor that say the Product Design department of each and every company will have organisational responsibility for an identical sub-set of these activities. What is intended, and required, is that the various activities associated with CIM can be defined in such a manner that they can be universally understood and related to.

The development of design rules for CIM systems has been approached from a simple but fundamental hypothesis that the basic or fundamental activities which need to go on within any manufacturing operation do not change - it is only the methods and technologies which are used to carry out these activities that change. The prehistoric designer of a megalithic monument such as Stonehenge may have used a stick and sand to construct a geometric representation of the product he wished to create; a mid-20th-century designer instead used a pen and drawing board, and a designer in 1985 might use a computer terminal for the same identical purpose. The essential or fundamental task has not changed during this time - only the methods and technologies which are used to carry out such tasks have changed.

New insights into basic activities also permit greater levels of sophistication and scientific precision within each activity, but do not change the fundamental list of activities which have been evolved over many decades of manufacturing experience. An early decision was therefore taken to base the required CIM systems structure upon these elemental manufacturing industry activities - rather than upon currently available products and offerings. There were several reasons for not allowing current IT offerings to dictate the required systems structure.

The first and most obvious reason is that different authorities and vendors have, to date, pursued dissimilar and often conflicting approaches to CIM. The second major difficulty is that many of even the most basic CIM related activities have yet to be addressed by the IT industry. Whilst manufacturing clients can frequently choose from a plethora of offerings to support certain of their activities, there are often no IT products of any kind to support other equally perplexed, equally costly, and equally time consuming activities - which need to be supported if CIM is to be effectively achieved.

Another very important reason for adopting this approach, was to identify a number of opportunity areas for new IT products. It is believed that this will be of benefit to both the IT industry, and to European manufacturing industry.

Before considering the detailed approach which is followed here it would perhaps be helpful to reiterate the essential aims and objectives of the project.

- Initially the principal, indeed the only, objective was to **identify** the sub-systems and interfaces which form essential parts of Computer Integrated Manufacturing. The development of Design Rules and Maxims followed as extensions to this undertaking.

Each sub-system has therefore been identified in terms of 'function', i. e. in terms of the business activities which are to be addressed and supported by each system. Design rules for sub-system functionality have been evolved to address two separate but very closely associated subjects:

1 The basic business activity which is to be addressed by the sub-system.
2 The minimum list of factors which must be used to condition or determine each business activity.

One example might serve to illustrate these two separate aspects of functionality. The first sub-system within Computer Aided Production Engineering(CAPE) concerns the evaluation of different proposed manufacturing technologies. A product, such as say a motor vehicle crankshaft, could be produced as a forging, as a spheroidal graphite casting, or possibly by some other manufacturing technology. The principal business function of the first CAPE sub-system is to decide (or to 'aid' a Production Engineer in deciding) which of all the available technologies should be chosen for the manufacture of a proposed new product. The minimum list of factors which must be used within this sub-system to condition or determine this decision would include

- Projected sales or manufacturing volumes.
- Details of all possible alternative technologies.
- Details of existing plant and equipment within each technology.
- Details of projected work load capacities on existing plant and equipment.
- Cost of procuring additional plant and equipment capacity.
- Rough processes, times and costs under each possible technology.

(Note: It will be appreciated that the above example has been greatly simplified in order to illustrate the difference between basic functions and conditioning functions. No inference should be made concerning the sub-system in question without a detailed study of the sub-system titled Evaluation of Design for Manufacturing Requirements of the CAPE section.)

CIM Design Rules do not instruct IT vendors on how systems have to be designed, or which technical standards have to be adopted. The above example of the first CAPE sub-system could, at the discretion of the vendor, be developed in any one of several different ways.

eg 1:
It could be developed as a conventional 'algorithmic' system, using traditional imperative computer programme coding techniques, against say a CODASYL type data base.

eg 2:
It could be developed by the use of knowledge representation and inferencing, by means of declarative type computer programming (ie. as an 'expert' system) against a Knowledge Base. Different vendors will be free to choose different technologies, and it is foreseen that alternative products will eventually become available to manufacturing clients.

For the development of Design Rules the following three steps were adopted.

Step 1   Flowcharts

The total CIM area was initially analysed into the five fundamental topics previously identified, and flowcharts were produced for each topic - detailing, in chronological sequence, the various activities and procedures which need to take place in order to take a product from the conceptual design stage, through to final manufacture.

Step 2   Sub-system tables

The next step was to group the very detailed 'tasks' shown on each chart into CIM 'activities' - or sub-systems. In grouping tasks together to form sub-systems it is necessary to take account of two quite separate factors - Functionality and Processability.

Functionality

In grouping a particular sub-set of tasks together to form an activity, cognisance has to be taken of 'logical procedural breaks'. Thus, the design of a product is seen to be a different procedural activity than say the development of an NC parts programme to manufacture that part. Several different factors may condition a logical procedural break. One reason might be that a person possessing a different kind of skill might be required to progress a procedure from a certain point. A procedural break may also be necessary if some time delay is unavoidable within a procedure - such as where a procedure generates a proposal which has to be approved before it is further acted upon.

Processability

Grouping of tasks to form activities also has to take cognisance of extreme differences in processing requirements. Thus computer aided draughting might be supportable by a relatively low-performance computer such as a micro-computer - whereas finite element analysis(FEA) would require much more powerful equipment. Many micro-computer vendors might therefore wish to develop a systems product to support draughting - but would be excluded from being able to do so if they were also required to provide support for FEA within the same product, in order to claim that their product complies with the Design Rules

It should be noted at this point that whilst technology was not allowed to influence the determination of the basic manufacturing tasks depicted on the Flowcharts, it has been allowed to influence the manner in which the basic tasks shown on the chart have been grouped together to form CIM sub-systems. However, as it is intended that the eventual rules will permit vendors to merge support for two or more sub-systems into a single product (subject to certain quite stringent conditions), changes in technology should not invalidate the structures which are proposed.

Step 3   Sub system interfaces

The third and final preparatory step to the creation of design rules or maxims, was to identify the way in which each sub-system relates to all other sub-systems. Essentially, this involved identifying the data inputs and outputs of each sub-system - and determining which sub-system has 'prime authorship' responsibility for each kind of data.

Following the above procedure, the various CIM sub-systems and their interconnections have been identified. A convenient way of representing such a comprehensive view of CIM is in the form of the 'round-table' chart which appears on the front cover. The applicability of CIM design rules to small firms was also investigated by the Department of Industrial Management, University of Dublin, Ireland. The results of this investigation are reported in an Appendix on 'CIM in the small firm'.

Chapter 3

Computer Aided Design

Introduction

The scope of Computer Aided Design (CAD) includes the use of a computer based system to assist all those tasks involved in the process of developing a concept for a product into a fully engineered design, described in sufficient detail to enable it to be manufactured.

The process starts with a Product Functional Specification, i.e. a statement of the parameters relating to the time, cost, size, weight, appearance, performance, durability etc. within which the design must be constrained. The process ends with the release from Engineering to Manufacturing of information describing the shapes of the constituent parts of the product, the materials from which they are to be made and the manufacturing processes and assembly instructions which may be mandatory to ensure the integrity of the design. The technical activities of engineering embraced by CAD can be divided into three broad categories :-

a   Design
b   Design Analysis
c   Engineering Test

Computer Aided Design is generally understood to embrace only the technical activities of engineering and does not usually include such things as manpower resource allocation systems, project control systems, parts usage and parts procurement systems, systems for accessing competitor and supplier information, administrative and accounting systems and other information systems necessary for the efficient running of an engineering department. While CAD does not embrace these non-technical activities, it should be borne in mind that as much as 75% of the time of technical staff is absorbed on non-technical work. However, since the use of CAD as part of a CIM environment requires both the management of large quantities of data and strict control of change, two particular administrative functions are regarded to be of sufficient fundamental importance to warrant inclusion within the scope of CAD for ESPRIT purposes. These two functions have been categorised as follows:-

d   CAD Administration
e   Design Modification and Engineering Change

Design

- Design is primarily concerned with establishing the geometrical shape of the parts which will make up the product, the materials from which they will be made and certain aspects of the manufacturing and assembly processes mandatory to ensure the integrity of the design.

Design can be divided broadly into three areas: Concept Design, Engineering Design and Detailed Design.

- Concept Design is concerned primarily with establishing the basic shape and appearance of the product resulting in a Design Proposal upon which a business decision to proceed can be based. It is also concerned with the broad evaluation of different ways in which the requirements of the Product Functional Specification could be satisfied. Establishing the basic shape of the product can potentially make much use of graphics systems capable of creating complex geometrical shapes; of displaying them in any view or of dynamic rotation of them; of displaying them in various colours, or combinations of colour, shading and surface texture to enable the designer to visualise already what the final appearance of the product will be. The database of geometric data established at

this stage will form the basis upon which more detailed design will follow.
- Engineering Design is the mainstream design activity. Based on the Design Proposal, it is concerned with the identification and design of assemblies, sub-assemblies and components. With reference to Design Analysis and Engineering Test, nominal geometric design will be carried out, materials identified, and the Bill of Materials and other part related lists will be initiated. Some of the techniques employed will be similar to those described in Concept Design. The end result of the activity will be pre-released design information, including nominal geometry, fairly comprehensive parts lists and materials details.
- Detailed Design is concerned with identifying an exhaustive list of all the parts to be incorporated in the product and with providing a complete specification of each one, in sufficient detail to enable it to be manufactured. Traditionally, specifications have consisted of sets of detailed drawings together with build and assembly instructions. It seems likely that the engineering drawing will remain in use for some time to come so that computer aided draughting systems will continue to be an important part of CAD. In the longer term, as greater use is made of computer readable CAD data as direct imput to control manufacturing plant and inspection equipment (CAM), the requirement for computer aided draughting will diminish.

Design Analysis
- Design Analysis is mainly concerned with the use of mathematical modelling techniques, either to predict theoretically during the design process the expected performance of the product, or else to understand the observed performance of an existing product under test or in service. It involves the use of techniques such as Finite Element analysis to determine stress and vibrational characteristics of a product and of many other calculations in such fields as heat transfer, thermodynamics, fluid dynamics, metal fatigue etc.

Engineering Test
- Engineering Test is concerned with the empirical determination of the performance and endurance characteristics of a designed product by physically testing a representative specimen or a prototype. Computers are used to control tests, by triggering actuators, for automatic data capture, by scanning of transducers and for data analysis. An Engineering Prototype Build activity will also be included.

CAD Administration
- This is the activity which will provide overall control of the design process and scheduling of design work. It may incorporate a CAD data management activity offering facilities to identify, locate and retrieve CAD data whilst controlling authorised access, maintaining data integrity and enforcing many company procedures.

Design Modification and Engineering Change
- This is the provision of a formal, controlling mechanism whereby requests for design modifications can be received from other CIM areas or sub-systems. Once a modification has been completed by a design area, all users or user sub-systems would be informed of the change.

Computer Aided Design is the integrated and systematic use of the above applications of computers in Engineering. Integration is achieved by ensuring that each step in the design process contributes data to a data base accessible to everyone else involved in the design, development and manufacturing process. It is also quite feasible to include in such a data base, lists and descriptions of standard parts or of parts which can be carried over from previous designs, and also to incorporate and enforce the use of engineering standards.

In this text attention has been focussed only on the technical areas of CAD (ie. Concept Design, Engineering Design and Detailed Design).

The inter-relationships between the major areas of CAD are summarised in the following diagram:-

INTER-RELATIONSHIPS BETWEEN MAJOR AREAS OF CAD

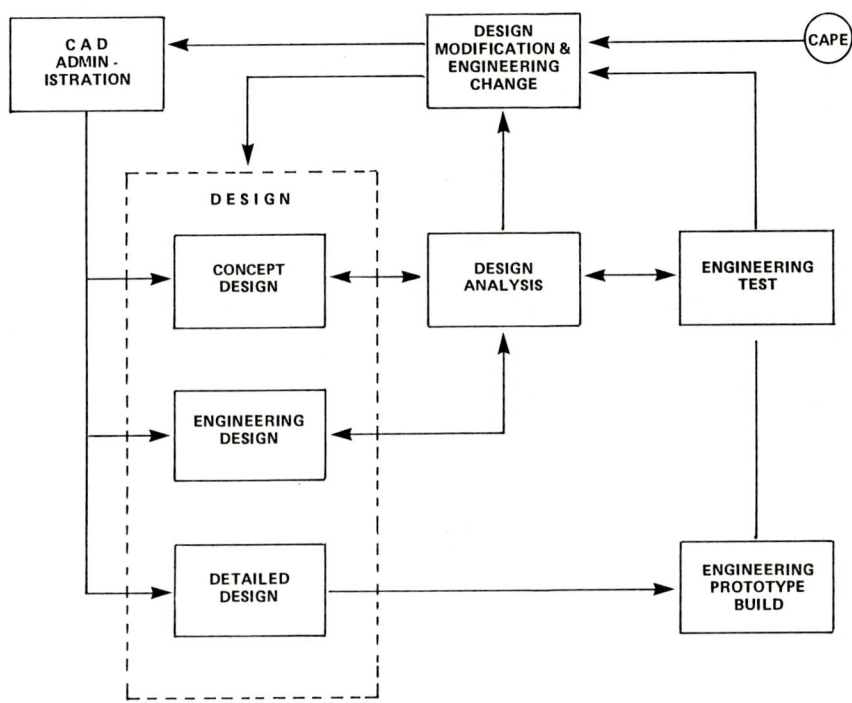

Computer Aided Design

Flowcharts

Computer Aided Design

COMPUTER AIDED DESIGN Sheet 1 of 5

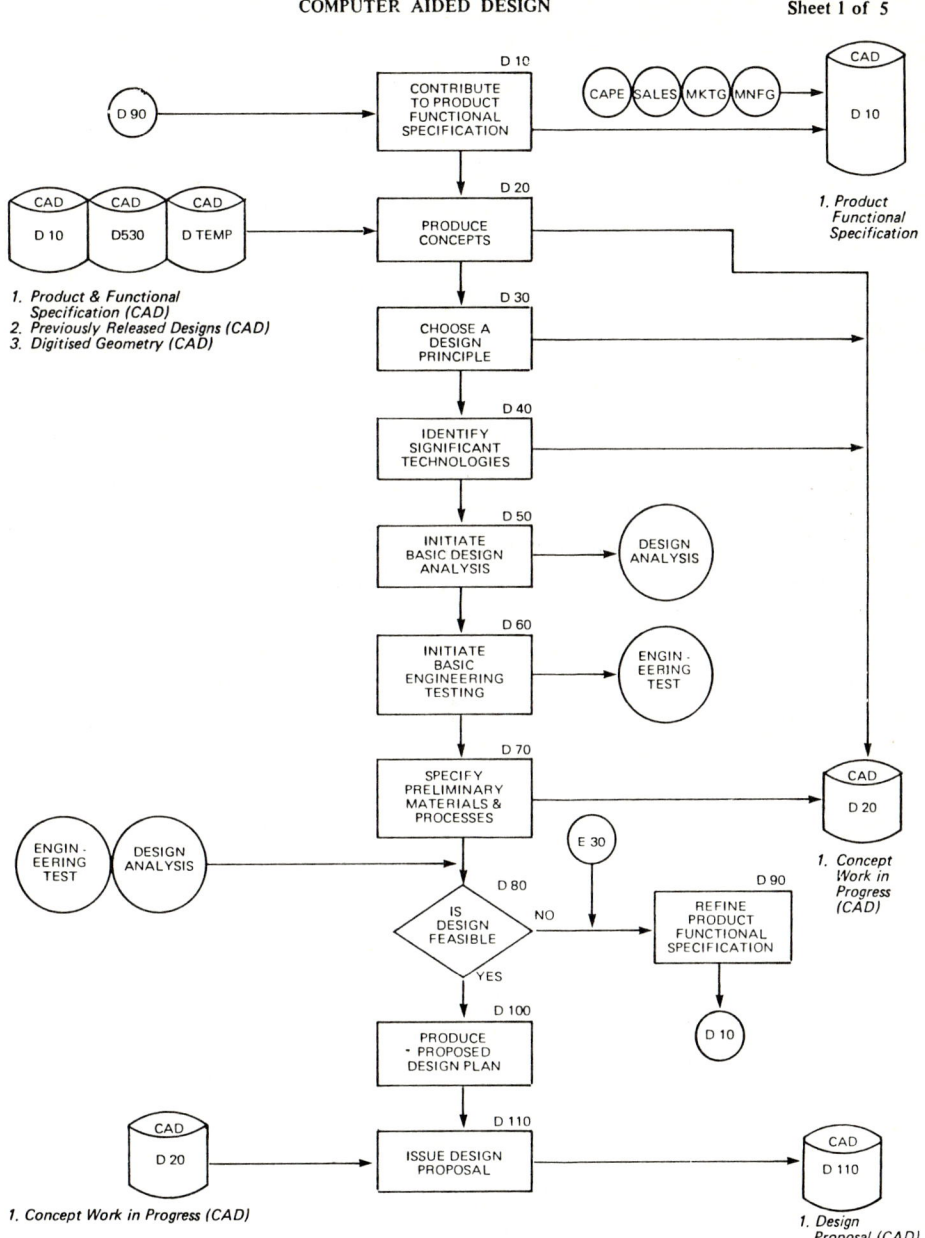

COMPUTER AIDED DESIGN

Sheet 2 of 5

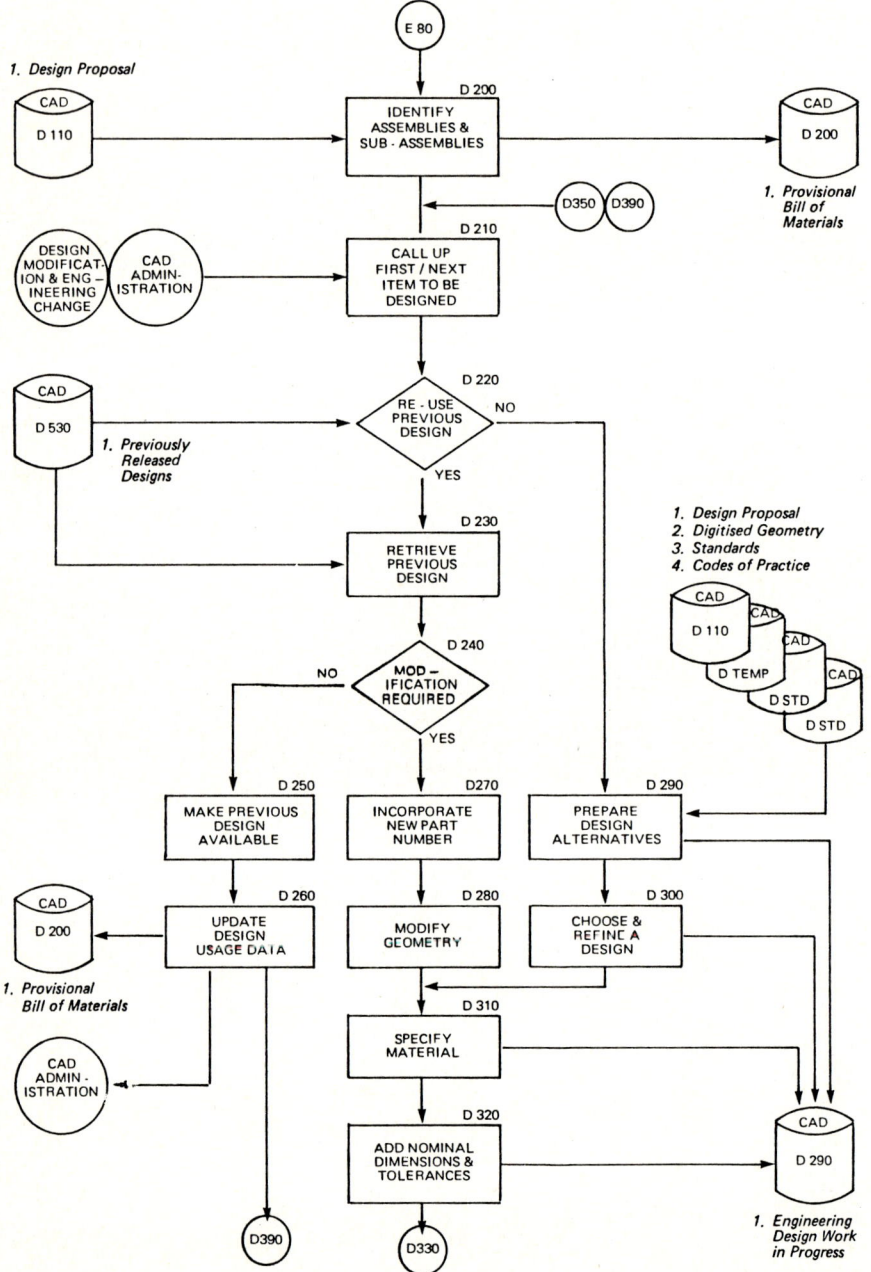

COMPUTER AIDED DESIGN

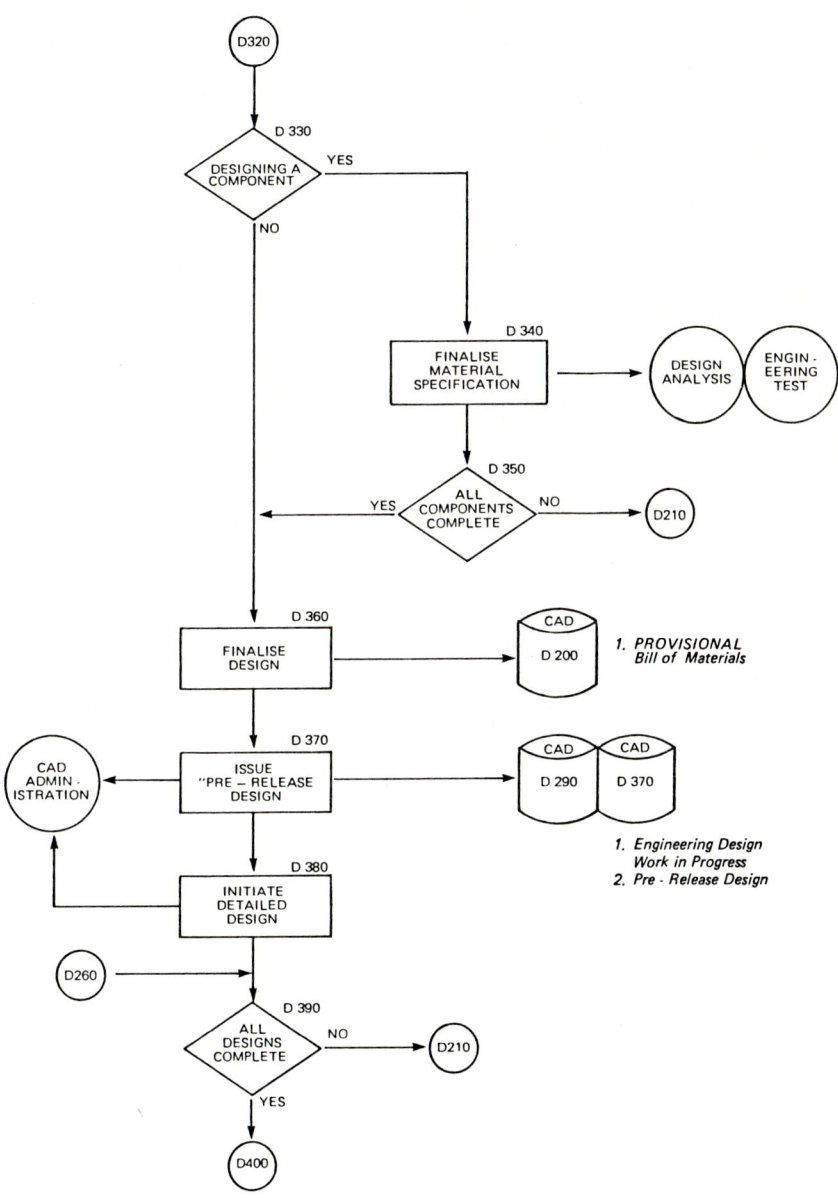

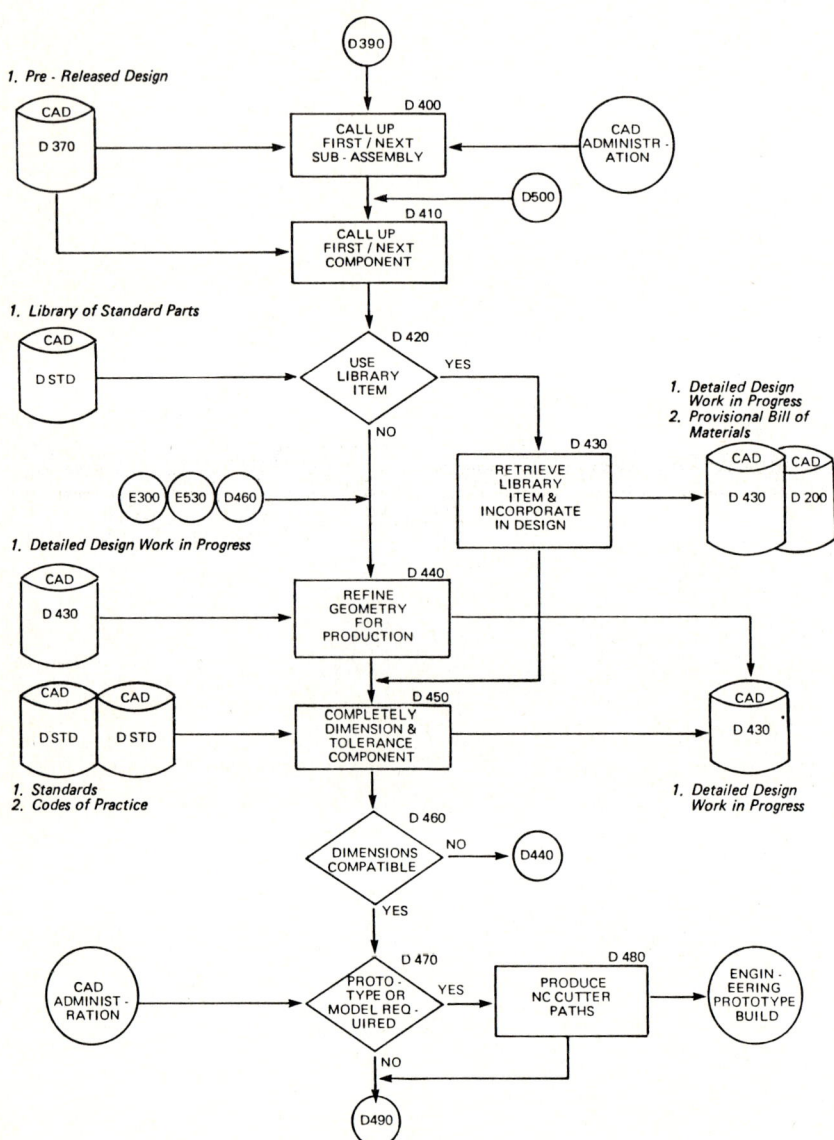

COMPUTER AIDED DESIGN

Sheet 5 of 5

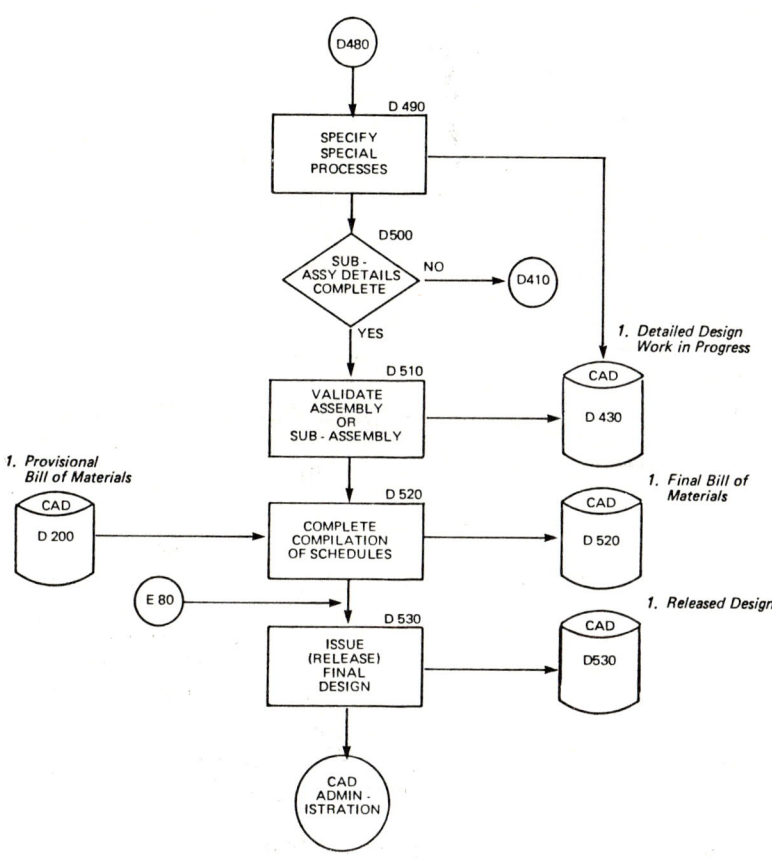

Narrative description of Computer Aided Design (CAD) network

D10   The Product Functional Specification will consist of a statement of the parameters relating to the time, cost, size, weight, appearance, performance, durability etc. within which the design must be constrained. Much of the input to the Specification will come from areas such as Sales, Marketing and Manufacturing (including CAPE). However, the designer, who is probably a senior designer at the Concept Design stage, will provide the majority of inspiration and original, inventive thought which will contribute to the formulation of the Specification.

Most companies in the machining industry will either be continually refining an existing product range or else repeatedly using the same basic design principles. Thus, the Functional Specification will invariably be based on what has been done before with some additions to account for, say, market trends or new technology. However, the fundamental nature of the product could undergo a significant change during the Concept Design stage based on a totally original idea by the designer for one particular detail. Due to this inspirational nature of Concept Design work, liaison with senior management will tend to be on an informal basis. Similarly, administrative control of the concept design process will tend not to be rigorous.

D20   Once the Product Functional Specification has been agreed as part of a business proposition at company level, conceptual work will commence in earnest. This will consist of the production of a number of broad concepts, artist's impressions and sketches in order to establish alternative approaches to designing the basic shape and appearance of the whole product rather than that of individual details. The activity will be informal in that detailed designs, with full dimensions and tolerances, will not be produced.

Typically, computer techniques used will be those geared towards three dimensional geometry manipulation and display, solid modelling, surfaces, surface texture, colour, shaded images, scaling, dynamic rotation of displays etc.

It may be a requirement to make reference to previous design work in order to minimise the proliferation of different types of manufactured detail. Such reference will require access to the database which contains all previously issued or released designs.

In the engineering design functions of some companies, it is normal practice to produce, either by Numerically Controlled machining or by manual methods, a clay, wooden or plastic model of the product or part thereof. In such cases, it may be necessary to provide a means whereby the geometry of the model can be digitised and the resultant data incorporated into the current design.

Concept work tends to be artistic and unstructured in nature dependent on the individual style of the designer, often relying on his inspiration and personalised style.

All work carried out at the Concept Design stage will contribute information to a Concept Work In Progress base of data which will be refined and added to as the Concept design task proceeds. This data will be protected from unauthorised access in that it will, at this stage, be private to the design function. It is only at the later stages of design that data tends to be made more widely available.

D30 Based on the initial concept work carried out in node D20, a particular design principle will be selected in order that more specific sketches and layout designs can be produced. The design principle chosen will form the basis of the Design Proposal which will be issued at the end of Concept Design in node D110. Similar techniques to those described for node D20 will be employed and again, the information produced will contribute to the Concept Work In Progress.

D40 If it is intended to employ any new technology, or to change the use of a current technology, it is vitally important that such changes are identified at as early a stage as is practically possible in order to maximise advance warning to those functions which may have to carry out a significant amount of work as a direct consequence. The changes will need to be highlighted when the Design Proposal is produced during the activities of node D110.

D50 Predictive calculations regarding the performance of the product will be made based on major assumptions regarding the product design. Such calculations will either confirm, or generate modifications to, those assumptions, thus guiding the design principles to be employed. Reference will continually be made to the original features and constraints contained in the Product Functional Specification.

D60 Some basic engineering testing will be carried out at an early stage of design for the same reasons as those given for node D50. For example, existing components, sub-assemblies or assemblies may be known to be included in the design, but under different operating conditions. In such cases, engineering test work can be carried out to evaluate product performance under those revised operating conditions. Similarly, minor modifications may be made to individual details, in a development mode, to explore the possibility of using existing designs with some modifications.

D70 The Design Proposal will go forward from Concept Design to CAPE in order that a business decision may be made. It is important, therefore, to include any information which is known at this early stage regarding the proposed use of materials and production processes.

Although CAPE will ultimately decide the way in which t he product will be manufactured, it is the responsibility of the Design function to nominate and describe those manufacturing processes which are mandatory to ensure the integrity of the design.

D80 Obviously, it is only sensible to put forward a Design Proposal which is feasible. If the overall Product Functional Specification and its associated constraints cannot be satisfied then a request to refine that Specification must be made. This assumes that the Concept Design work of nodes D20 to D70 inclusive has been exhaustive.

D90 Following the decision in node D80, a mechanism for modifying or requesting a change to be made to the Product Functional Specification is required. This is necessarily formal since it is not the Design function alone which contributes to the Specification.

This is also the node at which requests from CAPE will be received if CAPE cannot satisfy the Design Proposal requirements.

D100 A detailed plan will be required for the engineering design and development aspects of the proposed new product. Estimates must be made of workloads (ie what is to be done), manpower required, facilities required, costs and timescales (lead times). Areas embraced will be design, design analysis, prototype build and engineering test. Typically, critical items will be identified for analysis or testing as will those with expected long lead times. Workload will be planned such that the design of major components, sub-assemblies or assemblies will be scheduled and allocated to specific designers or design groups based on availability or expertise.

Similarly, the plan will include programmes for design analysis and prototype testing.

The plan will serve two main purposes. Firstly, it will give broad indications of costs and timescales which will need to be known to enable the business decision to be made. Secondly, it will form the basis of a more detailed plan to enable Engineering Design and Detailed Design to proceed efficiently once the Design Proposal has gained formal acceptance.

D110　Once it has been agreed that the design is feasible, then a full Design Proposal will be produced which will include **some** of the sketches and models produced in nodes D20 and D30. Not all Concept Work In Progress data will be included since some will be confidential to the Concept function. Although those data items taken from Concept Work In Progress will be mainly geometric in nature with associated pictorial or graphical representations, much of the Design Proposal itself will consist of text.

The Design Proposal will form part of the information required, at a senior level in the company, on which a business decision will be based as to whether to proceed or not. CAPE will play a significant role in this decision process. Further design activities will not proceed until the Design Proposal has been accepted and authorisation to proceed has been received.

D200　Once the Design Proposal has been accepted, the plans produced in node D100 may be implemented. Initial Engineering Design work will involve the refinement of any selection of assemblies and sub-assemblies so that design work can be broken down into manageable tasks.

Each sub-assembly and each of its constituent components will require unique identification by, for example, part number, in order to identify it at subsequent stages of design (and manufacture). Once identified, initial entries may be made to part related lists such as Bill of Materials etc. At this stage, such text information will be provisional since each component within the product will not have been fully identified.

D210　In order to follow the design process in detail, each item to be designed must be considered in turn (be it assembly, sub-assembly or component). The selection of the item to be designed will be governed by one of two factors. Firstly, CAD Administration will dictate design priority based on the plans produced in node D100 together with other information. Secondly, requests processed by a Design Modification and Engineering Change sub-system may cause a change in design priorities.

The procedure followed by nodes D210 to D390 inclusive will be similar for each design level for the product. Initially, design will be at the assembly level and the process described in nodes D210 to D390 will be followed. Subsequently, when sub-assemblies have been identified, design will proceed at that more detailed level but will again proceed through the same nodes.

D220　It may be possible, or even desirable, to re-use a component or sub-assembly which was designed on a previous occasion. In order to assess the possibility of such re-usability, a means of locating and viewing the previous design will be required. Probably, this design data management function will be provided by CAD Administration. Identification of the design may be by means of a unique reference number, such as a traditional drawing number, or by means of a coding and classification system which enables the identification of those components with similar or pre-defined characteristics.

Previous designs which have been issued for manufacture will be held in a "view only" form in that direct modification of those designs will be prohibited. The designer will require sufficient access to previous design work to enable him to decide whether he can use the previous design completely or whether some modifications would be suitable. Failing such re-use of previous work, a new design will be required.

D230 Once a previous design has been selected, a copy will be made on which further work may be carried out. This copy will become the working copy for subsequent tasks.

D240 It must be decided whether the previous design can be used in its currently copied form or whether modifications will be required.

D250 If no modification of the previous design is required, then the working copy of the design must be made available for incorporation into the overall product design. This will be necessary since it will be a requirement to produce a fully assembled design of the whole product, or major assemblies thereof, for validation purposes in node D510.

D260 Previous designs of sub-assemblies or components which are to be incorporated, without modification, into the current product design will require relevant changes to be made to any files containing data regarding the use of that previous design. For example, if a company employs a method of producing "where used" or traceability reports (that is, a means whereby the occurrence of a particular detail in **any** company product may be identified) then such usage data will be updated to reflect the new use of the previous design. Similarly, the Provisional Bill of Materials for the current design will be updated.

D270 In the case of a modification to a previous design, a new part number or unique reference will be assigned to the working copy of the previous design since, after modification, the part has essentially become a new one.

D280 The facilities offered by computer based design techniques enable geometrical representations to be modified by means of deletion, movement or addition of graphical entities and text. In this node, the designer will carry out such modification ensuring that all of the necessary changes have been made.

D290 In the case of a completely new design, a number of design alternatives will be prepared for consideration. These will be at a more detailed level of design than those which led to the issue of the Design Proposal in node D110 but will still be conceptual in nature. Thus the computer techniques used will be those geared towards three dimensional geometry manipulation and display, solid modelling, surfaces, surface texture, colour, shaded images, scaling, dynamic rotation of displays etc.

In the engineering design functions of some companies, it is normal practice to produce, either by Numerically Controlled machining or by manual methods, a clay, wooden or plastic model of the product or part thereof. In such cases, it may be necessary to provide a means whereby the geometry of the model can be digitised and the resultant data incorporated into the current design.

Each design alternative will be created in accordance with Codes of Practice and Design Standards which will have been nominated by the company. For example, it may be a legislative requirement that a product has a certain feature or performs to a particular recognised standard. In order that the designer may account for such a requirement, it will be necessary to make the relevant Code of Practice or Standard available to that designer.

D300   A single design will be selected from the various design alternatives prepared in node D290 following which a refinement of the geometrical representation of that design will be carried out. This will be necessary since the conceptual work of node D290 will only have been sufficient to enable the selection to be made. An example of refinement is where, in CAD terms, a conceptual sketch is converted into a three dimensional geometrical model.

D310   In the case of design at the sub-assembly level, a broad statement of material specification will be made for the constituent components. The precise material for each component will be finalised later in node D340.

D320   During the Engineering Design phase, particularly in node D300, the geometry will have been designed to nominal dimensions. In this node, the designer will add textual information to the geometry relating to these nominal dimensions and their associated nominal tolerances.

D330   It was stated in the narrative for node D210 that the procedure followed by nodes D210 to D390 inclusive will be similar for each design level of the product. However, this node (D330) allows a deviation from the procedure in order to cater for activities which are specifically required in the case of individual component design.

D340   At the component design level, the material specification selected in node D310 must be finalised. This may require confirmation from Design Analysis, Engineering Test or both.

D350   This section of the CAD network (nodes D340 and D350) specifically concerns individual component design and if all of the components which comprise the current sub-assembly have not yet been designed, then the next one must be selected by returning to node D210. If the current sub-assembly is complete, however, then the activity of node D360 will occur next.

D360   All relevant information regarding the current assembly or sub-assembly must be gathered or organised and the Bill Of Materials updated ready for node D370. Any missing information must be checked for and provided as necessary.

D370   At this stage of design, there is no further Engineering Design to be carried out since the next phase will involve Detailed Design. Thus the Work In Progress for the current design may be considered to have reached a significant stage at which other areas of Engineering may regard the design as fixed but it has not yet been authorised for manufacture. The design at this stage is often referred to as the "pre-released" design in that it is a provisional design made available before the full authorising procedure for issue or release has occurred. Typically, Design Analysis or Engineering Test will require access to the design as will any advanced manufacturing engineering function (e.g. CAPE).

D380   Detailed Design will usually be carried out by designers who are skilled in that particular activity (often referred to as "Detailers"). Thus, groups of components will be assigned to individual detailers as each provisional design becomes available.

D390   In parallel with the Detailed Design task, Engineering Design of further assemblies or sub-assemblies may be required. If this is the case, then it will be necessary to return to node D210 in order to continue.

D400   In order to follow the Detailed Design process, each sub-assembly of the pre-released design data will be considered in turn. The selection of the item to be designed will be governed by one of two factors. Firstly, CAD Administration will dictate design priority based on the plans produced in node D100 together with other information. Secondly, requests processed by a Design Modification and

Engineering Change sub-system may cause a change in design priorities.

D410 Similarly, as in node D400, within a particular sub-assembly, each component to be detailed will be considered in turn.

D420 Often, it will be possible to include a standard part in the design. Typically, items such as nuts, bolts, washers etc. will have been fully detailed previously and placed on a CAD library for repetitive use. It is essential to maximise the use of such library parts since the re-use of previously or currently manufactured items will inevitably reduce manufacturing and other costs.

D430 Once it has been decided to use a library item, then it must be retrieved from the library and incorporated in the current design. This will involve placing the library part in the correct geometrical orientation compared with the design under consideration and, depending on the library item, scaled accordingly. The Bill Of Materials will be updated as necessary.

D440 At this stage, the geometry produced to date must be refined, where necessary, to take full account of the intended production method. It is essential that the manufacture of the designed item is both practically achievable and efficient. The Detailer will liaise closely with both the manufacturing function and external suppliers in order to obtain sufficient information to enable the activity to be completed.

D450 Engineering Design was to nominal dimensions and tolerances only. The Detailer will be responsible for completing full detail dimensions and tolerances of the component to company and other standards.

D460 Once the dimensions and tolerances have been fully defined, a check must be made to see if they are compatible with:

(a) the material from which the component is to be manufactured

(b) the probable method of manufacture (which will be confirmed by CAPE)

Again, in order to complete this activity, the Detailer will liaise closely with the manufacturing function or external supplier as appropriate.

D470 The CAD Administration System will indicate if a prototype or model is required. If so, then node D480 will be necessary. Such a prototype or model may be required for a variety of purposes such as engineering testing, visual appreciation, submission to suppliers, etc.

D480 Close links will exist between Detailed Design and the Engineering Prototype Build functions in order to use the geometry of the component design as the basis to generate Numerical Control cutter paths. It is a function of Engineering Prototype Build to produce the actual NC machine instructions. (It is possible that CAM may employ the same Detailed Design sub-system to produce NC information)

D490 The design will require textual information to be added in those cases where special manufacturing processes must be carried out to ensure functional performance of the component.

D500 If the details are not complete for the current sub-assembly, then further work will be required.

D510 When all components have been detail designed, it will be necessary to assemble each of those detail designs into the relevant sub-assembly or assembly for validation purposes. For example, cumulative tolerances and clash detection are typical areas of concern. Once assembled, this total design will supersede the previously produced sub-assembly design which was "pre-released" in node D370.

D520   The Bill Of Materials and other parts list based schedules will have been compiled in provisional form at the start of the design process and additions and refinements made throughout. Before final release of the design, it must be checked that all schedules which will be included as part of the release have been completed satisfactorily.

D530   Permission or authorisation must be obtained to officially release or issue the design from Engineering. Once given, the design information will be placed on a "released database" such that further modifications are prohibited.

Computer Aided Design (CAD) sub-systems

The Flowchart for the Computer Aided Design activities in the ESPRIT Pilot Study has been broken down into sub-systems:

1. Concept Design
    (Nodes D10 to D110)

2. Engineering Design
    (Nodes D200 to D390)

3. Detailed Design
    (Nodes D400 to D530)

Concept Design sub-system (D1)

D1.1 **Sub-System Description**

The concept design stage of engineering is concerned with two aspects:

(a) Concept Design of the proposed product in order to establish its basic shape and appearance

(b) The Business Case or Proposition justifying the introduction of the proposed product

In order for these two phases to commence, a broad Functional Specification of the proposed product is drawn up. It will consist of product features and constraints based on information such as sales department forecasts, marketing, target dates, pricing, trends, fashions, technology, competition, etc.

Concept Design is then concerned with the broad evaluation of different ways in which the requirements of the Product Functional Specification could be satisfied. Establishing the basic shape of the product can potentially make much use of graphics systems capable of creating complex geometrical shapes; of displaying them in any view or of dynamic rotation of them; of displaying them in various colours, or combinations of colour, shading and surface texture to enable the designer to visualise already what the final appearance of the product will be. The database of geometric data established at this stage will form the basis upon which more detailed design will follow.

Concept Design Sub-System

D1.2 Flowchart

D1.2. SUB SYSTEM FLOWCHART

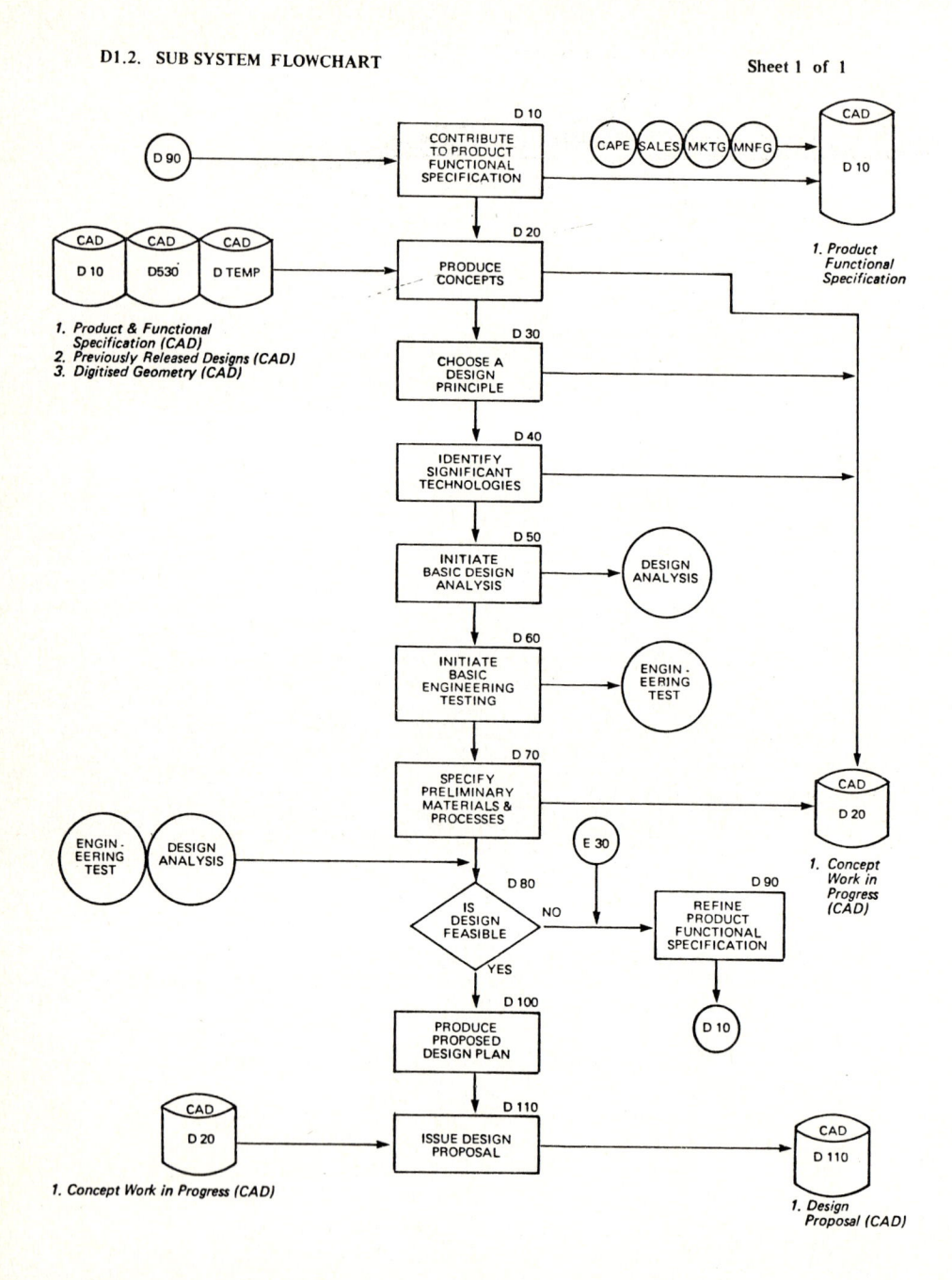

Sheet 1 of 1

D1.3 **Design Rules - Concept Design**

D1.3.1 **Input Data**

The sub-system must be capable of accepting the minimum data items listed below:

Data is required relating to:

- Previously Released Designs (Section D1.3.1.1)
- Digitised Geometry (Section D1.3.1.2)

D1.3.1.1 **Information relating to previously released designs**

- **Unique Identification**
  A unique identification, allocated or machine generated, which distinguishes each design from all other designs. A part number or traditional drawing number may be considered sufficient.

- **Geometric Data**
  A representation of the shape of a part in terms of a group of geometric entities. For example, a group of points, lines, curve functions, surfaces, etc. It is assumed that each type of entity will be used in a disciplined and identified manner.

- **Attribute Data**
  Additional attributes which would appear on a traditional engineering drawing such as linear or angular dimensions, diameters, surface finishes, tolerances, etc. The information would include only that which is necessary to enable the part to be manufactured within the constraints of the design. Some of the information will be implied by the geometric data but will not have been explicitly defined. It is assumed that each type of attribute will be used in a disciplined and identified manner.

- **Text Data**
  This refers to descriptive information which forms a fundamental part of the design. For example, special processes, assembly instructions, design title (eg. Connecting Rod), design notes, etc. It is assumed that, where any text item is related to a specific geometric entity, the associativity will be implemented in a disciplined and identified manner.

  It is also assumed that general textual descriptions relating to a whole design, rather than to a geometric entity within a design, will be associated with the unique identification of that design.

- **View Data**
  A collection of the information necessary to present a "picture" of the designed part to a human in an understandable and recognisable form. For example, general arrangements, layouts, technical illustrations, exploded views of assemblies, sections, isometric views. Such "pictures" may be in hard copy form (such as a pen plot) or for presentation on a CAD computer graphics display unit.

  View data is not a mandatory design data item. However, where such a facility is provided, it must be clearly stated whether or not there is associativity between the geometric data and the view data. Since the view data is a "pictorial" representation of the designed part, associativity would imply that a modification to the geometric data would automatically cause the changes to be accounted for in the view data.

D1.3.1.2 **Information related to digitised geometry**

The sub-system should be capable of accepting digitised co-ordinates. Such data will typically be derived from three dimensional models or from two dimensional drawings or sketches. It may be considered sufficient if the sub-system takes the digitised co-ordinates and converts them into geometric point entities. However, where the sub-system offers facilities to mathematically fit, for example, curves or surfaces to the digitised co-ordinates, thus providing a more sophisticated representation of the geometry, it must do so in a disciplined and identified manner.

D1.3.2 **Input reformatting**

The sub-system must be capable of reformatting, as required, input data into the form required by the sub-system. It is assumed that the input data will be structured in a disciplined and identified manner.

D1.3.3 **Unique identification**

A unique identity, allocated or machine generated, which distinguishes the current design from all other designs on the same sub-system and from all previous designs (which may reside on other sub-systems). A part number or traditional drawing number may be considered sufficient.

D1.3.4 **Restriction of access**

The sub-system must be capable of restricting the access to specified design information. Restriction must be applicable to individual sub-system users, groups of those users, and other sub-systems. A form of hierarchical password protection may be considered sufficient.

D1.3.5 **Output data**

The sub-system must be capable of producing data to the minimum levels specified in this section It must be in a form which will facilitate reformatting by a subsequent module, ie. all items must be structured in a disciplined and identified manner. Data is required relating to:

- Design Proposal (Section D1.3.5.1)

D1.3.5.1 **Information relating to design proposal**

- **Unique Identification**
  A unique identification, allocated or machine generated, which distinguishes the proposed design from all other designs. A part number or traditional drawing number may be considered sufficient.

- **Geometric Data**
  A representation of the shape of the proposed part in terms of a group of geometric entities. For example, a group of points, lines, curve functions, surfaces, etc. It is assumed that each type of entity will be used in a disciplined and identified manner.

  The Concept Design function will have produced a Work In Progress base of geometric data which will contain design information at various levels of sophistication. This is due to the computing techniques which may be used for conceptual work as described in the narrative for Concept Design. Thus, the geometric data will be biassed towards those techniques. In the case where the data is to be used by a less sophisticated sub-system (as is most likely), then the Concept Design sub-system must be capable of manipulating the geometric data into a suitable output form.

- **Attribute Data**

    Additional attributes which would appear on a traditional engineering drawing such as linear or angular dimensions, diameters, surface finishes, tolerances, etc. The information would include only that which is necessary to enable the part to be manufactured within the constraints of the design. Some of the information will be implied by the geometric data but will not have been explicitly defined. It is assumed that each type of attribute will be used in a disciplined and identified manner.

    Depending on company requirements, such attribute data may not be appropriate in all cases of Concept Design sub-system output.

- **Text Data**

    This refers to descriptive information which forms a fundamental part of the design. For example, special processes, assembly instructions, design title (eg. Connecting Rod), design notes, etc. It is assumed that, where any text item is related to a specific geometric entity, the associativity will be implemented in a disciplined and identified manner.

    It is also assumed that general textual descriptions relating to a whole design, rather than to a geometric entity within a design, will be associated with the unique identification of that design.

- **View Data**

    A collection of the information necessary to present a "picture" of the designed part to a human in an understandable and recognisable form. For example, general arrangements, layouts, technical illustrations, exploded views of assemblies, sections, isometric views. Such "pictures" may be in hard copy form (such as a pen plot) or for presentation on a CAD computer graphics display unit.

    View data is not a mandatory design data item. However, where such a facility is provided, it must be clearly stated whether or not there is associativity between the geometric data and the view data. Since the view data is a "pictorial" representation of the designed part, associativity would imply that a modification to the geometric data would automatically cause the changes to be accounted for in the view data.

Engineering design sub-system (D2)

D2.1 **Sub-system description**

Engineering Design is the mainstream design activity. Based on the Design Proposal, it is concerned with the identification and design of assemblies, sub-assemblies and components. With reference to Design Analysis and Engineering Test, nominal geometric design will be carried out, materials identified, and the Bill of Materials and other part related lists will be initiated. Some of the techniques employed will be similar to those described in Concept Design. The end result of the activity will be pre-released design information, including nominal geometry, fairly comprehensive parts lists and materials details.

The activities described in the narrative may be applied to all sizes of engineering operation, in that the design of complex products consisting of a multitude of components will involve more iterations within the sub-system than will the design of single details. However, the basic activities described will remain unchanged in each case.

Engineering design sub-system

D2.2 Flowchart

D2.2 SUB SYSTEM FLOWCHART

Sheet 1 of 2

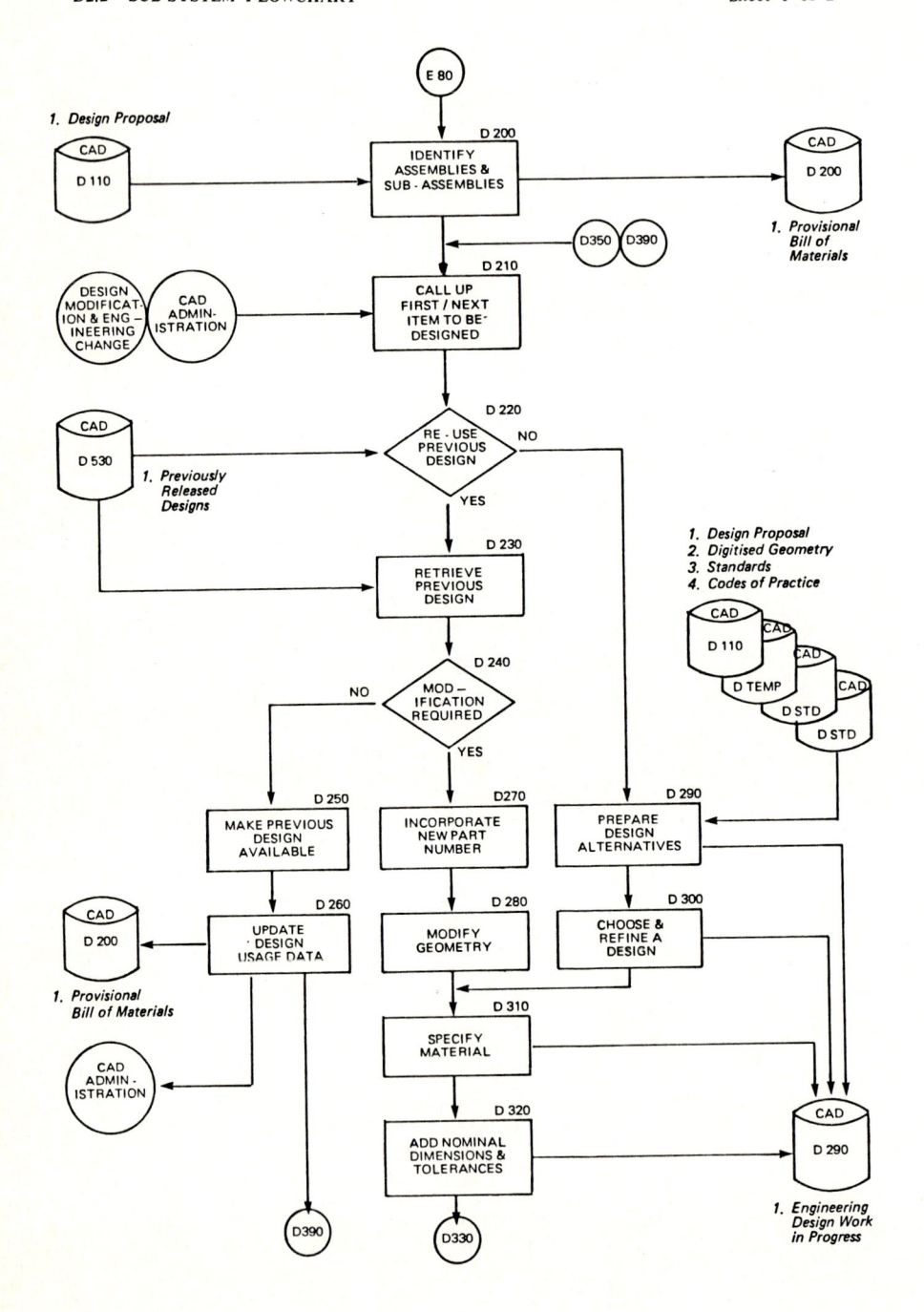

D2.2 SUB SYSTEM FLOWCHART

Sheet 2 of 2

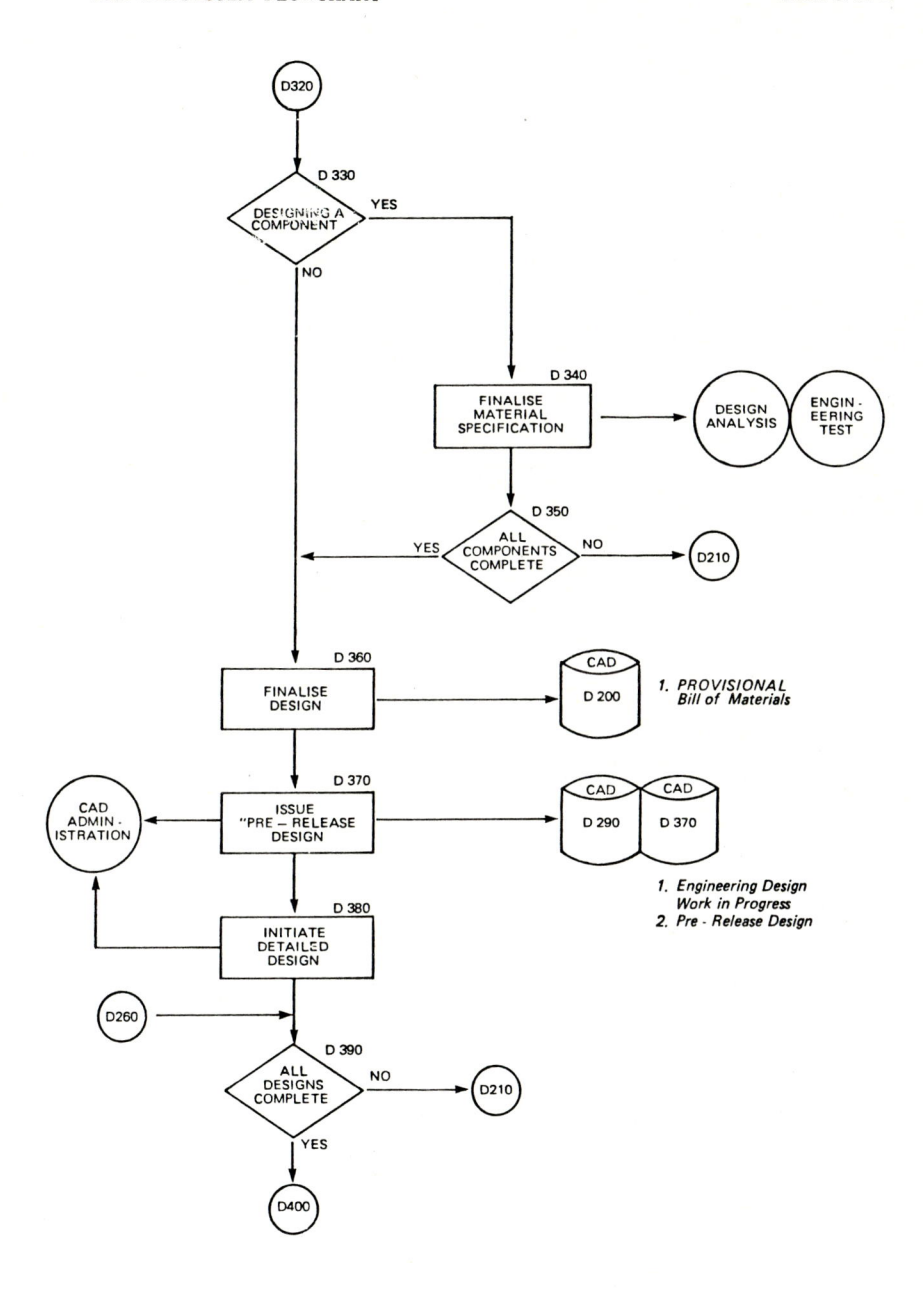

D2.3 Design Rules - Engineering Design

D2.3.1 Input Data

The sub-system must be capable of accepting the minimum data items listed below:

Data is required relating to:

- Design Proposal (Section D2.3.1.1)
- Previously Released Designs (Section D2.3.1.2)
- Digitised Geometry (Section D2.3.1.3)

D2.3.1.1 Information relating to design proposal

The Design Proposal will have similar data characteristics as those for previously released designs as described below in Section D2.3.1.2.

D2.3.1.2 Information relating to previous designs

- **Unique Identification**

    A unique identification, allocated or machine generated, which distinguishes the design from all other designs. A part number or traditional drawing number may be considered sufficient.

- **Geometric Data**

    A representation of the shape of a part in terms of a group of geometric entities. For example, a group of points, lines, curve functions, surfaces, etc. It is assumed that each type of entity will be used in a disciplined and identified manner.

- **Attribute Data**

    Additional attributes which would appear on a traditional engineering drawing such as linear or angular dimensions, diameters, surface finishes, tolerances, etc. The information would include only that which is necessary to enable the part to be manufactured within the constraints of the design. Some of the information will be implied by the geometric data but will not have been explicitly defined. It is assumed that each type of attribute will be used in a disciplined and identified manner.

- **Text Data**

    This refers to descriptive information which forms a fundamental part of the design. For example, special processes, assembly instructions, design title (eg. Connecting Rod), design notes, etc. It is assumed that, where any text item is related to a specific geometric entity, the associativity will be implemented in a disciplined and identified manner.

    It is also assumed that general textual descriptions relating to a whole design, rather than to a geometric entity within a design, will be associated with the unique identification of that design.

- **View Data**

    A collection of the information necessary to present a "picture" of the designed part to a human in an understandable and recognisable form. For example, general arrangements, layouts, technical illustrations, exploded views of assemblies, sections, isometric views. Such "pictures" may be in hard copy form (such as a pen plot) or for presentation on a CAD computer graphics display unit.

View data is not a mandatory design data item. However, where such a facility is provided, it must be clearly stated whether or not there is associativity between the geometric data and the view data. Since the view data is a "pictorial" representation of the designed part, associativity would imply that a modification to the geometric data would automatically cause the changes to be accounted for in the view data.

D2.3.1.3 **Information related to digitised geometry**

The sub-system should be capable of accepting digitised co-ordinates. Such data will typically be derived from three dimensional models or from two dimensional drawings or sketches. It may be considered sufficient if the sub-system takes the digitised co-ordinates and converts them into geometric point entities. However, where the sub-system offers facilities to mathematically fit, for example, curves or surfaces to the digitised co-ordinates, thus providing a more sophisticated representation of the geometry, it must do so in a disciplined and identified manner.

D2.3.2 **Input reformatting**

The sub-system must be capable of reformatting, as required, input data into the form required by the sub-system. It is assumed that the input data will be structured in a disciplined and identified manner.

D2.3.3 **Unique identification**

A unique identity, allocated or machine generated, which distinguishes the current design from all other designs on the same sub-system and from all previous designs (which may reside on other sub-systems). A part number or traditional drawing number may be considered sufficient.

D2.3.4 **Restriction of access**

The sub-system must be capable of restricting the access to specified design information. Restriction must be applicable to individual sub-system users, groups of those users, and other sub-systems. A form of hierarchical password protection may be considered sufficient.

D2.3.5 **Output data**

The sub-system must be capable of producing data to the minimum levels specified in this section

It must be in a form which will facilitate reformatting by a subsequent module, ie. all items must be structured in a disciplined and identified manner.

Data is required relating to:

- Bill Of Materials (Section D2.3.5.1)
- Pre-released design data (Section D2.3.5.2)

D2.3.5.1 **Information relating to bill of materials**

- **Product Identity**

    The unique identity of each product being considered for manufacture

- **Product Description**

    A title or description of the product to enable humans to recognise the entry in the bill of materials.

- **Constituent Parts**

The unique identities of all the details required to construct the product being considered for manufacture.

- **Part Descriptions**

    A title or description of each of the constituent parts to enable humans to recognise the entries in the bill of materials.

- **Quantity Required**

    For each detail within a product, the quantity of that detail required to manufacture one manufactured product.

D2.3.5.2 **Information relating to pre-released design data**

- **Unique Identification**

    A unique identification, allocated or machine generated, which distinguishes the design from all other designs. A part number or traditional drawing number may be considered sufficient.

- **Geometric Data**

    A representation of the shape of a part in terms of a group of geometric entities. For example, a group of points, lines, curve functions, surfaces, etc. It is assumed that each type of entity will be used in a disciplined and identified manner.

- **Attribute Data**

    Additional attributes which would appear on a traditional engineering drawing such as linear or angular dimensions, diameters, surface finishes, tolerances, etc. The information would include only that which is necessary to enable the part to be manufactured within the constraints of the design. Some of the information will be implied by the geometric data but will not have been explicitly defined. It is assumed that each type of attribute will be used in a disciplined and identified manner.

- **Text Data**

    This refers to descriptive information which forms a fundamental part of the design. For example, special processes, assembly instructions, design title (eg. Connecting Rod), design notes, etc. It is assumed that, where any text item is related to a specific geometric entity, the associativity will be implemented in a disciplined and identified manner.

    It is also assumed that general textual descriptions relating to a whole design, rather than to a geometric entity within a design, will be associated with the unique identification of that design.

- **View Data**

    A collection of the information necessary to present a "picture" of the designed part to a human in an understandable and recognisable form. For example, general arrangements, layouts, technical illustrations, exploded views of assemblies, sections, isometric views. Such "pictures" may be in hard copy form (such as a pen plot) or for presentation on a CAD computer graphics display unit.

    View data is not a mandatory design data item. However, where such a facility is provided, it must be clearly stated whether or not there is associativity between the geometric data and the view data. Since the view data is a "pictorial" representation of the designed part, associativity would imply that a modification to the geometric data would automatically cause the changes to be accounted for in the view data.

Detailed design sub-system (D3)

D3.1 **Sub-system description**

Detailed Design is concerned with identifying an exhaustive list of all the parts to be incorporated in the product and with providing a complete specification of each one, in sufficient detail to enable it to be manufactured.

Traditionally, specifications have consisted of sets of detailed drawings together with build and assembly instructions. It seems likely that the engineering drawing will remain in use for some time to come so that computer aided draughting systems will continue to be an important part of CAD. In the longer term, as greater use is made of computer readable CAD data as direct imput to control manufacturing plant and inspection equipment (CAM), the requirement for computer aided draughting will diminish. However, the activities described for Detailed Design can be applied in either of these short and long term cases.

The sub-system commences with the pre-released design which is a provisional design to nominal dimensions. Use is made of standard parts where possible and the design refined to account for possible production methods. Each detail is completely dimensioned and toleranced followed by specification of those production processes necessary for the integrity of the design.

The final output of the sub-system is a fully released design on which manufacture will be based together with a complete Bill of Materials.

Detailed design sub-system

D3.2 Flowchart

D3.2. SUB SYSTEM FLOWCHART

Sheet 1 Of 2

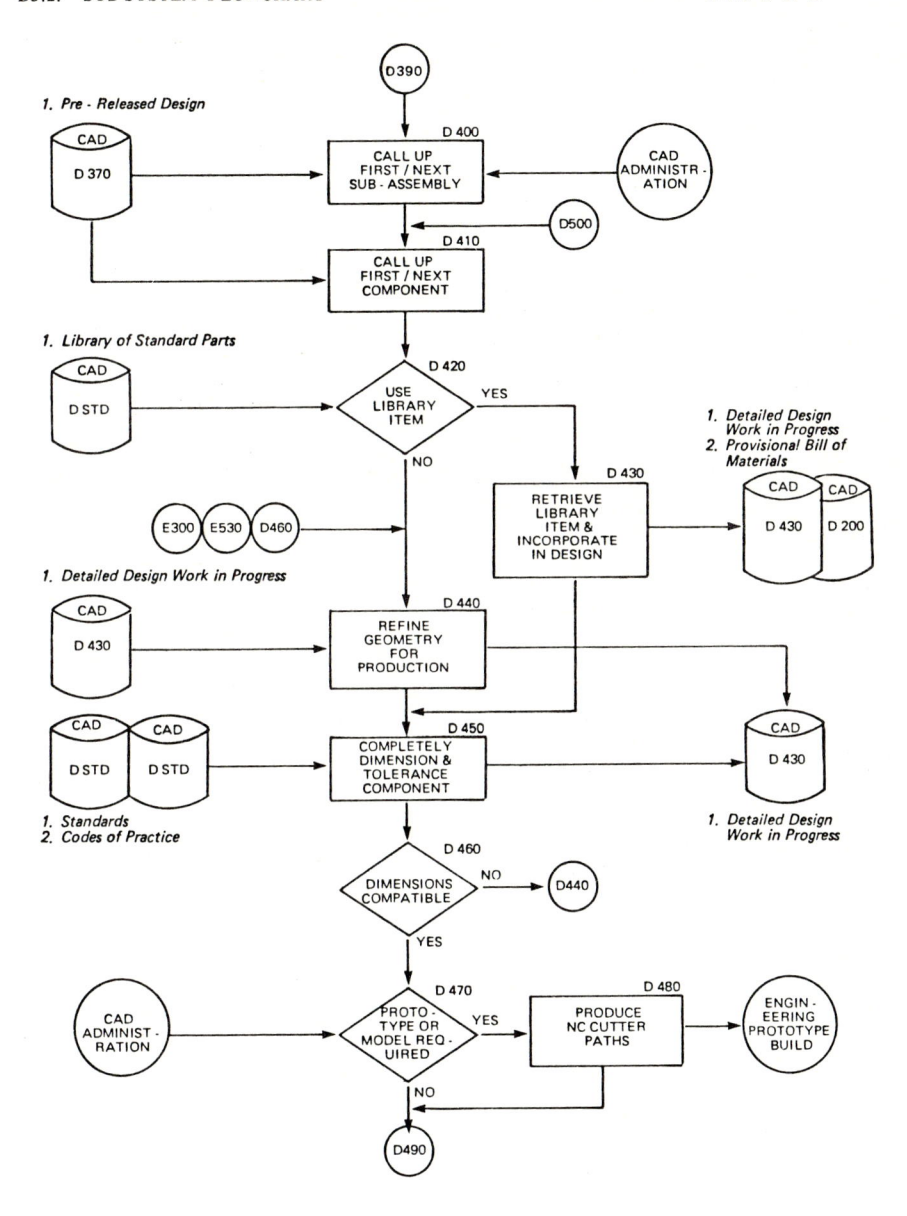

Computer Aided Design

D3.2. SUB SYSTEM FLOWCHART Sheet 2 of 2

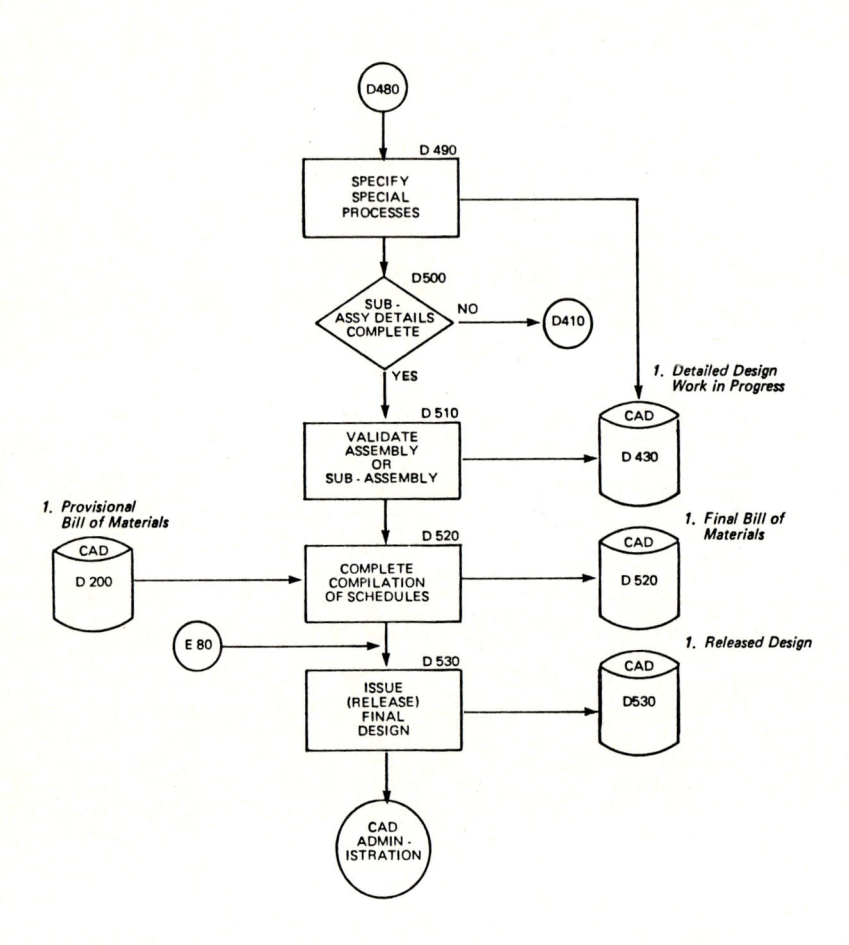

D3.3 Design rules - detailed design

D3.3.1 Input data

The sub-system must be capable of accepting the minimum data items listed below:

Data is required relating to:

- Pre-Released Designs (Section D3.3.1.1)

D3.3.1.1 Information relating to pre-released design data

- **Unique Identification**

    A unique identification, allocated or machine generated, which distinguishes the design from all other designs. A part number or traditional drawing number may be considered sufficient.

- **Geometric Data**

    A representation of the shape of a part in terms of a group of geometric entities. For example, a group of points, lines, curve functions, surfaces, etc. It is assumed that each type of entity will be used in a disciplined and identified manner.

- **Attribute Data**

    Additional attributes which would appear on a traditional engineering drawing such as linear or angular dimensions, diameters, surface finishes, tolerances, etc. The information would include only that which is necessary to enable the part to be manufactured within the constraints of the design. Some of the information will be implied by the geometric data but will not have been explicitly defined. It is assumed that each type of attribute will be used in a disciplined and identified manner.

- **Text Data**

    This refers to descriptive information which forms a fundamental part of the design. For example, special processes, assembly instructions, design title (eg. Connecting Rod), design notes, etc. It is assumed that, where any text item is related to a specific geometric entity, the associativity will be implemented in a disciplined and identified manner.

    It is also assumed that general textual descriptions relating to a whole design, rather than to a geometric entity within a design, will be associated with the unique identification of that design.

- **View Data**

    A collection of the information necessary to present a "picture" of the designed part to a human in an understandable and recognisable form. For example, general arrangements, layouts, technical illustrations, exploded views of assemblies, sections, isometric views. Such "pictures" may be in hard copy form (such as a pen plot) or for presentation on a CAD computer graphics display unit.

    View data is not a mandatory design data item. However, where such a facility is provided, it must be clearly stated whether or not there is associativity between the geometric data and the view data. Since the view data is a "pictorial" representation of the designed part, associativity would imply that a modification to the geometric data would automatically cause the changes to be accounted for in the view data.

D3.3.2 **Input reformatting**

The sub-system must be capable of reformatting, as required, input data into the form required by the sub-system. It is assumed that the input data will be structured in a disciplined and identified manner.

D3.3.3 **Unique identification**

A unique identity, allocated or machine generated, which distinguishes the current design from all other designs on the same sub-system and from all previous designs (which may reside on other sub-systems). A part number or traditional drawing number may be considered sufficient.

D3.3.4 **Restriction of access**

The sub-system must be capable of restricting the access to specified design information. Restriction must be applicable to individual sub-system users, groups of those users, and other sub-systems. A form of hierarchical password protection may be considered sufficient.

D3.3.5 **Dimensions and tolerances**

The sub-system must be capable of carrying out full dimensioning and tolerancing to stated, identified standards.

D3.3.6 **Output data**

The sub-system must be capable of producing data to the minimum levels specified in this section

It must be in a form which will facilitate reformatting by a subsequent module, ie. all items must be structured in a disciplined and identified manner.

Data is required relating to:

- Final Bill Of Materials (Section D3.3.6.1)
- The Issued (Released) Design (Section D3.3.6.2)
- NC Cutter Paths (Section D3.3.6.3)

D3.3.6.1 **Information relating to bill of materials**

- **Product Identity**

    The unique identity of each product being considered for manufacture

- **Product Description**

    A title or description of the product to enable humans to recognise the entry in the bill of materials.

- **Constituent Parts**

    The unique identities of all the details required to construct the product being considered for manufacture.

- **Part Descriptions**

    A title or description of each of the constituent parts to enable humans to recognise the entries in the bill of materials.

- **Quantity Required**

    For each detail within a product, the quantity of that detail required to manufacture one manufactured product.

D3.3.6.2 **Information relating to issued (released) design data**

Computer Aided Design

- **Unique Identification**

    A unique identification, allocated or machine generated, which distinguishes the design from all other designs. A part number or traditional drawing number may be considered sufficient.

- **Geometric Data**

    A representation of the shape of a part in terms of a group of geometric entities. For example, a group of points, lines, curve functions, surfaces, etc. It is assumed that each type of entity will be used in a disciplined and identified manner.

- **Attribute Data**

    Additional attributes which would appear on a traditional engineering drawing such as linear or angular dimensions, diameters, surface finishes, tolerances, etc. The information would include only that which is necessary to enable the part to be manufactured within the constraints of the design. Some of the information will be implied by the geometric data but will not have been explicitly defined. It is assumed that each type of attribute will be used in a disciplined and identified manner.

- **Text Data**

    This refers to descriptive information which forms a fundamental part of the design. For example, special processes, assembly instructions, design title (eg. Connecting Rod), design notes, etc. It is assumed that, where any text item is related to a specific geometric entity, the associativity will be implemented in a disciplined and identified manner.

    It is also assumed that general textual descriptions relating to a whole design, rather than to a geometric entity within a design, will be associated with the unique identification of that design.

- **View Data**

    A collection of the information necessary to present a "picture" of the designed part to a human in an understandable and recognisable form. For example, general arrangements, layouts, technical illustrations, exploded views of assemblies, sections, isometric views. Such "pictures" may be in hard copy form (such as a pen plot) or for presentation on a CAD computer graphics display unit.

    View data is not a mandatory design data item. However, where such a facility is provided, it must be clearly stated whether or not there is associativity between the geometric data and the view data. Since the view data is a "pictorial" representation of the designed part, associativity would imply that a modification to the geometric data would automatically cause the changes to be accounted for in the view data.

D3.3.6.3   **Information relating to NC cutter paths**

    Numerical Control cutter location data should be output to conform to recognised, stated standards.

Chapter 4

Computer Aided Production Engineering

Introduction

Computer Aided Production Engineering is the process of deciding how to manufacture a product that has been specified by the design function.

The broad activities that the production engineering function embraces include the following main areas:

a) Process Planning
b) Plant Layout
c) Part-Programming
d) Production Tool and Fixture Design
e) Material Handling

Process Planning

- This is the activity of deciding which manufacturing processes and machines should be used to perform the various operations necessary to produce a component, and the sequence that the processes should follow.

Plant Layout

- This is the activity of ensuring that the manufacturing facilities required in a Company are sufficient to produce the volume of articles required and that they are positioned in such a manner that the flow and rate of production is optimised. The use of simulation techniques are highly desirable in proving the efficiency of a plant layout proposal.

Part-Programming

- This is the activity concerned with producing the instruction set required by NC machines to machine a component in the prescribed manner determined by the production engineering function.

Production Tool and Fixture Design

- This is the activity concerned with designing and producing all of the machines, cutting tools, holding jigs and fixtures, and other ancillary equipment required to manufacture the components to the required design specification.

Material Handling

- This is the activity concerned with the design, manufacture and control of the equipment required to store, transport, load and unload manufacturing processes. This includes computer controlled systems such as warehouses, automatic guided vehicles, and robotic pick and place systems.

    For the one-year pilot study within the ESPRIT project, it was necessary to consider and develop design rules for only a sub-set of the activities described above, because of the amount of work involved. For this reason, the Computer Aided Production Engineering (CAPE) Section that follows has selected and concentrated on the Process Planning, and Production Tool and Fixture Design activities. The remaining areas should form part of a future ESPRIT Project.

Computer Aided Production Engineering

Flowcharts

COMPUTER AIDED PRODUCTION ENGINEERING

Sheet 1 of 7

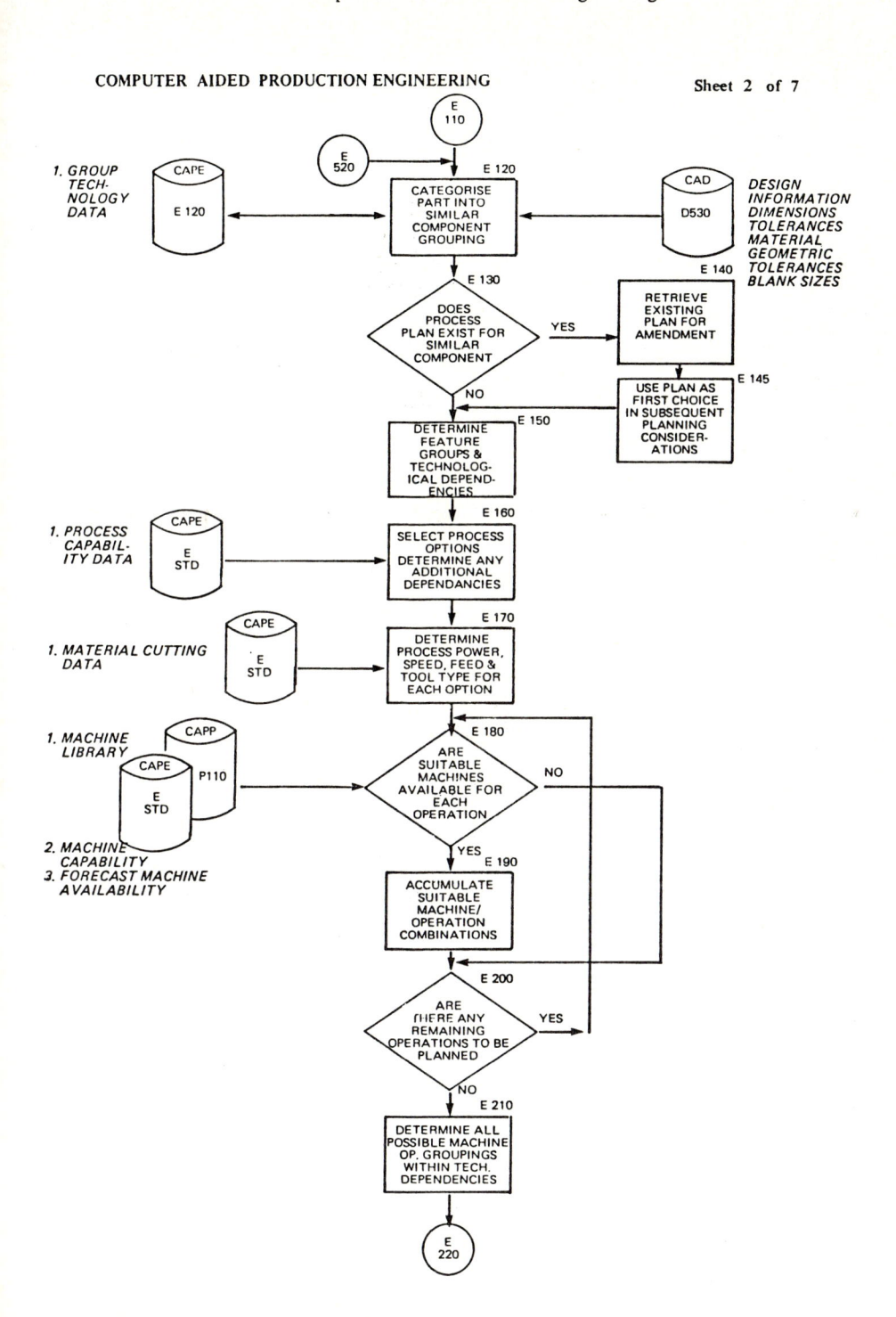

Computer Aided Production Engineering

COMPUTER AIDED PRODUCTION ENGINEERING　　　　　　　　　　Sheet 2 of 7

```
                              ( E 110 )
                                 │
                    ( E 520 ) ──→│
                                 ↓  E 120
    1. GROUP         ┌─CAPE─┐   ┌──────────────┐   ┌─CAD─┐   DESIGN
       TECH-         │      │   │  CATEGORISE  │   │     │   INFORMATION
       NOLOGY        │ E 120│──→│   PART INTO  │←──│ D530│   DIMENSIONS
       DATA          │      │   │    SIMILAR   │   │     │   TOLERANCES
                     └──────┘   │  COMPONENT   │   └─────┘   MATERIAL
                                │   GROUPING   │                GEOMETRIC
                                └──────────────┘       E 140  TOLERANCES
                                       │                      BLANK SIZES
                                       ↓  E 130
                                  ╱─────────╲         ┌──────────────┐
                                 ╱   DOES    ╲        │   RETRIEVE   │
                                ╱   PROCESS   ╲ YES   │   EXISTING   │
                               ⟨ PLAN EXIST FOR⟩─────→│    PLAN FOR  │
                                ╲   SIMILAR   ╱       │  AMENDMENT   │
                                 ╲ COMPONENT ╱        └──────────────┘
                                  ╲─────────╱                │
                                     │ NO                    │ E 145
                                     │           ┌──────────────────┐
                                     │           │   USE PLAN AS    │
                                     │    E 150  │   FIRST CHOICE   │
                                     │←──────────│  IN SUBSEQUENT   │
                                     ↓           │    PLANNING      │
                            ┌─────────────────┐  │  CONSIDER-       │
                            │   DETERMINE     │  │    ATIONS        │
                            │    FEATURE      │  └──────────────────┘
                            │   GROUPS &      │
                            │  TECHNOLOG-     │
                            │  ICAL DEPEND-   │
                            │    ENCIES       │
                            └─────────────────┘
                                     │
                                     ↓  E 160
    1. PROCESS       ┌─CAPE─┐   ┌──────────────┐
       CAPABIL-      │  E   │   │SELECT PROCESS│
       ITY DATA      │ STD  │──→│   OPTIONS    │
                     └──────┘   │DETERMINE ANY │
                                │  ADDITIONAL  │
                                │ DEPENDANCIES │
                                └──────────────┘
                                     │
                                     ↓  E 170
    1. MATERIAL      ┌─CAPE─┐   ┌──────────────┐
       CUTTING       │  E   │   │  DETERMINE   │
       DATA          │ STD  │──→│PROCESS POWER,│
                     └──────┘   │ SPEED, FEED &│
                                │ TOOL TYPE FOR│
                                │ EACH OPTION  │
                                └──────────────┘
                                     │
                                     ↓  E 180
    1. MACHINE     ┌─CAPP─┐         ╱─────────╲
       LIBRARY     │      │        ╱    ARE    ╲
                 ┌─CAPE─┐ │       ╱   SUITABLE  ╲    NO
                 │  E   │P110──→ ⟨   MACHINES    ⟩────→
                 │ STD  │ │       ╲ AVAILABLE FOR╱
                 └──────┘─┘        ╲    EACH    ╱
                                    ╲OPERATION ╱
    2. MACHINE                       ╲────────╱
       CAPABILITY                       │ YES
    3. FORECAST MACHINE                 ↓  E 190
       AVAILABILITY              ┌──────────────┐
                                 │  ACCUMULATE  │
                                 │   SUITABLE   │
                                 │   MACHINE/   │
                                 │  OPERATION   │
                                 │ COMBINATIONS │
                                 └──────────────┘
                                        │
                                        ↓  E 200
                                   ╱─────────╲
                                  ╱    ARE    ╲   YES
                                 ╱  THERE ANY  ╲─────→
                                ⟨   REMAINING   ⟩
                                 ╲OPERATIONS TO╱
                                  ╲ BE PLANNED╱
                                   ╲─────────╱
                                        │ NO
                                        ↓  E 210
                                 ┌──────────────┐
                                 │DETERMINE ALL │
                                 │POSSIBLE MACH.│
                                 │ OP. GROUPINGS│
                                 │ WITHIN TECH. │
                                 │ DEPENDENCIES │
                                 └──────────────┘
                                        │
                                        ↓
                                     ( E 220 )
```

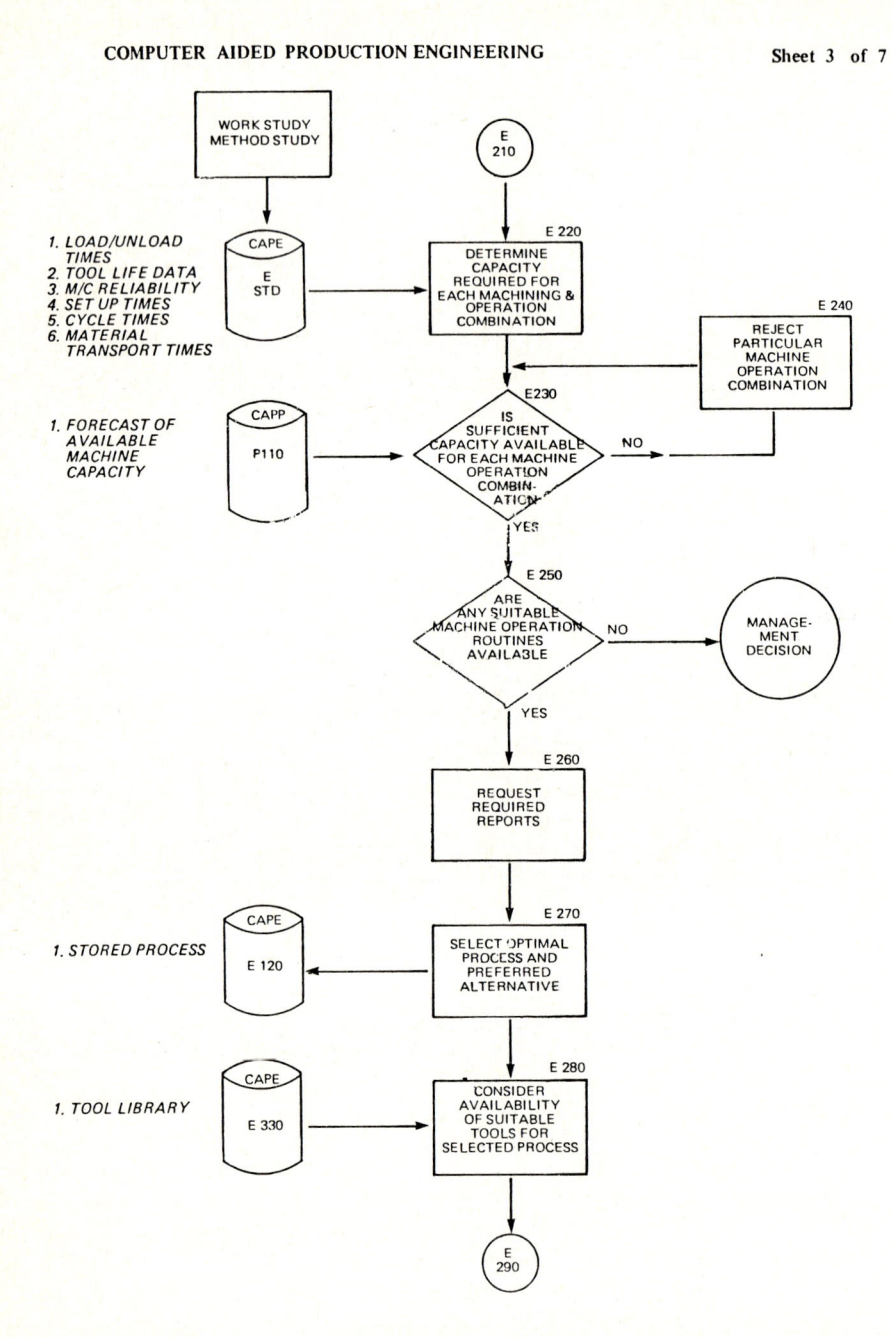

COMPUTER AIDED PRODUCTION ENGINEERING   Sheet 3 of 7

- WORK STUDY METHOD STUDY
- E 210

1. LOAD/UNLOAD TIMES
2. TOOL LIFE DATA
3. M/C RELIABILITY
4. SET UP TIMES
5. CYCLE TIMES
6. MATERIAL TRANSPORT TIMES

CAPE — E STD

E 220 — DETERMINE CAPACITY REQUIRED FOR EACH MACHINING & OPERATION COMBINATION

E 240 — REJECT PARTICULAR MACHINE OPERATION COMBINATION

1. FORECAST OF AVAILABLE MACHINE CAPACITY

CAPP — P110

E 230 — IS SUFFICIENT CAPACITY AVAILABLE FOR EACH MACHINE OPERATION COMBINATION — NO / YES

E 250 — ARE ANY SUITABLE MACHINE OPERATION ROUTINES AVAILABLE — NO → MANAGEMENT DECISION / YES

E 260 — REQUEST REQUIRED REPORTS

1. STORED PROCESS

CAPE — E 120

E 270 — SELECT OPTIMAL PROCESS AND PREFERRED ALTERNATIVE

1. TOOL LIBRARY

CAPE — E 330

E 280 — CONSIDER AVAILABILITY OF SUITABLE TOOLS FOR SELECTED PROCESS

E 290

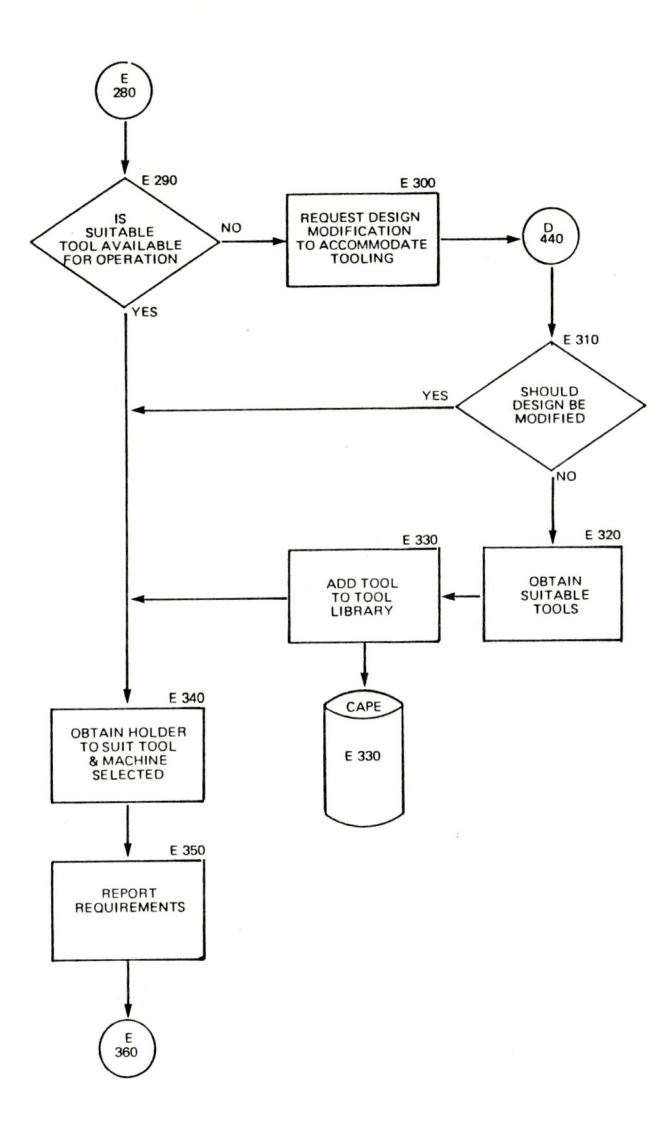

COMPUTER AIDED PRODUCTION ENGINEERING Sheet 4 of 7

COMPUTER AIDED PRODUCTION ENGINEERING

Sheet 5 of 7

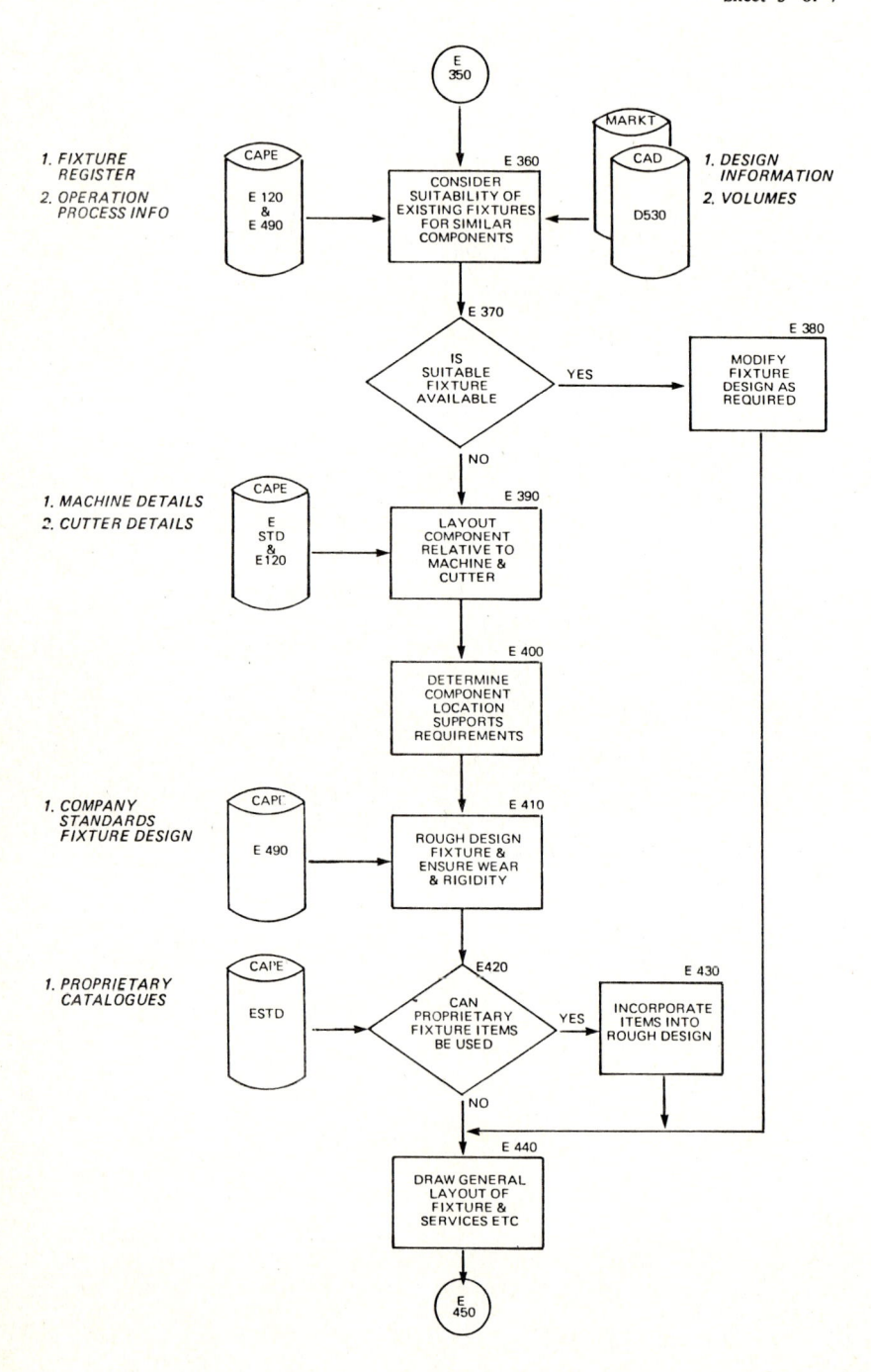

```
                              ( E 350 )
                                 │
                                 ▼
  1. FIXTURE                 ┌─────────┐          ┌─────────┐
     REGISTER                │  CAPE   │   E 360  │  MARKT  │   1. DESIGN
  2. OPERATION               │  E 120  │──┐    ┌──│   CAD   │      INFORMATION
     PROCESS INFO            │    &    │  ▼    ▼  │  D530   │   2. VOLUMES
                             │  E 490  │  CONSIDER│         │
                             └─────────┘  SUITABILITY OF
                                          EXISTING FIXTURES
                                          FOR SIMILAR
                                          COMPONENTS
                                                 │
                                               E 370
                                                 ▼
                                           ╱ IS       ╲              ┌──────────┐
                                          ╱ SUITABLE   ╲    YES      │  E 380   │
                                          ╲ FIXTURE    ╱ ──────────► │ MODIFY   │
                                           ╲ AVAILABLE╱              │ FIXTURE  │
                                                │                    │ DESIGN AS│
                                                │ NO                 │ REQUIRED │
                                                ▼                    └──────────┘
  1. MACHINE DETAILS        ┌─────────┐   E 390
  2. CUTTER DETAILS         │  CAPE   │   LAYOUT
                            │    E    │   COMPONENT
                            │   STD   │──►RELATIVE TO
                            │    &    │   MACHINE &
                            │  E120   │   CUTTER
                            └─────────┘     │
                                          E 400
                                            ▼
                                       DETERMINE
                                       COMPONENT
                                       LOCATION
                                       SUPPORTS
                                       REQUIREMENTS
                                            │
  1. COMPANY               ┌─────────┐    E 410
     STANDARDS             │  CAPE   │
     FIXTURE DESIGN        │  E 490  │──► ROUGH DESIGN
                           └─────────┘    FIXTURE &
                                          ENSURE WEAR
                                          & RIGIDITY
                                            │
  1. PROPRIETARY           ┌─────────┐    E 420                   E 430
     CATALOGUES            │  CAPE   │     ╱  CAN  ╲         ┌──────────┐
                           │  ESTD   │──► ╱ PROPRIETARY╲ YES │INCORPORATE│
                           └─────────┘    ╲ FIXTURE ITEMS╱──►│ ITEMS INTO│
                                           ╲ BE USED  ╱     │ROUGH DESIGN│
                                              │NO           └──────────┘
                                              ▼                   │
                                            E 440 ◄───────────────┘
                                          DRAW GENERAL
                                          LAYOUT OF
                                          FIXTURE &
                                          SERVICES ETC
                                              │
                                              ▼
                                           ( E 450 )
```

COMPUTER AIDED PRODUCTION ENGINEERING

Sheet 6 of 7

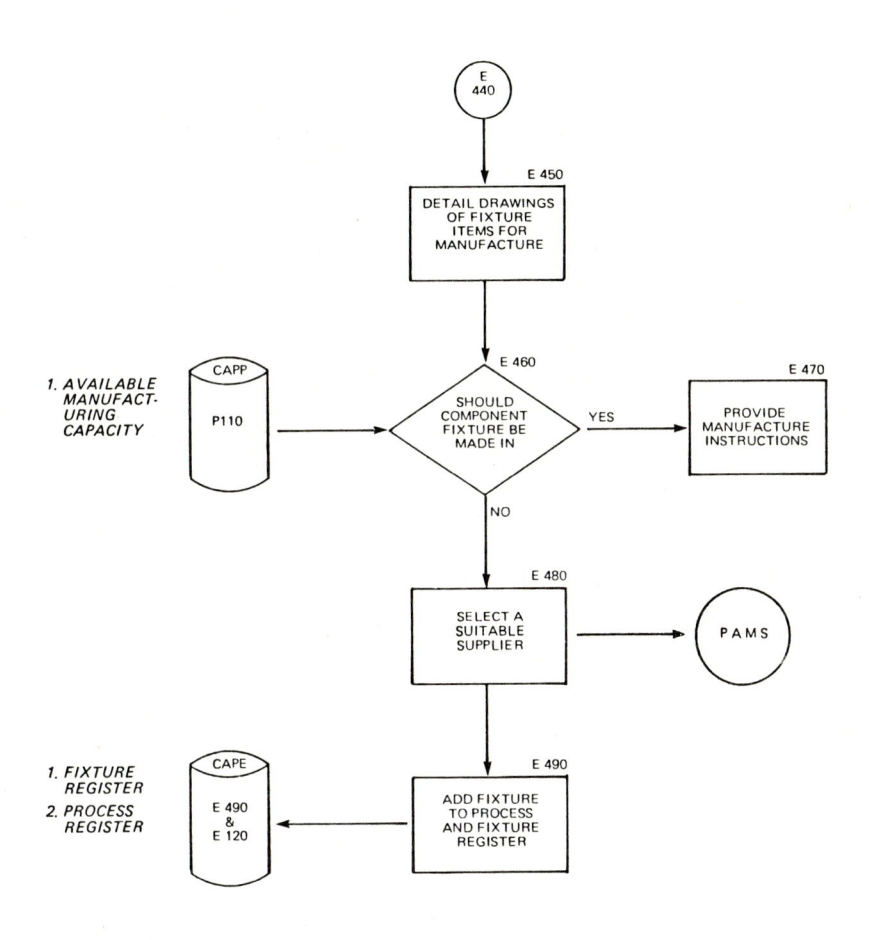

```
                            ( E 440 )
                               │
                               ▼                E 450
                      ┌──────────────────┐
                      │ DETAIL DRAWINGS  │
                      │    OF FIXTURE    │
                      │   ITEMS FOR      │
                      │   MANUFACTURE    │
                      └──────────────────┘
                               │
                               ▼                E 460
                    ╱╲
   1. AVAILABLE   (CAPP)      SHOULD                        E 470
      MANUFACT-   ┌────┐    COMPONENT      YES      ┌──────────────┐
      URING       │P110│──▶  FIXTURE BE   ─────▶   │   PROVIDE    │
      CAPACITY    └────┘     MADE IN                │ MANUFACTURE  │
                              ╲╱                    │ INSTRUCTIONS │
                               │ NO                 └──────────────┘
                               ▼                E 480
                      ┌──────────────────┐
                      │    SELECT A      │           ╱      ╲
                      │    SUITABLE      │─────▶   (  PAMS  )
                      │    SUPPLIER      │           ╲      ╱
                      └──────────────────┘
                               │
   1. FIXTURE    (CAPE)        ▼                E 490
      REGISTER   ┌────┐  ┌──────────────────┐
   2. PROCESS    │E 490│◀─│  ADD FIXTURE    │
      REGISTER   │  &  │  │  TO PROCESS     │
                 │E 120│  │  AND FIXTURE    │
                 └────┘  │    REGISTER     │
                          └──────────────────┘
```

COMPUTER AIDED PRODUCTION ENGINEERING

Sheet 7 of 7

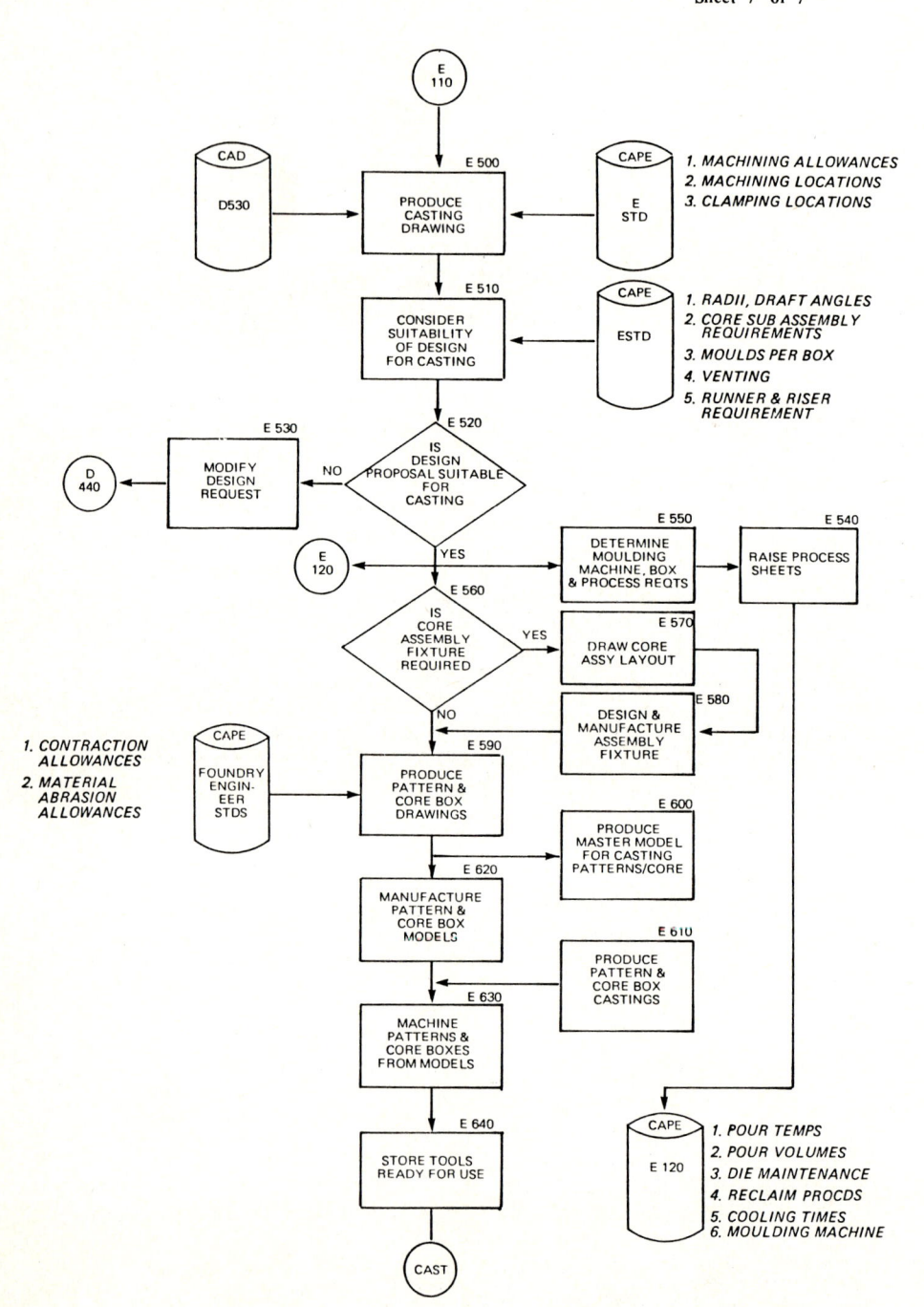

Narrative description of
Computer Aided Production Engineering (CAPE) network

E10   Very often an Engineering design is handed to the Manufacturing functions of a company with no consideration being given to the manufacturing difficulties that the design may create. It is considered necessary that a design proposal should be evaluated by the Manufacturing function for two purposes; namely:-

    (a)   To review the practical alternative technologies that may be used by a company to manufacture the component.

    (b)   To remove design attributes that will create manufacturing difficulties, which, after discussion with the Design function, are considered to be unimportant and changeable.

    The information used to review the alternative manufacturing technologies considered practical by the company will include a knowledge of the existing machines, facilities and equipment already available. Obviously, however, the existence of machines does not necessarily mean that sufficient capacity will be available for the production of the required volume of components. It is therefore necessary for a knowledge of the spare capacity to be made available by the Production Planning function, over the planned time horizon for the component.

E20   The decision process is concerned with agreeing with the Design function that a feature will be redesigned. If it is agreed that the design is to be modified, then the request will be passed to a formal procedure within the Design function and further evaluation of that design will only continue with a formal response.

E30   To prevent the possibility of various company functions, including Production Engineering, working on design information that may have been changed by the Design function, then a formal 'raise of modification' procedure must be adopted by the Design function and used in the notification of amendment information.

E40   When a design is considered to be acceptable regarding a particular manufacturing technology, then an evaluation of the effect upon the company by adopting that approach must be made. This will include such factors as the cost of new plant and facilities, the time required to commence production of the component, the manpower implications, etc.

E50   The evaluation process carried out in node E40 must be repeated for any other possible manufacturing technologies that are considered to be sensible options by the company.

E60   Once an evaluation of the effects and implications to the company has been made for each sensible option, then these must be reported for comparison and consideration by the company management in order to decide which manufacturing technology should be adopted.

E70   Once the best option has been chosen according to the decision criteria considered important by the company, then action can start to be taken to finalise the design details and to start to order new facilities as required. The subject of the Design Rules required for the plant layout activity will not be addressed by this Esprit pilot project.

E100   The pilot study within the Esprit project was confined to the production of design rules for the activities concerned with the manufacture of 'machined' components. Other methods of manufacture should be addressed by future Esprit projects.

E110	The activities concerned with the manufacture of 'machined' components may be carried out on a preformed item that is required to be produced, rather than using standard raw material. For the sake of completeness, the network has been produced to use casting as an example of the preformed component. It is recognised that other processes may be employed to produce the preformed item and these should be the subject of future Esprit projects.
E120	It is obviously desirable to use previous experience in planning the manufacture of components that are similar to the detail under consideration. However it may be quite impossible to trace the documentation providing that experience from the use of keys such as description, or part number of the new item. The increasingly popular solution to this difficulty is to code the item into a category of parts that have similar manufacturing or design characteristics. The retrieval of past experience for similar components then becomes a relatively simple activity.
E130	Using the classification code generated in Node E120, a search can be made to see if documentation exists that may provide useful previous experience.
E140	If a previous plan exists for a similar component, this will be retrieved and used as an indication of how the new item might be processed. Practical constraints existing on the machine shop, as well as design variations between the components, may require the plan to be amended. In particular, there may not be sufficient spare machine capacity on the machines used by a previous plan to allow the process to be feasible for the new item. For this reason, the previous experience should only be used as a guide for the subsequent planning processes.
E150	A preliminary task in determining a process plan is to identify the feature groups that are to be machined (see Glossary of Terms for Feature Groups). For example a tapped hole may be regarded as a feature group which comprises two features of a drilled hole, and a subsequent dependent feature of a thread. Alternatively, each of the features could be regarded as separate feature groups in their own right.
	In addition to the feature groups, it is necessary to specify technological dependencies between the feature groups. This will ensure that the suggested processes that are to be used to produce the detail are correctly ordered. For example, it may be necessary to produce a datum before a subsequent feature is produced that has a geometric tolerance dependency upon the datum. Similarly, it is necessary to produce a hole before it can be tapped.
E160	In addition to the feature group description, a complete set of information describing the attributes of the feature is also required. This information will be matched against the process capability of various processes, and a selection of options made. The options are required (if available) to allow for an alternative process to be considered if the prime selection is not available for some reason when specific machines are being assigned. It is sensible for the options to be presented in a preferred ranked order according to some criteria considered important by the company. For example, a hole to a given accuracy might be formed by either drilling or the much more accurate process of jig boring. The preferred selection might be the cheaper process of drilling, but if that machinery is fully utilised with no spare capacity, then jig boring might become a very sensible option.
E170	Once a process and its options have been selected and ordered, it is possible from standard cutting data to determine the upper limit of machining conditions including tool type, depth of cut, cutting speeds, etc. This information together with the ranked process derived in node E160 can now be passed to a Machine Selection Routine commencing at node E180.

E180   A process which has been determined in E160 will be selected and matched against the process capability in the machine tool library.

Each process list selected will carry with it a "desirability factor" (generated at E160) which depends upon its suitability to perform the particular operation. By this means it becomes possible to explore all the ways of producing a feature (or feature group) and to rank them in respect of stated criteria. An example of the value of this approach is the identification that a hole may be produced on either a drilling machine or a jig borer. Under normal circumstances the cheaper drilling machine will be chosen, but in the absence of spare capacity the jig borer can be used.

If none of the listed machine tools has the capacity of performing the selected process a warning must be issued

E190   Each process with its associated list of capable machine tools will be accumulated at this stage. This activity will yield the following information:

- Detail identity
- Feature groups identity
- Process
- List of all machine tools which are capable of performing the selected process.

E200   The next machining process will be selected and the above procedure repeated. The iteration will continue until all processes have been completed.

E210   A process may be performed on more than one machine so it is necessary at this stage to collect the information generated at E190 into a set of combinations of the following form:

Operation/Capable machine tools

Operations will then be ordered into an acceptable sequence in accordance with manufacturing requirements and practice. The presence of Technological Dependencies (specified in E150) will govern the sequence of some operations.

Having established the sequence of operations, all the possible combinations of machines which can perform those operations will be assembled to form a set of possible sequences. Each combination of machines will be presented with an associated overall desirability factor derived from the individual desirability factors generated in E160.

E220   It will be necessary at this stage to calculate the time required to perform an operation on a particular machine tool; and this must be done for every combination of operation/machine tool.

The operation time will be calculated from the following information:

- Workpiece material and condition
- Cutting tool material and type
- Machine tool
- Work study synthetic times
- Standard cutting speed, feed and depth of cut for operation/material/tool combinations.

NOTE: This facility would also be used by stage E170 in order to derive an initial approximation of the cutting speed, feed and power requirements for use in the machine selection procedure.

The individual operation times will be summed to establish the total time that a detail would spend on a particular machine tool. This value will be multiplied by the expected volume of production from CAPP/P110 to determine the machine capacity requirement.

The process is repeated for all the machine tools involved to provide a forecast of machine capacity required to produce the selected detail.

Additional information which can be generated at this stage may include:
- Forecast of cutting tool life and consumption
- Forecast of material transport requirements and times
- Estimated machine tool maintenance requirements

E230 The machine capacity requirement for each combination of operations will be compared with the forecast of spare machine capacity (from CAPP/P110) in the following way:

Operation:	Capable Machine Tools:
1	M1, M2, M5, M8
2	M1, M2, M8
3	M4, M7
4	M3
5	M1, M2, M6

Sum of Operations   Machine Combinations Possible:
produces a detail
     M1, M2, M5)
     M1, M1, M1)
     M4, M4, M4)------etc.
     M3, M3, M3)
     M1, M1, M1)

Sequence No: (1) (2) (3) (4) (5) (6) (7)...(N)

Consider Sequence No.1:

The capacity requirement of each machine is compared with the forecast available capacity. A sequence in which all machines have sufficient available capacity will be listed under the heading of "Suitable Sequence".

E240 If any machine has insufficient available capacity the sequence being considered will be rejected by E230 and listed under the heading of "Rejected Sequence".

E250 All "Suitable Sequences" are ranked in the order of their overall desirability factors.

If a suitable sequence does not exist a warning will be reported. Such a situation can arise in either of two ways:
1   None of the machine tools held in the library is capable of performing a particular operation. It will therefore be necessary for management to decide if the purchase of a machine tool to perform that operation is justified.

2   If one or more of the machines in a sequence does not have forecasted available capacity to satisfy the requirements of that sequence.

E260   Under normal circumstances a ranked list of Suitable Sequences will be reported.

In the absence of any Suitable Sequence owing to lack of machine capacity, it is possible to recall the ranked list of Rejected Sequences. From this list it may be possible, for example, to upgrade a sequence from Rejected to Suitable by overtime working on one or more machines.

E270   From the list of Suitable Sequences generated in E260 the highest ranked sequence is considered the optimum solution. A number of sequences of lesser rank will be chosen as "Preferred Alternative Sequences".

All these sequences are then transferred to the E120 information file for use by the CAM function.

Preferred Alternative Sequences are included as a safeguard against the breakdown of machine tools or the shortages of tooling and other unpredictable transient events.

E280   Once a machine routing has been selected, the next operation is to select suitable cutting tools from the Company Library. The type of tools required will have been specified in the process selection sub-system (E150 - E170) for the alternative process options being proposed, and finalised once a specific machine is selected

E290   The type of tool required for an operation is compared with the library of standard tools held by the Company. If a tool exists that satisfies the cutting characteristics but which varies dimensionally from the required tool, then it is possible that a modification to the size of the component will allow the standard tool to be used rather than incur Company expense in the purchase of new tools.

E300   If a possible dimensional change to the component will allow a standard Company tool to be used, then a modification request will be sent to the Design function.

E310   If a decision is taken to modify the design of the component to accommodate the use of standard tools, then those tools will be assigned to the process plan for the operation being considered.

E320   If a design cannot be modified, then tools have to be purchased.

E330   New purchased tools must be added to the Company Tool Library.

E340   Once a tool has been selected for a particular operation, then this information together with a knowledge of the machine tool to be used will allow the appropriate tool holder to be selected from the Company inventory and assigned to the Process Plan.

E350   Once the tools and holder have been selected to suit the planned machinery operations, these are added to the Process Plan ready for use by other Company functions.

E360   Once a Process Plan has been derived to route a component through a series of machinery operations, it is necessary to design fixtures to hold the component in the correct machining position. The first activity in this process should be to determine from previous experience whether a fixture had been manufactured for a similar component that may aid the design of the newly required fixture. The most efficient method of retrieving experience is to code the component into a category of parts that have similar characteristics, and to examine the fixtures previously designed for that category.

E370	Once previous fixture designs have been retrieved, these can be examined to determine whether an existing fixture can be utilised.
E380	If an existing fixture can be modified to accommodate the component, then the time required to design the fixture may be substantially reduced.
	Substantial savings may also be made in the total operating efficiency of the manufacturing plant, if the fixture can be designed and modified to be universally usable for a whole family of components using the same machines.
E390	If it is necessary to design a new fixture, the first task is to layout the component in its correct relevant position on the selected machine bed, in its correct attitude to the selected cutting tool. The purpose of this is to provide the fixture designer with the working space that is available for the fixture to occupy.
E400	Once the component has been correctly positioned, it is necessary to decide how to support the item. This will obey the principles of restraining the six degrees of freedom.
E410	The fixture can be designed to fit into the available space generated by node E390 and meeting the support requirements considered by node E400.
E420	To minimise the eventual task of manufacturing the fixture, it may be desirable to utilise proprietary fixture items already purchased by the Company. These will be matched against the rough design concept generated by node E410 and wherever usable, will be allocated to the final fixture design.
E440	In addition to the holding and clamping aspects of the fixture, it may be necessary to provide services, such as air lines, to operate the fixture. To ensure that no foul situation exists, it is necessary to draw a general layout showing the position of the services.
E460	Once the fixture has been designed it is then necessary to decide whether the item should be manufactured within the Company or sub-contracted. The decision to manufacture internally will depend upon the available machine capacity of the production plant.
E490	Once the fixture has been designed it will be recorded on a fixture register and also added to the Process Plan for the component.
E500 to E640	This section of the flowchart is outside the terms of reference of the ESPRIT first-year pilot study, which have confined the work to machining activities. The chart has been included in the report to illustrate that other activities, such as casting considerations, must be taken into account before machine process planning can commence.

Computer Aided Production Engineering (CAPE) sub-systems

The Flowchart for the Production Engineering activities in the ESPRIT Pilot Study has been broken down into sub-systems:

1 Evaluation of 'design for manufacturing' requirements
    (Nodes E10 to E80)

2 Product coding/classification and retrieval
    (Nodes E120 to E145)

3 Process selection
    (Nodes E150 to E170)

4 Machine and routing selection
    (Nodes E180 to E270)

5 Cutting tool selection
    (Nodes E280 to E350)

6 Fixture design
    (Nodes E360 to E490)

Design for manufacturing evaluation sub-system (E1)

E1.1 **Sub-system description**

Once a Company has decided to design and produce a product that has been recognised by the Marketing function as saleable, then it becomes necessary to design the product and determine how it should be manufactured. However, the cost of manufacturing the product can be greatly influenced by the detailed design, and in order to ensure that the best course of action is taken by the Company, it is necessary for the Design function and the Production Engineering function to jointly evaluate the 'Design for Manufacture' concepts from initial rough design proposals through to finished detail design.

This sub-system is concerned with the formal evaluation of the costs of alternative manufacturing technologies that might be considered usable by the Company. It will keep track of the design modifications requested by the manufacturing function and the cost implications resulting from the amendments. The successful design modifications will of course feed back to provide design principles for the Company.

Once the design concepts are satisfactory, the sub-system will allow estimates to be made by the Production Engineering function of the times required to manufacture the proposal through each process of each of the technologies being considered. These times, together with a knowledge of the intended marketing volumes, will allow estimates to be made of the required capacity which will be compared with available capacity of existing facilities, over the forecast production time horizon of the product. The capacities considered will not only be those of machinery, but will also cover other factors such as manpower, skill, etc., considered important by the user.

Finally, the sub-system will allow the accumulation of costs, timing information, and subjective factors in a manner considered important by the user to produce a capital and operating cost parameter that will allow comparison of the merits of the alternative design proposals and technologies being considered.

Design for manufacturing evaluation sub-system

E1.2 Flowchart

E 1.2 SUB - SYSTEM FLOWCHART   Sheet 1 of 1

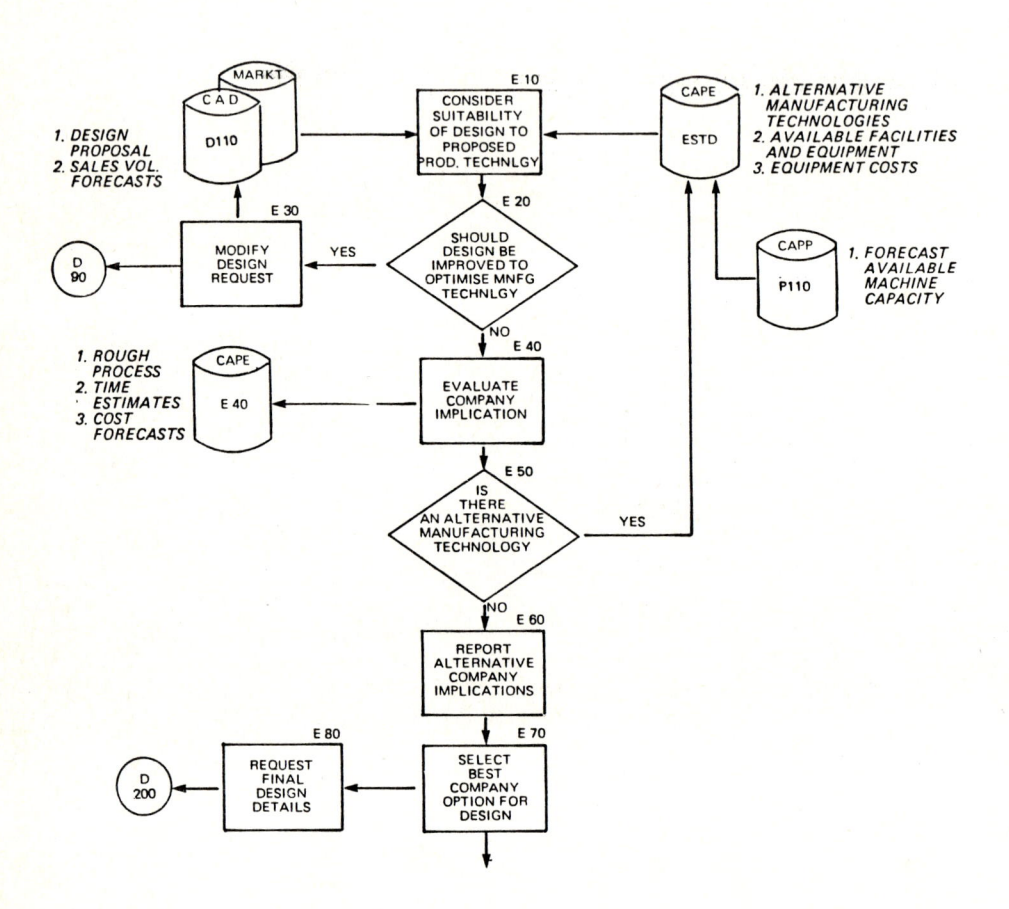

E1.3 **Design rules - design for manufacturing evaluation**

E1.3.1 **Subjective Factors**

The sub-system must provide a mechanism to allow an evaluation to be made of subjective factors that are to be considered in the decision making process. The system must allow the user to specify his own subjective factors, and their relative importance (eg. health hazard, prestige, political possibilities, competitiveness).

E1.3.2 **Proposal Identification**

The sub-system must be capable of evaluating the set of factors and their alternative values for a particular design and manufacturing proposal which must be uniquely identified and recognised by the system. The unique identifier must incorporate a means of distinguishing between different levels of design proposal such as modification issues within a specific design approval.

E1.3.3 **Capacity Evaluation**

The sub-system must be capable of predicting manufacturing capacity requirements for each potential technology and of comparing this with available capacity in order to evaluate required additional capacity. The manufacturing capacity considerations must include basic factors of plant, machinery and equipment, manpower, storage and internal transportation facilities. The sub-system should be capable of considering further manufacturing capacity factors considered important by the user and specified by him. For example, such factors as power services, external delivery transportation, data communications, etc.

E1.3.4 **Proposal Comparison Factor**

The sub-system must be capable of evaluating all the quantifiable factors considered important by the user, and of producing a resultant composite cost that will allow a direct comparison of the alternative strategies. The sub-system must allow the user to specify the algorithms by which the quantified parameters are combined to evaluate the cost.

E1.3.5 **Automatic Comparison**

It must be possible for the user of the sub-system to directly compare the results of the different strategies considered or to compare a sub-set of these results which he has nominated.

E1.3.6 **Proposal Variations**

The sub-system must have the ability to retain all of the input data to be considered for an option, and to allow re-evaluation and comparison of alternative specifications involving a variation of algorithms, parameters, or subjective factor weightings already recorded.

E1.3.7 **Responsibility Identification**

The sub-system must be capable of allowing an identification of the author of the individual inputs of information for responsibility and authority traceability purposes.

E1.3.8 **Permanent Algorithm Definition**

The sub-system should allow for the retention of algorithms nominated by the user for use in future comparison exercises.

Product coding/classification sub-system (E2)

E2.1 **Sub-system description**

This sub-system is concerned with the task of allowing the coding of characteristics of a product in such a manner that a retrieval can be made of information relating to existing products that will assist Design/Production Engineers with their work on the new product.

Obviously the information of interest to a manufacturing engineer will be very different to the information required by a Designer and the system must be capable of meeting the specific requirements of either or both functions. Both functions will need to input the type of characteristic information that will be of specific use, but recognition should be given to the possibility in the future of dimensional information being automatically generated from CAD systems.

Once similar component information has been retrieved, this may be amended as required to meet the variation requirements of the new component thereby substantially reducing the effort required. The broad areas of code commonality between parts will allow family groupings of parts to be achieved and defined.

Product coding/classification sub-system

E2.2 Flowchart

E 2.2 SUB-SYSTEM FLOWCHART　　　　　　　　　　　Sheet 1 of 1

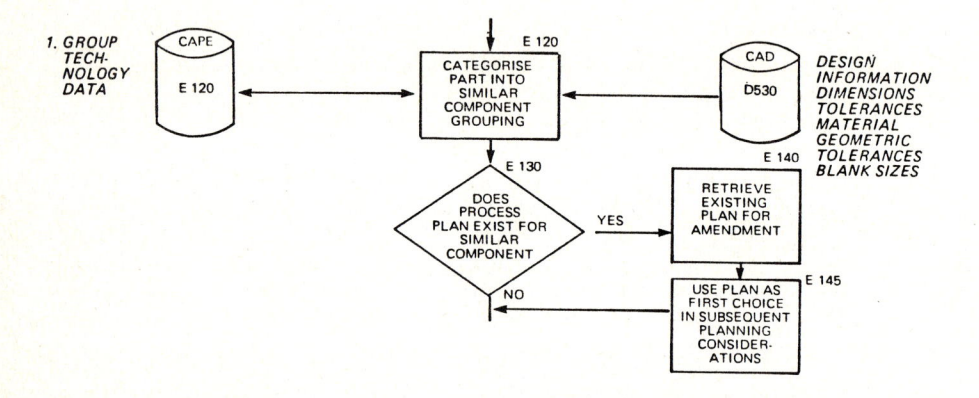

1. GROUP TECH-NOLOGY DATA

CAPE — E 120

E 120 CATEGORISE PART INTO SIMILAR COMPONENT GROUPING

CAD — D530

E 140 *DESIGN INFORMATION DIMENSIONS TOLERANCES MATERIAL GEOMETRIC TOLERANCES BLANK SIZES*

E 130 DOES PROCESS PLAN EXIST FOR SIMILAR COMPONENT — YES → RETRIEVE EXISTING PLAN FOR AMENDMENT

NO

E 145 USE PLAN AS FIRST CHOICE IN SUBSEQUENT PLANNING CONSIDERATIONS

E2.3 Design rules - product coding/classification

E2.3.1 Attribute category definition

The sub-system should allow the user to define a category for any particular attribute value, any range of attribute values, or any combination of attribute values according to a user defined algorithm, and there should be no limit to the number of categories that can be defined:

    eg. Material = mild steel.
          1 Kg LT. Weight LT. 2 Kg
          Length GT. 3 (Diameter)

The above requirements should not preclude a sub-system from incorporating predefined standard categories.

E2.3.2 Category sub grouping

The sub-system should allow the user to define 'minor' categories that form a sub set of an existing defined major category and it must be possible to successively group or subdivide to create broader or narrower categories. Any limit to the number of levels should be user defined.

    eg. 1 Kg LT. Weight LT. 2 Kg could be a major category comprising 10 minor categories each covering a distinct 100gm weight range within the major category.

E2.3.3 Code format

The sub-system should allow the user to define the length and format of codes assignable to categories used, in addition to any predefined standard codes incorporated into the system for predefined categories. The sub-system must allow the incorporation of newly assigned category codes into the overall entity code in a manner defined by the user.

E2.3.4 Category redefinition

The sub-system should allow the redefinition of the eligibility rules which determine the membership of categories and allow automatic reassignment of the total entity code of existing entities as a result of the restatement of the reassigned code variables:

    eg. It might be required to redefine the ratio of the diameter to the length of the component from 1/3 to 1/4.

E2.3.5 Entity coding

The sub-system should not require a user to code necessarily all defined categories for a component thereby allowing the retrieval of information that may assist in category definition.

E2.3.6 Retrieval category definition

The sub-system should allow the user to define any sub set of categories against which retrieval of similar information will be made. The system should be able to identify where similarities in code exist with the exception of categories that have been added to the system since the original entities were coded.

Process selection sub-system (E3)

E3.1   **Sub-system description**

This sub-system is concerned with the task of examining a detail to decide which processes would be suitable to convert the initial component blank into the finished article.

The first task is to examine the physical attributes of the finished detail, such as dimension, surface finish, geometric tolerance relationships, etc., and to compare these with the initial condition of the blank to determine features that are to be created by the machining activities. The sub-system then examines all the available processes and selects those that can produce the features to the desired accuracy. As there may be more than one process capable of creating particular features, optional process lists are produced which are ordered to take account of process precedence needs.

Finally the system determines for each process option, the recommended metal cutting conditions including such factors as speeds, feeds, machine power and tool type required.

Process selection sub-system

E3.2 Flowchart

E 3.2 SUB-SYSTEM FLOWCHART　　　　　　　　　　Sheet 1 of 1

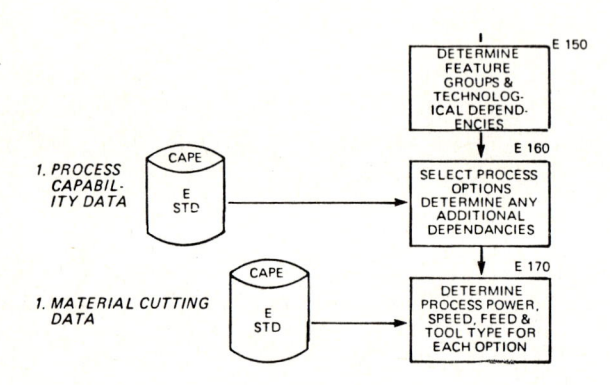

```
                                            ┌──────────────┐ E 150
                                            │  DETERMINE   │
                                            │   FEATURE    │
                                            │   GROUPS &   │
                                            │  TECHNOLOG-  │
                                            │ ICAL DEPEND- │
                                            │    ENCIES    │
                                            └──────┬───────┘
                     ╭─────╮                       │  E 160
                     │CAPE │                ┌──────▼───────┐
  1. PROCESS         │  E  │                │SELECT PROCESS│
     CAPABIL-        │ STD │───────────────▶│   OPTIONS    │
     ITY DATA        ╰─────╯                │DETERMINE ANY │
                                            │  ADDITIONAL  │
                                            │ DEPENDANCIES │
                                            └──────┬───────┘
                     ╭─────╮                       │  E 170
                     │CAPE │                ┌──────▼───────┐
  1. MATERIAL CUTTING│  E  │                │  DETERMINE   │
     DATA            │ STD │───────────────▶│PROCESS POWER,│
                     ╰─────╯                │SPEED, FEED & │
                                            │TOOL TYPE FOR │
                                            │ EACH OPTION  │
                                            └──────────────┘
```

E3.3 Design rules - process selection

E3.3.1 Input data

The system must be capable of accepting the minimum data items listed below, for any detail:

- **Identity**

    The identity must be a unique reference by which the detail can be selected - a part number is a common example.

- **Dimensioning**

    The dimensions, tolerances, geometric tolerances and datum points for both the finished machined detail and the unmachined blank. This information must be quoted in suitable form to allow absolute determination of the precise shape, form, size and position of every feature.

- **Surface Finish**

    The surface finish or texture requirements of any element of the detail.

- **Material**

    The type of material from which the detail is to be made. Where appropriate this should include the grade or composition of the material, eg. "high carbon steel" rather than "steel".

- **Material State**

    Any material treatment, (eg. heat treatment, or surface hardening) that may affect the process selection.

- **Weight**

    The approximate weight of the unmachined blank, which at extremes may condition process selection.

- **Planned Production Volume**

    An estimate of the volume, or rate of production, of the detail. This provides a guide, in some cases, to the type of production facilities likely to be needed or most appropriate (eg. a slow process may be appropriate for a small production volume).

- **Known Similar Parts**

    If group technology data has been used to derive similar details (E120 - E145), these should be identified to enable their existing data to be retrieved as a basis for subsequent processing in this sub-system. This may enable some functions of this sub-system to be carried out on a "same as, except" basis.

E3.3.2 Input re-formatting

Data will be received in a Batch format (see Section 7). The sub-system must be capable of re-formatting, as required, input data into the form required by the sub-system. It is assumed that the input data will be structured in a disciplined and identified manner.

E3.3.3 Feature group identification

The sub-system must carry out (with manual assistance if required) the identification of a complete set of feature groups.

A feature group is a user defined set of machining features which are considered as a group when performing process selection.

These feature groups represent, in total, the complete transformation of the detail blank into a fully machined detail :

eg. Feature Group 001 - End Face
Feature Group 002 - Tapped Hole
Feature Group 003 - Keyway

NB: Each feature group must have a unique identity which distinguishes that group from all others forming part of the same detail.

A feature group may contain one or more features which will require subsequent machining - the level of content may vary according to local requirements.

E3.3.4 **Dimensions**

There must be the capability to hold (for each identified feature group) sufficient dimension and tolerance data to absolutely define the nature, characteristics, sizes and locations of all elements of a feature group, and any dimensional relationships between feature groups.

E3.3.5 **Surface texture**

Provision must be made to hold required surface texture or finish information for any relevant elements of a feature group, since process selection will be affected by this.

E3.3.6 **Precedence between feature groups**

Where a mandatory precedence exists between feature groups this must be recorded:

eg. A location face must be produced before any feature group requiring this face as a datum point can be produced.

E3.3.7 **Process list creation**

The sub-system must carry out the breakdown of each feature group into a set of processes capable of producing the feature group in its required state. This set of processes is a process list. It will be dependent on local process capabilities. If more than one set of processes is capable of producing a particular feature group, each set will constitute a separate process list:

eg. Feature Group 009 - Tapped Hole

Process List 1   - Drill   Process List 2   - Jig Bore
                 - Tap   - Tap

NB: A process list may contain only one process or several processes.

E3.3.8 **Process list identity**

Each process list must have a unique identification within its feature group for reference purposes. This will apply even when only one process list exists for a feature group.

E3.3.9 **Process list desirability**

Each process list must have the capability of carrying a desirability rating, assessed on the basis of local factors deemed appropriate. This rating will be used in the subsequent selection of the best machines and operations to be used in manufacture. It may also be used to restrict, if required, the number of process lists ultimately related to a feature group.

E3.3.10 **Process identification**

Each process in a process list must have a unique identification within its process list. This may be allocated or machine generated. This will apply even when only one process forms a process list:

    eg: Feature Group 002   -   Internal Spline

Process List 1	Process List 2
Desirability 7	Desirability 4
Process 01 - Drill	Process 01 - Spark Erode
Process 02 - Broach	

E3.3.11 **Precedence between processes**

Where a process list consists of more than one process, and a mandatory precedence exists between any of these processes, this must be recorded.

E3.3.12 **Tool characteristics**

For each identified process, each characteristic of the tool required must be described, quantified where necessary, and uniquely identified. For example:

Characteristic 01: Tool type	- Twist Drill.
Characteristic 02: Tool material	- Tungsten Carbide.
Characteristic 03: Maximum length	- 200 mm.
Characteristic 04: Minimum length	- 150 mm.
Characteristic 05: Diameter	- 25 mm.

E3.3.13 **Machine characteristics**

For each process all characteristics of the machine required must be described, approximately quantified, and uniquely identified.

For example, for process 002, process list 001, feature group 008:

Speed	10m/min.
Feed	0.4mm/rev.
Depth of Cut	2mm
Power	40 Kw

E3.3.14 **Output data**

The sub-system must be capable of producing data to the minimum levels shown below. It must be in a form which will facilitate re-formatting by a subsequent module, ie. all items must be structured in a disciplined and identified manner.

Structure of data items

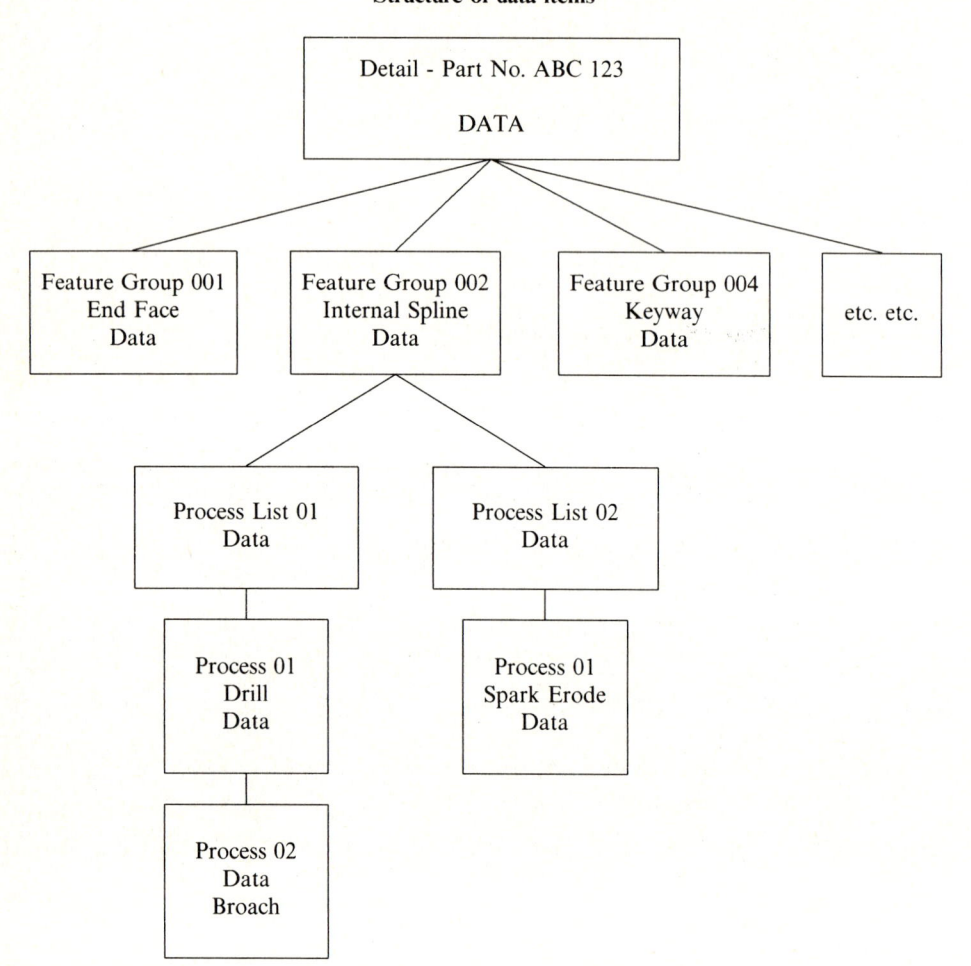

```
                    ┌─────────────────────────────┐
                    │  Detail - Part No. ABC 123  │
                    │           DATA              │
                    └─────────────────────────────┘
          ┌──────────────┬──────────────┬──────────────┐
┌─────────────────┐ ┌─────────────────┐ ┌─────────────────┐ ┌──────────┐
│ Feature Group 001│ │Feature Group 002│ │Feature Group 004│ │ etc. etc.│
│    End Face     │ │ Internal Spline │ │     Keyway      │ │          │
│      Data       │ │      Data       │ │      Data       │ │          │
└─────────────────┘ └─────────────────┘ └─────────────────┘ └──────────┘
                       ┌───────┴───────┐
              ┌─────────────────┐ ┌─────────────────┐
              │ Process List 01 │ │ Process List 02 │
              │      Data       │ │      Data       │
              └─────────────────┘ └─────────────────┘
                       │                   │
              ┌─────────────────┐ ┌─────────────────┐
              │   Process 01    │ │   Process 01    │
              │      Drill      │ │   Spark Erode   │
              │      Data       │ │      Data       │
              └─────────────────┘ └─────────────────┘
                       │
              ┌─────────────────┐
              │   Process 02    │
              │      Data       │
              │     Broach      │
              └─────────────────┘
```

Description of data items

Data Relating to Detail

- **Identity**

    The identity must be a unique reference by which the detail can be selected - a part number is a common example.

- **Material**

    The type of material from which the detail is to be made. Where appropriate this should include the grade or composition of the material, eg. "high carbon steel" rather than "steel".

- **Material State**

    This refers to any material treatment, eg. heat treatment, or surface hardening.

- **Weight**

    The approximate weight of the unmachined blank, which at extremes may condition process selection.

- **Planned Production Volume**

    An estimate of the volume, or rate of production, of the detail. This provides a guide, in some cases, to the type of production facilities likely to be needed or most appropriate (eg. a slow process may be appropriate for a small production volume).

- **Known Similar Parts**

    If group technology data has been used to derive similar details (E120 - E145), these should be identified to enable their existing data to be retrieved as a basis for processing in subsequent sub-systems.

Data Relating to Feature Groups

The following data form a minimum for each feature group within a detail:

- **Feature Group Identity**

    A unique identity which distinguishes any feature group from all others forming part of the same detail. A number, eg. "Feature Group 001" may be considered sufficient.

- **Dimensioning**

    Sufficient dimension and tolerance data must be present to absolutely define the nature characteristics, sizes and locations of all elements of each feature group, and of any dimensional relationships with any other feature group.

- **Surface Finish**

    Any surface texture or finish information relevant to any element of the feature group must be shown.

- **Precedence**

    Where a mandatory precedence exists between feature groups (eg. a location face which must be produced before a feature group requiring this face as a datum point can be produced) the feature groups requiring prior production must be identified.

Data Relating to Process Lists

The following data form a minimum for each process list within a feature group:

- **Process List Identity**

A unique identity which distinguishes any process list from all others which are production options for a particular feature group. A number, eg. "Process List 001" may be considered sufficient. If only one process list has been identified as viable to produce a particular feature group it must still carry an identity.

- **Process List Desirability Factor**

    A weighting factor must be provided for optional process lists to indicate their relative desirability in a particular manufacturing environment. It must be provided for even when only one process list is available for the feature group.

Data Relating to Processes

The following data form a minimum for each process within a process list:

- **Process Identity**

    A unique identity which distinguishes any individual process in a process list, eg. process 001. If a particular process list contains only one process, this process must still have a unique identity.

- **Process Type**

    The type of process to be carried out should be identified, possibly by a coding system meeting specific implementation requirements, eg:

    CG = Centreless Grinding
    RM = Reaming
    JB = Jig Boring
    DR = Drilling

- **Tool Characteristics**

    Each characteristic of the tool required for the process must be described, quantified where necessary and uniquely identified in a form capable of being re-formatted subsequently for tool selection purposes.

    For example, for process 001, process list 002, feature group 004, process type DR:

Characteristic 01: Tool type	- Twist Drill.
Characteristic 02: Tool material	- Tungsten Carbide.
Characteristic 03: Maximum length	- 200 mm.
Characteristic 04: Minimum length	- 150 mm.
Characteristic 05: Diameter	- 25 mm.

- **Machine Characteristics**

    For each process, the minimum characteristics required of the machine performing the process must be identified and listed in a form capable of being re-formatted subsequently for machine selection purposes.

    For example, for process 002, process list 001, feature group 008:

Speed	10m/min.
Feed	0.4mm/rev.
Depth of Cut	2mm
Power	40 Kw

- **Precedence**

    Where a mandatory precedence exists between processes within a process list (eg. a hole must be drilled before it can be tapped) the processes requiring prior performance must be identified.

Machine selection sub-system (E4)

E4.1 **Sub-system description**

Once a detail has been analysed into the various Feature Groups that are to be machined, and the process possibilities have been determined for the creation of these Feature Groups, it then becomes necessary to decide which specific machine tools are to be used for each machining operation and which route through the various machines will be adopted.

The Machine Selection Sub-System carries out these tasks from information previously evaluated by the Process Selection Sub-System, together with additional information concerned with the Machine Capability and available unused capacity.

The sub-system considers, for each operation to be performed, which machines are available that satisfy the process requirement and which have available spare capacity to be utilised. Once this has been completed, the sub-system then determines all of the combinations of machine and operations that could be regarded as possible routes for the detail being considered.

For each of the possible machine routes determined, the next task is to calculate, from synthetic data, the floor-to-floor time for each group of operations assigned to each machine and to use this information to calculate the required capacity.

The required capacity for each group of operation/machine combinations can then be compared with the known available capacity for each machine and all feasible process routings extracted.

Finally, the sub-system has to decide from the feasible routings generated, which one should be selected as the prime routing, and which of the remaining feasible routes should be retained as preferred alternatives to be used in the event of a machine shop problem.

The selection of the prime routing can either be judged by some ranking based upon the preferred use of machines set by the user, or by evaluating the options through an optimising package (such as a simulation package) that will judge each option according to some user criteria.

Machine selection sub-system

E4.2 Flowchart

E 4.2 SUB-SYSTEM FLOWCHART

Sheet 1 of 1

Inputs (left side):
1. MACHINE LIBRARY
2. MACHINE CAPABILITY
3. FORECAST MACHINE AVAILABILITY

Database: CAPP P110 / CAPE E STD

E 180 — ARE SUITABLE MACHINES AVAILABLE FOR EACH OPERATION — NO (loop back)
↓ YES

E 190 — ACCUMULATE SUITABLE MACHINE/OPERATION COMBINATIONS

E 200 — ARE THERE ANY REMAINING OPERATIONS TO BE PLANNED — YES (loop back)
↓ NO

E 210 — DETERMINE ALL POSSIBLE MACHINE OP. GROUPINGS WITHIN TECH DEPENDENCIES

Input: WORK STUDY METHOD STUDY

Inputs:
1. LOAD/UNLOAD TIMES
2. TOOL LIFE DATA
3. M/C RELIABILITY
4. SET UP TIMES
5. CYCLE TIMES
6. MATERIAL TRANSPORT TIMES

Database: CAPE E STD

E 220 — DETERMINE CAPACITY REQUIRED FOR EACH MACHINING & OPERATION COMBINATION

E 240 — REJECT PARTICULAR MACHINE OPERATION COMBINATION

Input:
1. FORECAST OF AVAILABLE MACHINE CAPACITY

Database: CAPP P110

E 230 — IS SUFFICIENT CAPACITY AVAILABLE FOR EACH MACHINE OPERATION COMBINATION — NO → E 240
↓ YES

E 250 — ARE ANY SUITABLE MACHINE OPERATION ROUTINES AVAILABLE — NO → MANAGEMENT DECISION
↓ YES

E 260 — REQUEST REQUIRED REPORTS

E 270 — SELECT OPTIMAL PROCESS AND PREFERRED ALTERNATIVE

Output: 1. STORED PROCESS — CAPE E 120

E4.3 **Design rules - machine selection**
E4.3.1 **Input data**

The sub-system must be capable of accepting the minimum data items listed below:

Data is required relating to:

- The detail to be manufactured (Section E4.3.1.1).
- The machines available in the manufacturing location (Section E4.3.1.2).
- The capacity available on those machines over the planning horizon (Section E4.3.1.3).

E4.3.1.1   **Information relating to detail**

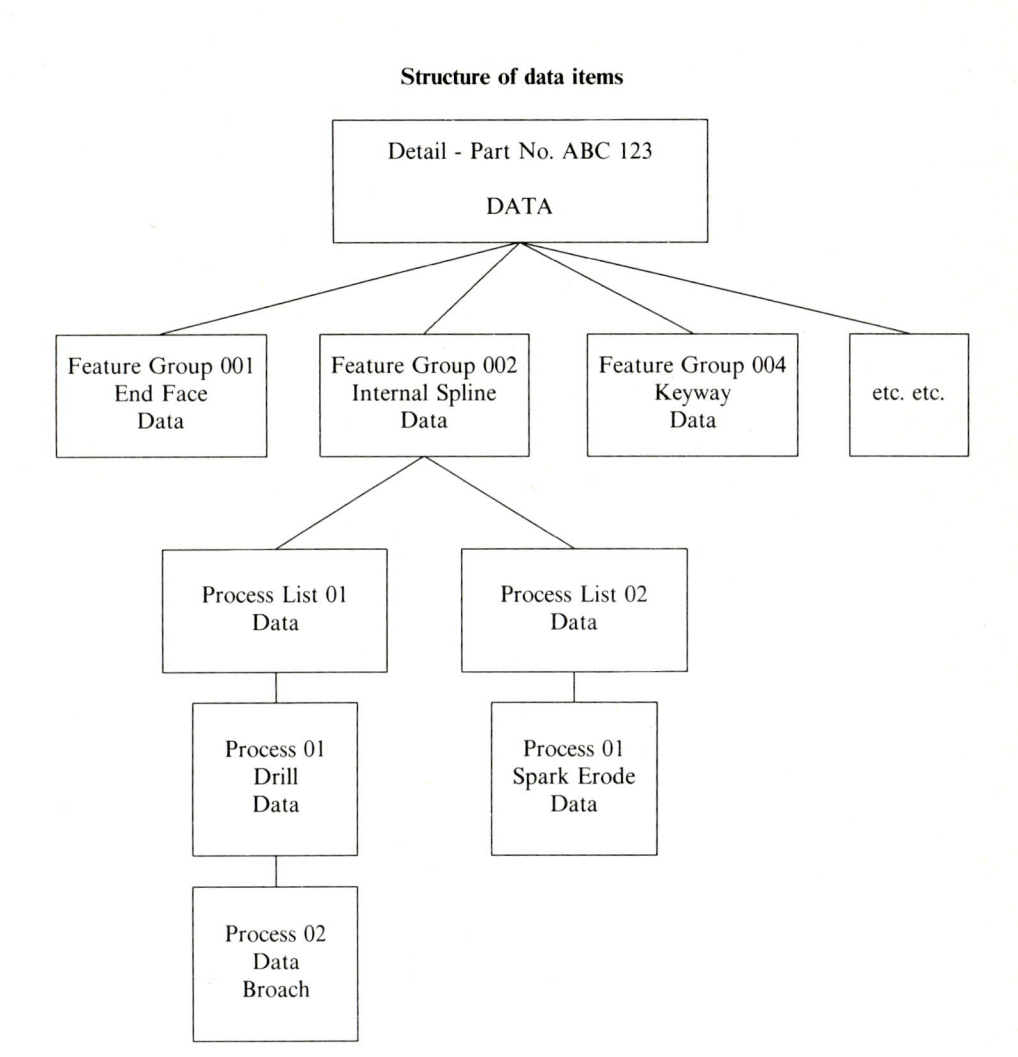

Structure of data items

```
                    ┌─────────────────────────┐
                    │ Detail - Part No. ABC 123│
                    │          DATA           │
                    └─────────────────────────┘
           ┌──────────┬──────────┴──────┬──────────┐
┌──────────────┐ ┌──────────────┐ ┌──────────────┐ ┌─────────┐
│Feature Group │ │Feature Group │ │Feature Group │ │         │
│    001       │ │    002       │ │    004       │ │etc. etc.│
│  End Face    │ │Internal Spline│ │   Keyway    │ │         │
│    Data      │ │    Data      │ │    Data      │ │         │
└──────────────┘ └──────────────┘ └──────────────┘ └─────────┘
                    ┌──────┴──────┐
            ┌──────────────┐ ┌──────────────┐
            │Process List 01│ │Process List 02│
            │    Data       │ │    Data       │
            └──────────────┘ └──────────────┘
                    │                │
            ┌──────────────┐ ┌──────────────┐
            │ Process 01    │ │ Process 01    │
            │    Drill      │ │ Spark Erode   │
            │    Data       │ │    Data       │
            └──────────────┘ └──────────────┘
                    │
            ┌──────────────┐
            │ Process 02    │
            │    Data       │
            │    Broach     │
            └──────────────┘
```

E4.3.1.1   **Descriptions of data items**

Data Relating to Detail

- **Identity**

    The identity must be a unique reference by which the detail can be selected - a part number is a common example.

- **Surface Finish**

    The surface finish or texture requirements of any element of the detail.

- **Material**

    The type of material from which the detail is to be made. Where appropriate this should include the grade or composition of the material, eg. "high carbon steel" rather than "steel".

- **Material State**

    This refers to any material treatment, eg. heat treatment, or surface hardening.

- **Weight**

    The approximate weight of the unmachined blank, which at extremes may condition process selection.

- Planned Production Volume

    This should be provided as a guide, in some cases, to the type of production facilities likely to be needed or most appropriate.

- **Known Similar Parts**

    If group technology data has been used to derive similar details (E120 - E145), these should be identified to enable their existing data to be retrieved as a basis for subsequent processing in this sub-system.

Data Relating to Feature Groups

The following data form a minimum for each feature group within a detail:

- **Feature Group Identity**

    A unique identity, allocated or machine generated, which distinguishes any feature group from all others forming part of the same detail. A number, eg. "Feature Group 001" may be considered sufficient.

- **Dimensioning**

    Sufficient dimension and tolerance data must be present to absolutely define the nature characteristics, sizes and locations of all elements of each feature group, and of any relationships with any other feature group.

- **Surface Finish**

    Any surface texture or finish information relevant to any element of the feature group must be shown.

- **Precedence**

    Where a mandatory precedence exists between feature groups (eg. a location face which must be produced before a feature group requiring this face as a datum point can be produced) the feature groups requiring prior production must be listed.

Data Relating to Process Lists

The following data form a minimum for each process list within a feature group:

- **Process List Identity**

    A unique identity, allocated or machine generated, which distinguishes any process list from all others which are production options for a particular feature group. A number, eg. "Process List 001" may be considered sufficient. If only one process list has been identified as viable to produce a particular feature group it must still carry an identity.

- **Process List Desirability Factor**

    A weighting factor must be provided for optional process lists to indicate their relative desirability in a particular manufacturing environment. It must be provided for even when only one process list is available for the feature group.

Data Relating to Processes

The following data form a minimum for each process within a process list:

- **Process Identity**

    A unique identity, allocated or machine generated, which distinguishes any individual process in a process list, eg. process 001. If a particular process list contains only one process, this process must still have a unique identity.

- **Process Type**

    The type of process to be carried out should be identified, possibly by a coding system meeting specific implementation requirements, eg:

    CG = Centreless Grinding
    RM = Reaming
    JB = Jig Boring
    DR = Drilling

- **Tool Characteristics**

    Each characteristic of the tool required for the process must be described, quantified where necessary and uniquely identified in a form capable of being re-formatted subsequently for tool selection purposes.

For example, for process 001, process list 002, feature group 004, process type DR:

Characteristic 01: Tool type	- Twist Drill.
Characteristic 02: Tool material	- Tungsten Carbide.
Characteristic 03: Maximum length	- 200 mm.
Characteristic 04: Minimum length	- 150 mm.
Characteristic 05: Diameter	- 25 mm.

- **Machine Characteristics**

    For each process, the minimum characteristics required of the machine performing the process must be identified and listed in a form capable of being re-formatted subsequently for machine selection purposes.

For example, for process 002, process list 001, feature group 008:

Speed	10m/min.
Feed	0.4mm/rev.
Depth of Cut	2mm
Power	40 Kw

- **Precedence**

    Where a mandatory precedence exists between processes within a process list (eg. a hole must be drilled before it can be tapped) the processes requiring prior performance must be listed.

E4.3.1.2 **Information relating to machines in the manufacturing location:**

- **Machine Identity**

    A unique identity (eg. asset number) by which a particular machine, or group of like machines, can be uniquely identified.

- **Machine Process Type**

    The processes which can be performed by the machine, eg. mill, drill, spotface, ream, tap.

- **Power/Speed Characteristics**

    The evaluated characteristics of the machine, eg. speed/power characteristics of the spindles, feedrate/power characteristics of the machine axes.

- **Accuracy**

    The limits of accuracy to which the machine can perform.

- **Handling Capacity**

    Maximum dimensions and weight of detail which can be mounted on the machine or transported by any transport system linked to the machine.

E4.3.1.3 **Information relating to capacity**

The system must be capable of considering the utilisation of any machine on already planned work. For each machine this utilisation must be available in a form which relates the total production time available on that machine with the time already allocated to the production of other details. It must be expressed in terms of elapsed times, not percentages. It must also be related to specific calendar periods, eg. by day, week, month, etc., as appropriate to the implementation.

eg: Machine Number 24597

Time Period	Total Available Hrs/Week	Allocated Hrs/Week	Free Hrs/Week
1st Quarter 1985	39	36	3
2nd Quarter 1985	39	34	5
3rd Quarter 1985	77	34	43
4th Quarter 1985	77	32	45

E4.3.2 Input re-formatting

Data will be received in a Batch format (see Section 7). The sub-system must be capable of re-formatting, as required, input data into the form required by the sub-system. It is assumed that the input data will be structured in a disciplined and identified manner.

E4.3.3 Processing scope

When potential machines and process routings are generated, the sub-system must have the capability of retaining and considering all options generated. If practical considerations impose a limit on the size of the range of options that may be handled, any restriction of the number of options should be carried out in accordance with the rules for "Selection Process". (Appendix 3)

E4.3.4 Pre-defined processes

The sub-system must be capable of receiving and operating on processes which have been previously defined, and must be able to respond to a ranking procedure associated with such processes.

E4.3.5 Identity of process

The system must be capable of relating individual processes to the identity of the detail and feature group.

E4.3.6 Machining cycle time

The sub-system must include a means of calculating the synthetic time required to perform a defined process on a defined machine tool (ie. a defined operation). Furthermore, it must include a means of using synthetic work-study information (directly input by the user or held in a library) to enable synthetic floor-to-floor production times to be calculated.

E4.3.7 Machine capacity

The system must be capable of recognising process capacity availability over the production time horizon of the detail being considered. This capacity must be capable of being used in a comparison with the capacity requirements of each machine tool selected.

E4.3.8 Machining capacity commitment

When a final allocation of operations to particular machines has been made, the additional capacity required of each machine must be made available for up-dating capacity planning data.

E4.3.9 Contingency reporting

In the absence of any suitable process/routings being generated, the system must allow the consideration of processes/routings which fail to meet the specified criteria (eg. insufficient process capacity).

E4.3.10 Machine routing optimisation

The system must be capable of selecting an optimum machine routing according to user defined and weighted parameters, or by use of an optimisation routine (eg. modelling/simulation package). Where one or more alternative routings have been identified, the capability must exist for these to be output in ranked preference in addition to the optimum routing.

E4.3.11 Grouping of like machines

The system must have the capability of considering a group of like machines as a single machining entity. This is to ensure spare capacity of like machines can be fully utilised.

E4.3.12 **Output data**

The sub-system must be capable of producing data to the minimum levels shown below. It must be in a form which will facilitate re-formatting by a subsequent module, ie. all items must be structured in a disciplined and identified manner.

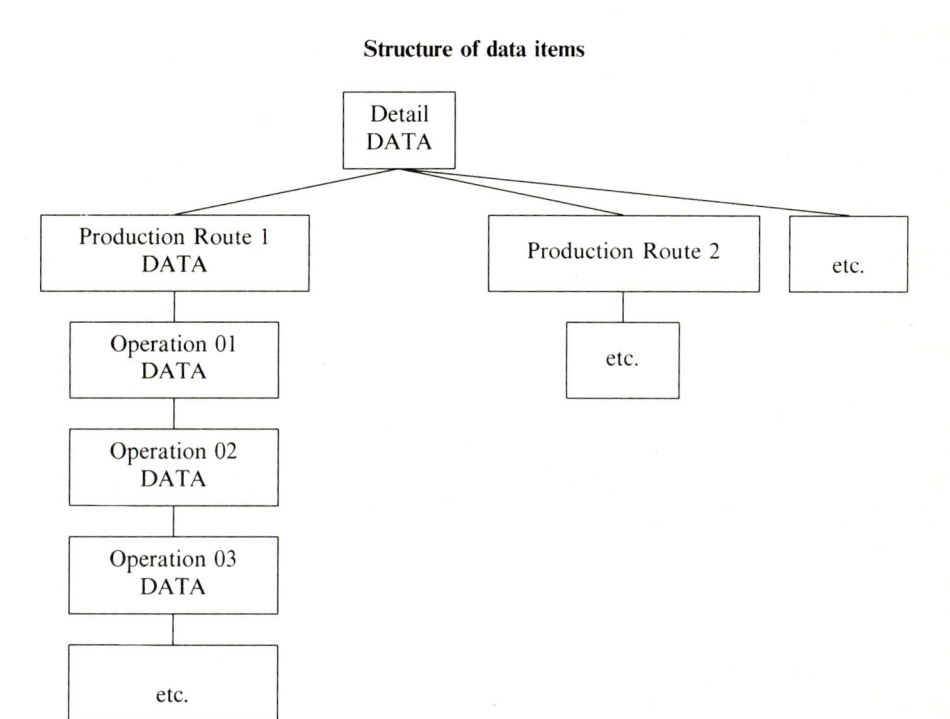

Structure of data items

```
                    ┌──────────┐
                    │  Detail  │
                    │   DATA   │
                    └────┬─────┘
         ┌───────────────┼───────────────┐
┌────────┴────────┐  ┌───┴──────────┐  ┌─┴────┐
│ Production Route 1│  │Production Route 2│ │ etc. │
│      DATA       │  │              │  │      │
└────────┬────────┘  └──────┬───────┘  └──────┘
┌────────┴────────┐         │
│  Operation 01   │     ┌───┴──┐
│     DATA        │     │ etc. │
└────────┬────────┘     └──────┘
┌────────┴────────┐
│  Operation 02   │
│     DATA        │
└────────┬────────┘
┌────────┴────────┐
│  Operation 03   │
│     DATA        │
└────────┬────────┘
┌────────┴────────┐
│      etc.       │
└─────────────────┘
```

Description of data items

Data Relating to Detail

- **Identity**

   The identity must be a unique reference by which the detail can be selected - a part number is a common example.

Data Relating to Production Route

- **Identity**

   Each potential production route which is valid for manufacturing the detail must have a unique reference by which it can be selected, eg. Production Route 01.

- **Routing Preference**

   Each production route should be given a preference indication, eg. Option 1.

Data Relating to Operation

- **Identity**

   Each operation must be given a unique reference within its production route. This reference should also indicate the sequence in which operations are to be carried out:

   eg: Operation 01
         Operation 02
         Operation 03
         etc.

- **Operation Description**

   Detail description, quantified where necessary, of the operation to be performed, eg: Drill Orthogonal Hole

   Diameter 12mm +/- 0.5mm
   Depth 4cm +/- 0.2mm
   Location   5cm from location face 001 +/- 0.2mm
                7cm from location face 003 +/- 0.2mm

- **Machine Settings**

   All relevant settings of the machine on which the operation is to be carried out must be given,

   eg: depth of cut   2mm
         speed   10m/min
         feed   0.4mm/rev

- **Tool Characteristics**

   Each characteristic of the tool(s) required for the operation must be described and quantified where necessary for tool selection purposes:

   eg: For detail ABC 123, Production Route 1, Operation 06

Tool type   = Twist Drill
Tool Material   = Tungsten Carbide
Maximum Length   = 200 mm
Minimum Length   = 150 mm
Diameter   = 25 mm

- **Timing Synthetic**

  Estimated operation time per detail.

- **Tool Life Synthetic**

  Estimated Tool Life before replacement.

Cutting tool selection sub-system (E5)

E5.1 **Sub-system description**

This sub-system is concerned with the task of examining the planned route of a detail through its various machining operations and of determining for each machine and operation, which cutting tool and holders should be used.

The sub-system will receive from a previous sub-system, information concerned with the operations and machines that are to be used for the production of the features of the detail. It will also receive information concerned with the overall required characteristics of the cutting tool, such as cutting tip material, the maximum and minimum dimensions, etc.

From a consideration of the cutting tool characteristics required, the system will retrieve the identity of suitable tools that will meet the requirements from a tool library. The library will hold information about all the tools that the Company wishes to consider available for consideration.

Once suitable tools have been selected, these will then be matched against suitable tool holders that will not only be capable of holding the selected tools, but also of fitting the holding characteristics of the machines being considered.

When the final selection of tools and holders has been made, the selection will then be used to update the information on the detail process planning file.

Cutting tool selection sub-system

E5.2 Flowchart

E 5.2 SUB-SYSTEM FLOWCHART　　　　　　　　Sheet 1 of 1

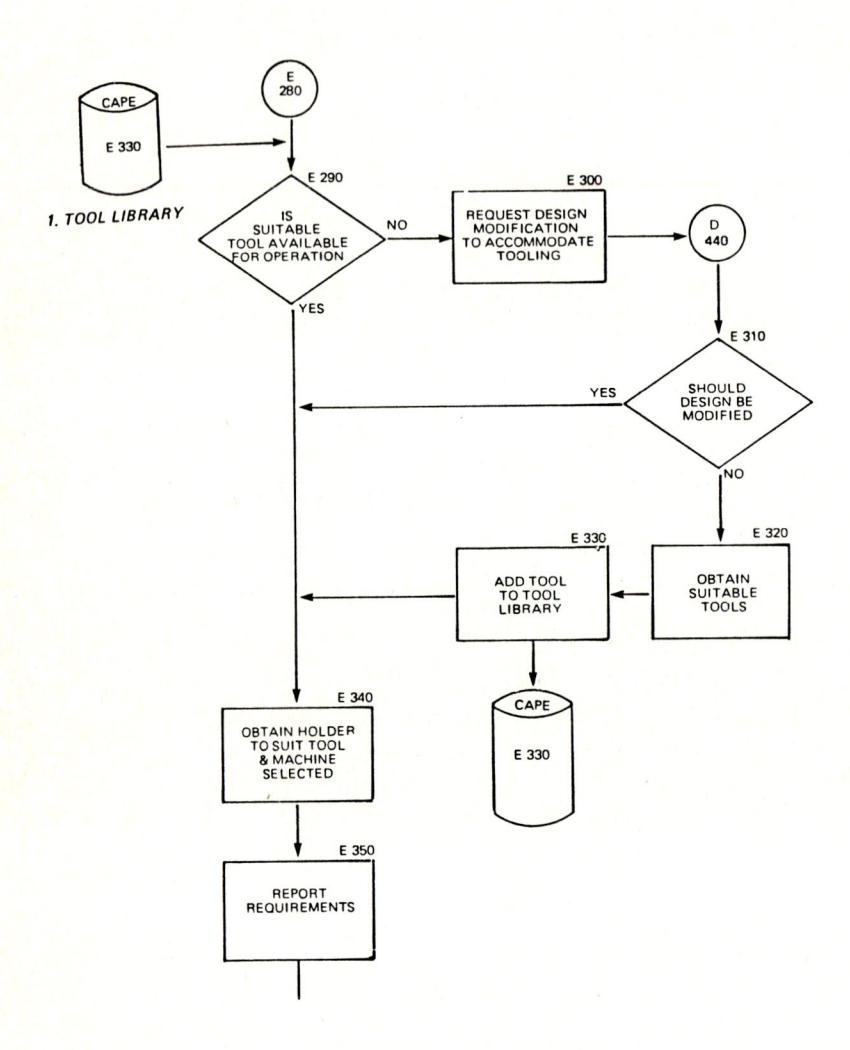

E5.3	**Design rules - cutting tool selection**
E5.3.1	**Input data**

The sub-system must be capable of accepting the minimum data items listed below:

- The data relating to the detail and its planned operation (E5.3.1.1).
- The data relating to the known cutting tools (E5.3.1.2).
- The data relating to tool holders (E5.3.1.3).

Structure of data items

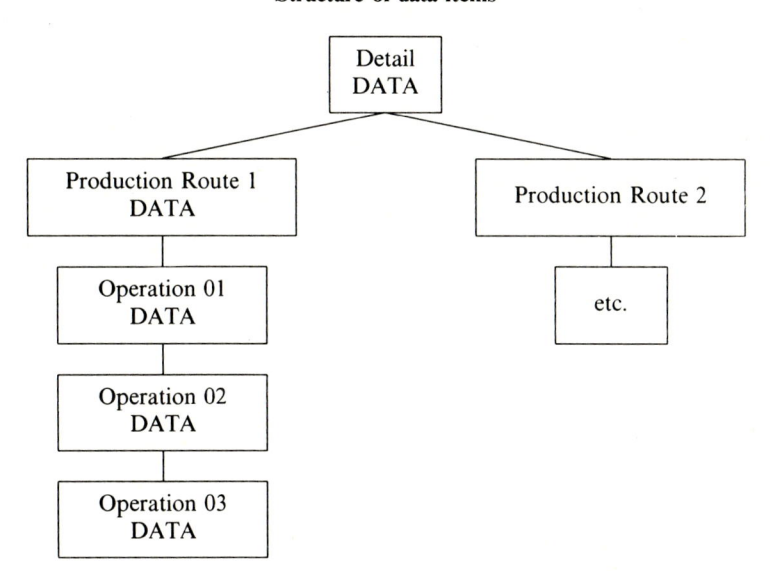

```
                    Detail
                    DATA
                   /      \
       Production Route 1    Production Route 2
            DATA                    
              |                     |
        Operation 01               etc.
           DATA
              |
        Operation 02
           DATA
              |
        Operation 03
           DATA
```

E5.3.1.1 **Information relating to detail and planned operations**

Data Relating to Detail:

- **Identity**

    The identity must be a unique reference by which the detail can be selected - a part number is a common example.

Data Relating to Production Route:

- **Identity**

    Each production route which is selected as either the prime route or preferred alternatives must have a unique reference by which it can be identified:

    eg. Production Route 01.

Data Relating to Operation:

- **Identity**

    Each operation must be given a unique reference within its production route. This reference should also indicate the sequence in which operations are to be carried out:

    eg. Operation 01
    Operation 02
    Operation 03
    etc.

- **Tool Characteristics**

    Each characteristic of the tool(s) required for the operation must be described and quantified where necessary:

    eg. for detail ABC 123, Production Route 1, Operation 06.

    | Tool type | = Twist Drill |
    | Tool Material | = Tungsten Carbide |
    | Maximum Length | = 200 mm |
    | Minimum Length | = 150 mm |
    | Diameter | = 25 mm |

- **Machine Identity**

    Each machine should have a unique reference which can be used to indicate which machines are being used to carry out individual operations. The reference should be in the format used by The Company Asset Register.

- **Machine Spindle Characteristics**

    Each characteristic of the machine spindle required for the purpose of tool holder selection should be described and quantified where necessary:

    eg: Taper   40 I.S.O.
    Clamping Method   Power draw bar/screw.

- **Tool Type**

    The type of tool should be identified possibly by a coding system to meet the specific requirements of the system user:
    eg. Drill, broach, ream, etc.

E5.3.1.2   **Information relating to known cutting tools**

- **Tool Identity**

    The cutting tools must be referenced by a unique identity that will allow an individual tool to be selected.

    It is conceivable that the tool type may be incorporated into the Tool Identity.

- **Tool Supplier**

    It must be possible to identify the source of a tool that is selected for subsequent purchasing action.

- **Tool Characteristics**

    Each characteristic of each tool must be described and quantified where necessary for comparison with the required tool characteristics for tool selection purposes:

    eg.   Length   175 mm
           Diameter   30 mm
           Tip Material   High Speed Steel
           etc.

- **Usage Indicator**

    An indication of whether a particular cutting tool is currently used.

E5.3.1.3   **Information relating to tool holder**

- **Identity**

    The identity must be a unique reference by which the holder may be selected.

- **Characteristic**

    Each characteristic necessary to match the suitability of the holder to fit a particular machine tool, and to hold a particular cutting tool:

    eg.

        Collet   25 mm +/- 0.5 mm
        Taper   40 I.S.O.

E5.3.2   **Input re-formatting**

Data will be received in a batch format. The sub-system must be capable of re-formatting, as required, input data into the form required by the sub-system. It is assumed that the input data will be structured in a disciplined and identified manner.

E5.3.3   **Cutting tool selection**

The sub-system must be capable of retrieving from a tool library the identification and associated properties of all cutting tools that will satisfy the required characteristics of cutting tools already determined by a previous machine routing and operation decision.

E5.3.4   **Tool holder selection**

The sub-system must be capable of retrieving the identity, with associated properties, of tool holders that satisfy the holding characteristics of the cutting tools that are selected as potentially usable for each particular operation. These must also satisfy the holding characteristics of the machine to be used.

E5.3.5 **Selection criteria**

The sub-system must be capable of allowing selection criteria to be used for the final selection of a tool when more than one suitable cutting tool has been identified.

The lack of a suitable tool holder will not preclude a tool from appearing in the 'suitable' list, although the existence of a suitable holder may form an important part of the selection criteria.

The selection process should be in accordance with the rules for Selection Process (Appendix 3).

After selection of the best tool, ranked alternative tooling should be recorded in accordance with the selection criteria already used. This will allow alternative tooling to be selected in the event of non-availability of the preferred tooling.

E5.3.6 **User defined tooling categories**

The sub-system must allow a user to nominate data categories for each type of cutting tool, and to define the characteristics comprised by a specification of a particular category according to his particular requirements. The sub-system must be able to determine the data required in a tool specification either from the manner in which it is recorded, or by associating a particular set of characteristics with the tools category:

eg. The sub-system might be able to determine that a specification for a tool in the category type 'drill' will comprise values for the following:

Diameter
Length
Tip Material
Tip Angle
Shank Type
etc.

if this is how a particular user wishes to specify his drills.

E5.3.7 **Partial tool specification**

The sub-system must be capable of operating with complete and partial tool specifications as its selection requirements, and must retrieve every tool from the tool library that satisfies a selection requirement:

eg. There might be three instances of 6mm diameter drills between 10cm and 15cm long, but only one with these characteristics and a carbide tip.

E5.3.8 **Total route tool selection**

The sub-system must be capable of accepting a complete set of input data relating to the production of a detail. The sub-system should also allow the input to include more than one routing plan for a detail where a prime routing and preferred alternatives have been selected.

Computer Aided Production Engineering   103

E5.3.9   **Output data**

The sub-system must be capable of producing data to the minimum levels shown below. It must be in a form which will facilitate re-formatting by a subsequent module, ie. all items must be structured in a disciplined and identified manner.

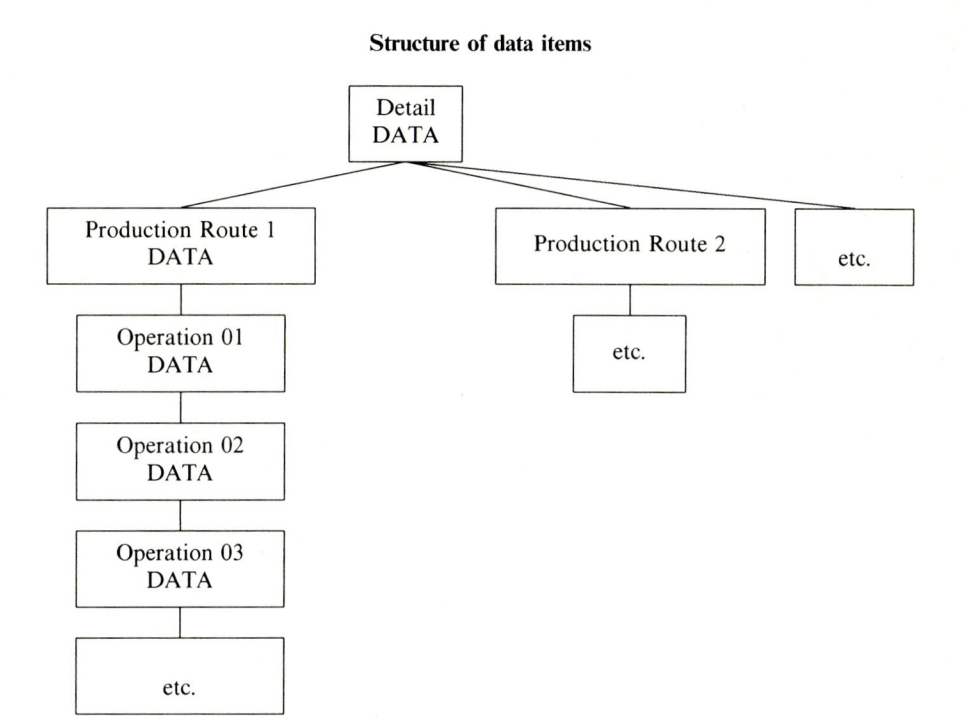

Structure of data items

```
                        Detail
                        DATA
          ┌───────────────┼───────────────┐
   Production Route 1   Production Route 2   etc.
        DATA                 DATA
          │                   │
   Operation 01              etc.
        DATA
          │
   Operation 02
        DATA
          │
   Operation 03
        DATA
          │
         etc.
```

Description of data items

Data Relating to Detail

- **Identity**

    The identity must be a unique reference by which the detail can be selected - a part number is a common example.

Data Relating to Production Route

- **Identity**

    The 'prime route' and the preferred alternatives must have a unique reference by which it can be selected, eg. Production Route 1.

Data Relating to Operation

- **Identity**

    Each operation must be given a unique reference within its production route. The reference should also indicate the sequence in which operations are to be carried out:

    eg. Operation 01
    Operation 02
    Operation 03
    etc.

- **Machine Identity**

    Each machine, or group of machines where like machine cells have been selected, should be referenced by a unique identity from which it can be located.

- **Tool Identity**

    Each tool selected to perform the considered operation must be referenced by its unique identity.

- **Tool Description**

    Detail description of the tool, quantified as required to show the characteristics of the tool for confirmation purposes.

- **Tool Holder Identity**

    Each tool holder selected to perform the considered operation must be referenced by unique identity.

- **Tool Holder Description**

    Detail description of the tool holder, quantified as required to show the characteristics of the holder for confirmation purposes.

Chapter 5

Computer Aided Production Planning

Introduction

Overview

The scope of this section will include the computer aided functions for the forecasting of long term demands, the planning of production requirements based on the dated demands for end products, the short term scheduling of orders for manufacture using associated resource constraint profiles, and the real time activity of selection and sequencing of the next manufacturing process when a required manufacturing resource becomes available.

a) **Long term forecasting**

Forecasting of long term fixed resource requirements involves the simulation of the manufacturing processes using data on forecast sales volumes or marketshare potential, product profiles, products' demands on resources, inventory holding policies and planned manufacturing facilities. Mathematical modelling may be used as an aid in developing this data.

The purpose of this exercise is to identify potential resource over or under loads within sufficient timescales to permit the advance planning of changes to the resources. It also aids decision making on future product ranges, mixes and rates. Executive decisions on the provision of resource changes provide the broad constraints under which the subsequent production planning will operate.

b) **Production Planning**

The actual customer demands for products in terms of firm orders, the projections of orders resulting from statistical forecasting techniques, and the inventory holding policies provide the basis for the planning of production.

The associated data requirements, such as the finished product profiles, the current and future product component structures and the product manufacture process methods are utilised to level the order demands over the production planning horizon. The process of levelling, either by manual interventions or using computerised techniques, is of necessity an interactive one to achieve a balance between the conflicts of meeting customer order due dates, seasonal variations in order demands, and the available resources.

c) **Production scheduling**

The segmentation of the Factory Production Plan into sub-elements over shorter time horizons, e.g. one week or one month, provides the basis for short term scheduling of the manufacturing areas.

Scheduling these sub-elements involves the process of optimising the manufacturing requirements against material, labour and facility constraints. The result of this optimisation is a detailed Production Schedule defining for each Work Centre the set of products to process and their respective quantity off and completion due date.

Changes in manufacturing requirements arising from whatever cause (eg. the introduction of additional urgent orders) may initiate the production scheduling process to provide revised manufacturing schedules if necessary.

d) **Production sequencing and monitoring**

To gain the full potential of computer aided manufacturing systems, it is necessary to utilise computer controls for the sequencing of manufacture. In this event, the Manufacturing Schedules are used to reflect the broad definition of manufacturing demand and the manufacturing process loads.

The selection of a particular work centre machine or a sequence of machines when they become available for more processing must be optimised with regard to material availability, priority on the Manufacturing Schedules, manufacturing route, machine tool changes etc. The selection of a particular machining process or processes would be conveyed to the computer aided manufacturing system computers to initiate transport of the required material, and the provision of the required tooling and part-programs to production machines.

At this stage considerable interaction with the Computer Aided Manufacturing and Computer Aided Storage and Transportation systems is necessary both to retrieve status information on resources and to advise required production sequences.

Computer Aided Production Planning

Flowcharts

COMPUTER AIDED PRODUCTION PLANNING

Sheet 1 of 5

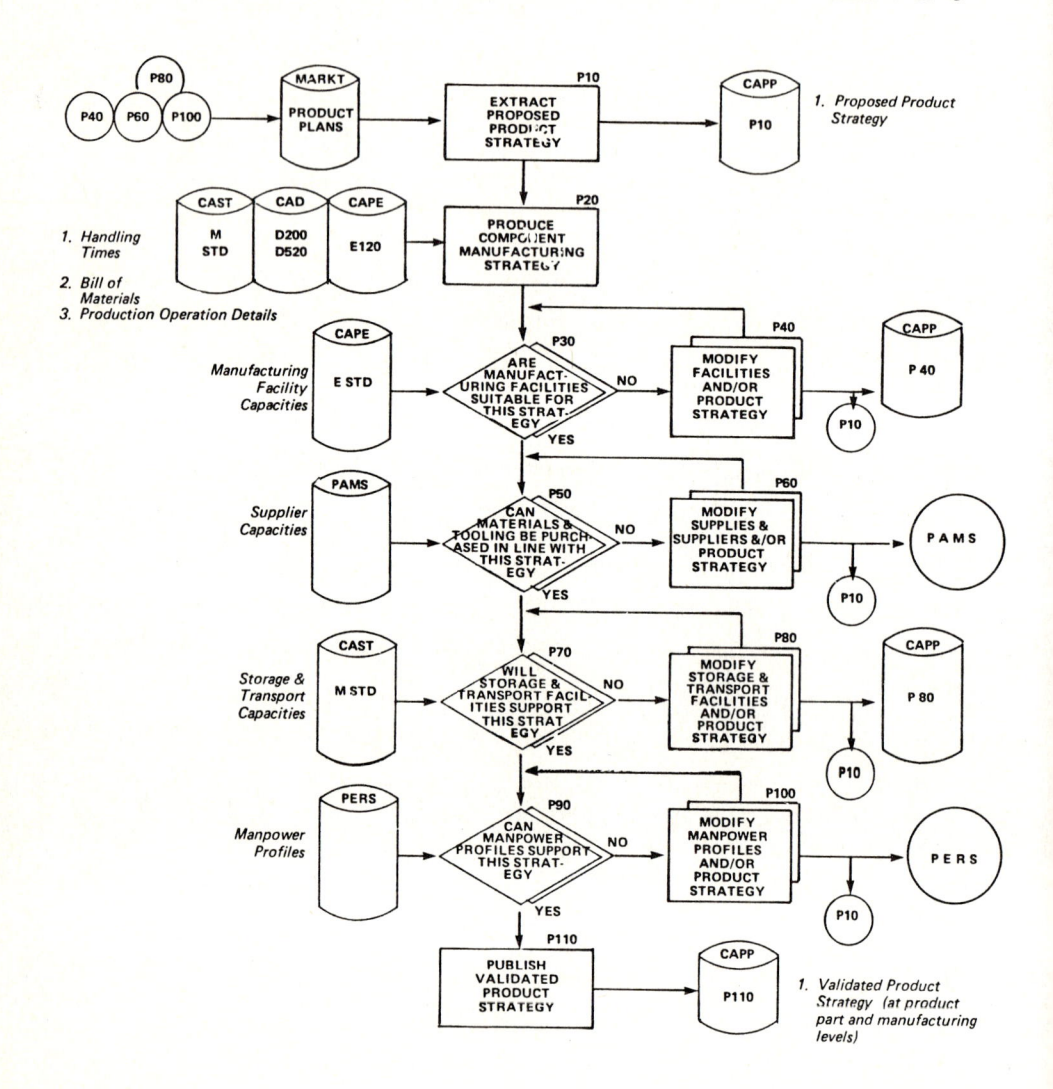

Flowchart:

- P40, P60, P80, P100 → **MARKT PRODUCT PLANS** → **P10 EXTRACT PROPOSED PRODUCT STRATEGY** → **CAPP P10**
  1. Proposed Product Strategy

- Inputs (CAST M STD, CAD D200/D520, CAPE E120) → **P20 PRODUCE COMPONENT MANUFACTURING STRATEGY**
  1. Handling Times
  2. Bill of Materials
  3. Production Operation Details

- *Manufacturing Facility Capacities* (CAPE E STD) → **P30 ARE MANUFACTURING FACILITIES SUITABLE FOR THIS STRATEGY** —NO→ **P40 MODIFY FACILITIES AND/OR PRODUCT STRATEGY** → **CAPP P40**, P10
  YES ↓

- *Supplier Capacities* (PAMS) → **P50 CAN MATERIALS & TOOLING BE PURCHASED IN LINE WITH THIS STRATEGY** —NO→ **P60 MODIFY SUPPLIES & SUPPLIERS &/OR PRODUCT STRATEGY** → **PAMS**, P10
  YES ↓

- *Storage & Transport Capacities* (CAST M STD) → **P70 WILL STORAGE & TRANSPORT FACILITIES SUPPORT THIS STRATEGY** —NO→ **P80 MODIFY STORAGE & TRANSPORT FACILITIES AND/OR PRODUCT STRATEGY** → **CAPP P80**, P10
  YES ↓

- *Manpower Profiles* (PERS) → **P90 CAN MANPOWER PROFILES SUPPORT THIS STRATEGY** —NO→ **P100 MODIFY MANPOWER PROFILES AND/OR PRODUCT STRATEGY** → **PERS**, P10
  YES ↓

- **P110 PUBLISH VALIDATED PRODUCT STRATEGY** → **CAPP P110**
  1. Validated Product Strategy (at product part and manufacturing levels)

COMPUTER AIDED PRODUCTION PLANNING

Sheet 2 of 5

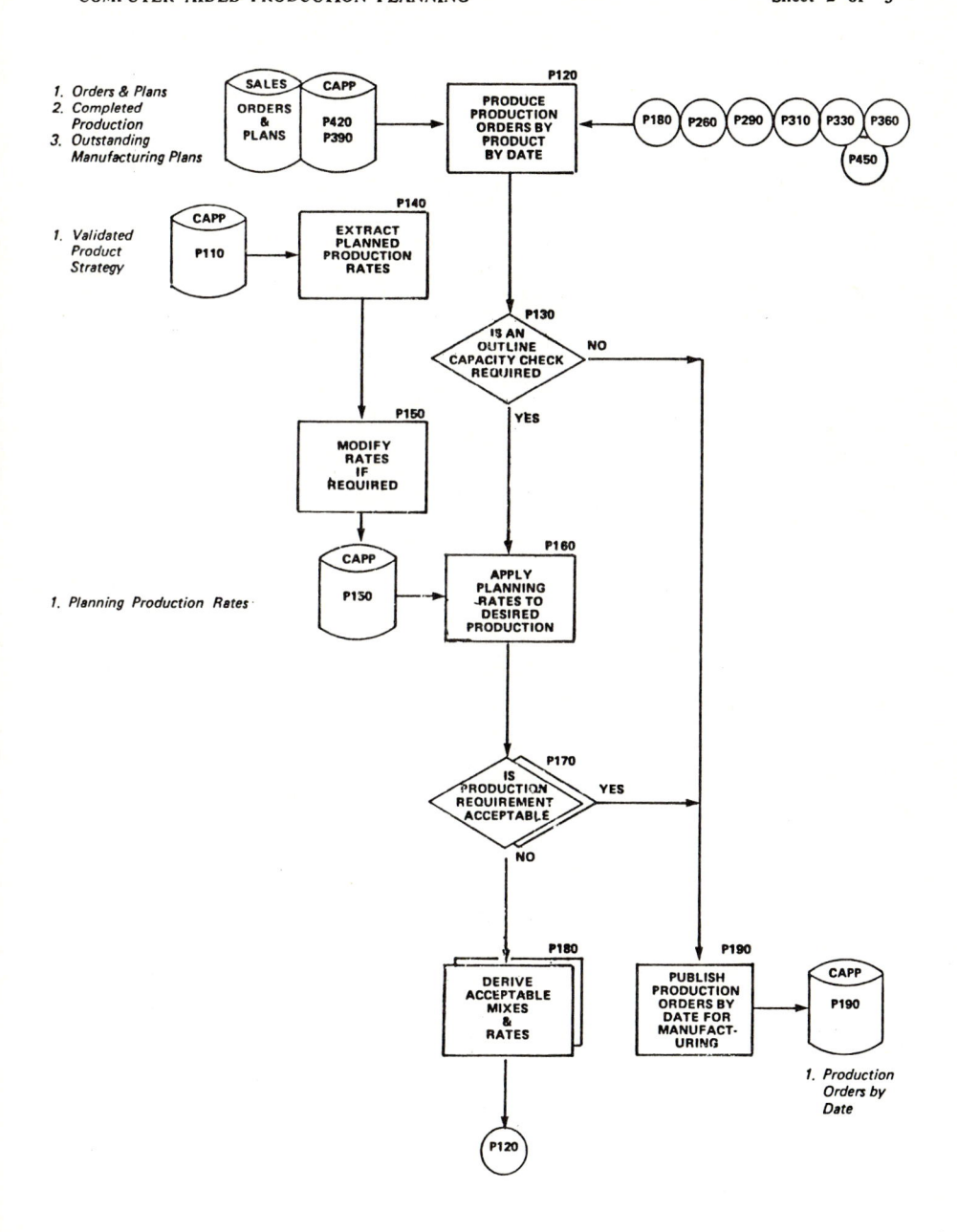

1. Orders & Plans
2. Completed Production
3. Outstanding Manufacturing Plans

1. Validated Product Strategy

1. Planning Production Rates

1. Production Orders by Date

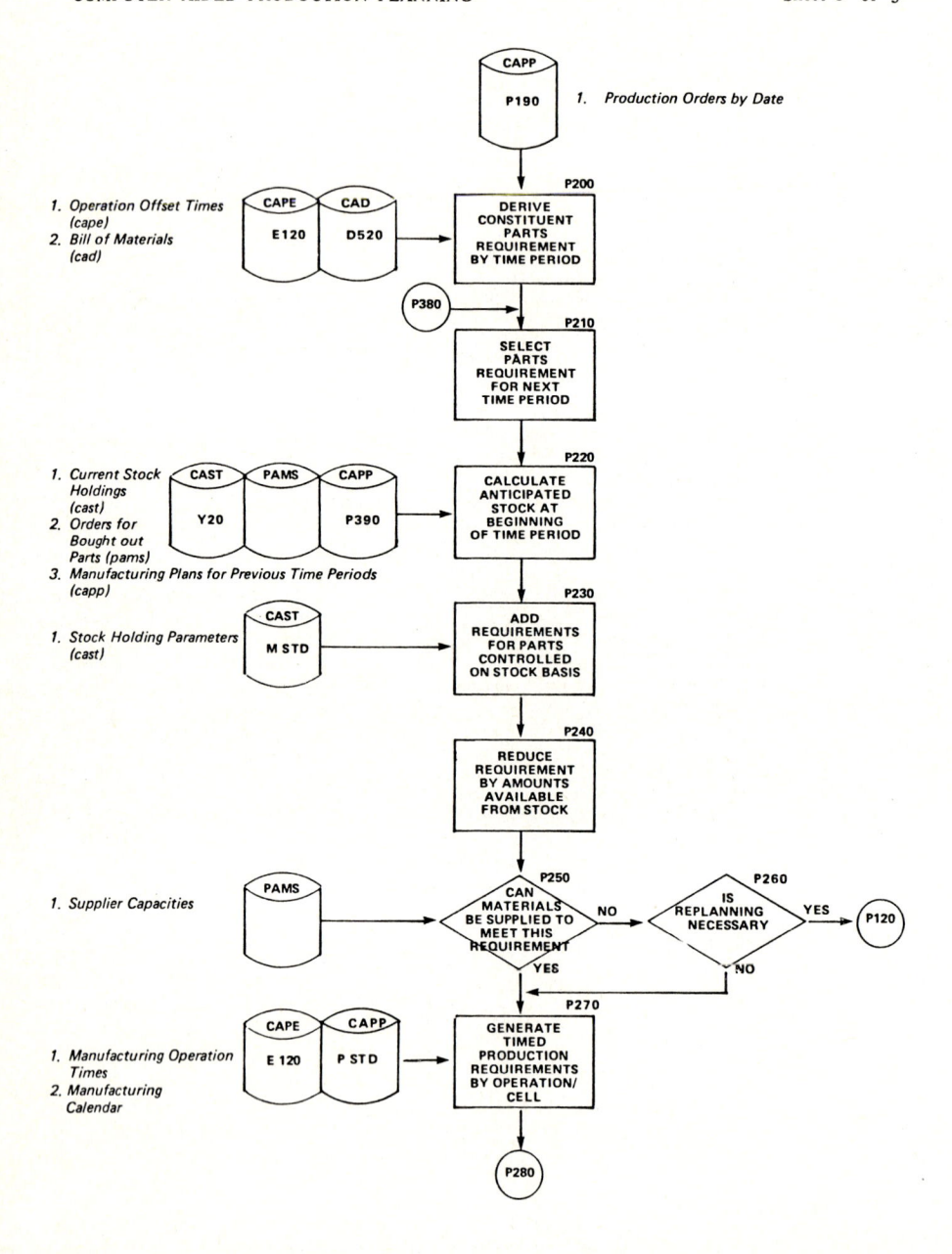

COMPUTER AIDED PRODUCTION PLANNING Sheet 3 of 5

```
                              ┌─────┐
                              │CAPP │
                              │P190 │   1. Production Orders by Date
                              └──┬──┘
                                 │
                                 ▼ P200
                          ┌──────────────┐
  1. Operation Offset Times   ┌────┐ ┌────┐   │   DERIVE     │
     (cape)              │CAPE│ │CAD │──▶│ CONSTITUENT  │
  2. Bill of Materials   │E120│ │D520│   │    PARTS     │
     (cad)               └────┘ └────┘   │ REQUIREMENT  │
                                         │ BY TIME PERIOD│
                                         └──────┬───────┘
                              ┌─────┐           │
                              │P380 │──────────▶│
                              └─────┘           ▼ P210
                                         ┌──────────────┐
                                         │   SELECT     │
                                         │    PARTS     │
                                         │ REQUIREMENT  │
                                         │   FOR NEXT   │
                                         │ TIME PERIOD  │
                                         └──────┬───────┘
                                                ▼ P220
  1. Current Stock    ┌────┐ ┌────┐ ┌────┐ ┌──────────────┐
     Holdings         │CAST│ │PAMS│ │CAPP│ │  CALCULATE   │
     (cast)           │ Y20│ │    │ │P390│▶│ ANTICIPATED  │
  2. Orders for       └────┘ └────┘ └────┘ │   STOCK AT   │
     Bought out                            │  BEGINNING   │
     Parts (pams)                          │ OF TIME PERIOD│
  3. Manufacturing Plans for Previous      └──────┬───────┘
     Time Periods (capp)                          │
                                                  ▼ P230
                              ┌────┐       ┌──────────────┐
  1. Stock Holding Parameters │CAST│       │     ADD      │
     (cast)                   │MSTD│──────▶│ REQUIREMENTS │
                              └────┘       │  FOR PARTS   │
                                           │  CONTROLLED  │
                                           │ ON STOCK BASIS│
                                           └──────┬───────┘
                                                  ▼ P240
                                           ┌──────────────┐
                                           │   REDUCE     │
                                           │ REQUIREMENT  │
                                           │  BY AMOUNTS  │
                                           │  AVAILABLE   │
                                           │  FROM STOCK  │
                                           └──────┬───────┘
                                                  ▼ P250         P260
                       ┌────┐             ◇ CAN ◇       ◇  IS   ◇
  1. Supplier Capacities│PAMS│────────────▶MATERIALS ──NO──▶ REPLANNING ──YES──▶(P120)
                       └────┘              BE SUPPLIED       NECESSARY
                                           TO MEET THIS
                                           REQUIREMENT         │NO
                                                  │YES          │
                                                  ◀─────────────┘
                                                  ▼ P270
                              ┌────┐ ┌────┐ ┌──────────────┐
  1. Manufacturing Operation  │CAPE│ │CAPP│ │   GENERATE   │
     Times                    │E120│ │PSTD│▶│    TIMED     │
  2. Manufacturing            └────┘ └────┘ │  PRODUCTION  │
     Calendar                              │ REQUIREMENTS │
                                           │ BY OPERATION/│
                                           │    CELL      │
                                           └──────┬───────┘
                                                  ▼
                                               (P280)
```

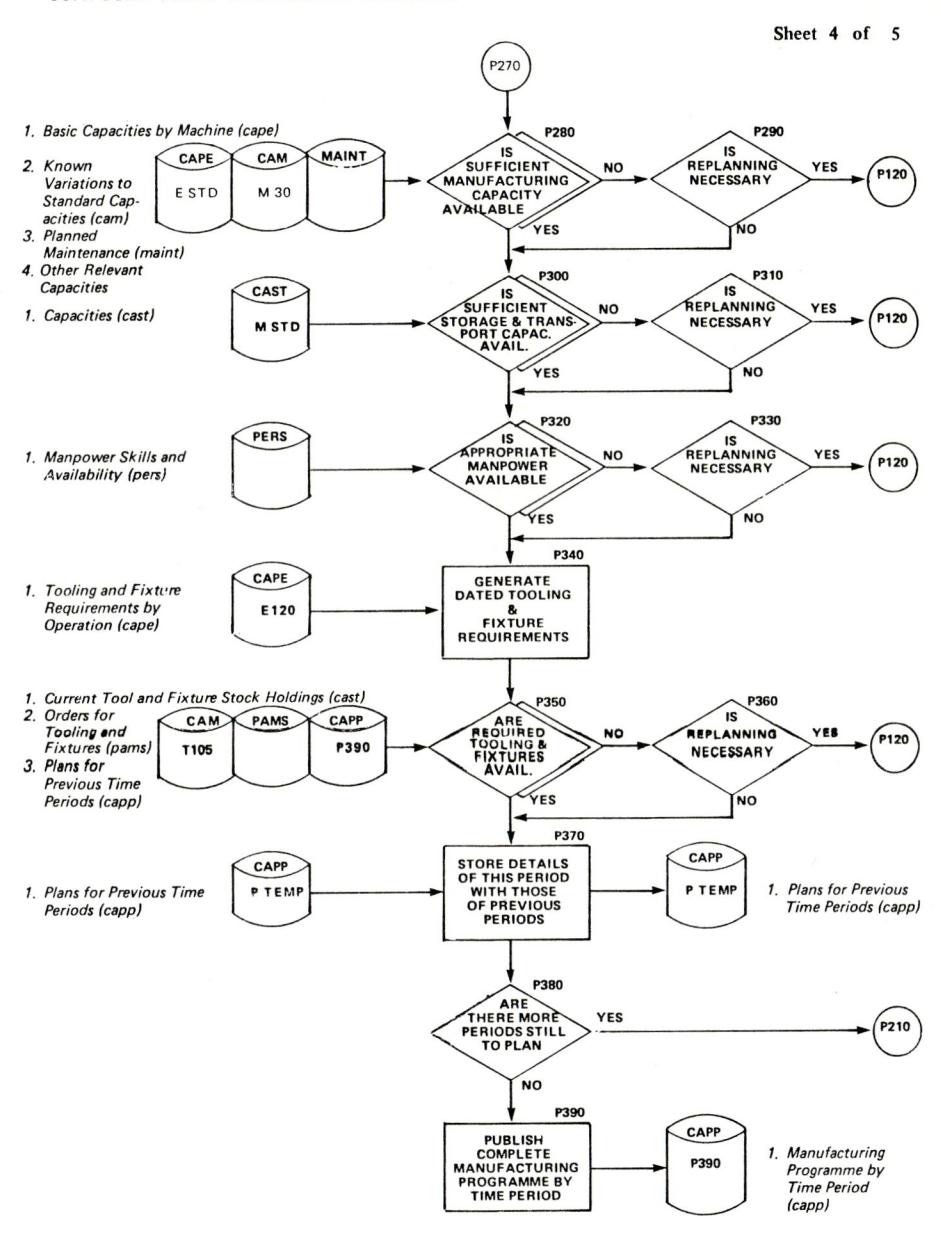

COMPUTER AIDED PRODUCTION PLANNING

Sheet 4 of 5

1. Basic Capacities by Machine (cape)
2. Known Variations to Standard Capacities (cam)
3. Planned Maintenance (maint)
4. Other Relevant Capacities

1. Capacities (cast)

1. Manpower Skills and Availability (pers)

1. Tooling and Fixture Requirements by Operation (cape)

1. Current Tool and Fixture Stock Holdings (cast)
2. Orders for Tooling and Fixtures (pams)
3. Plans for Previous Time Periods (capp)

1. Plans for Previous Time Periods (capp)

1. Plans for Previous Time Periods (capp)

1. Manufacturing Programme by Time Period (capp)

COMPUTER AIDED PRODUCTION PLANNING

Sheet 5 of 5

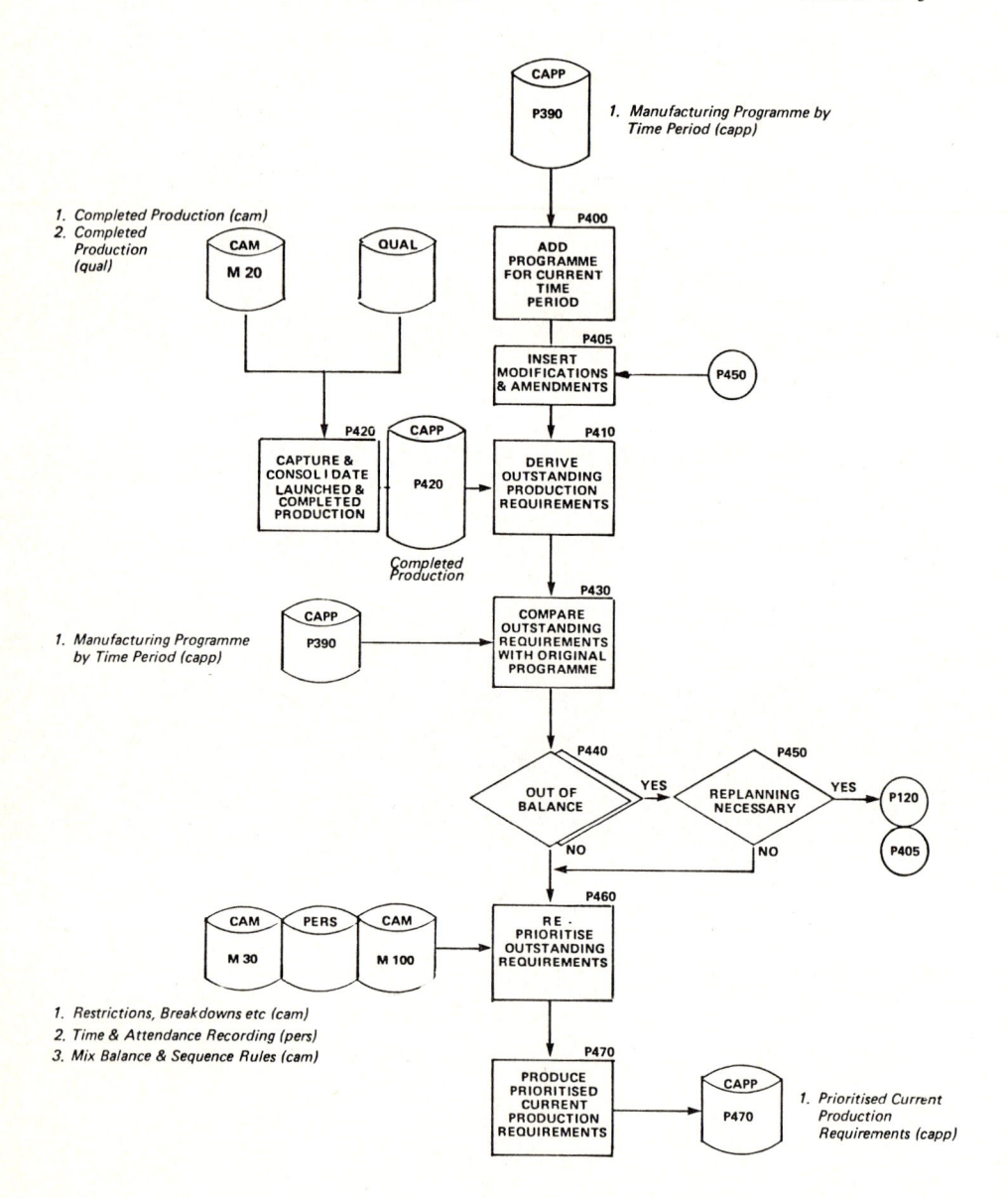

1. Manufacturing Programme by Time Period (capp) — CAPP P390

1. Completed Production (cam) — CAM M20
2. Completed Production (qual) — QUAL

P400 ADD PROGRAMME FOR CURRENT TIME PERIOD

P405 INSERT MODIFICATIONS & AMENDMENTS ← P450

P420 CAPTURE & CONSOLIDATE LAUNCHED & COMPLETED PRODUCTION → CAPP P420 Completed Production

1. Manufacturing Programme by Time Period (capp) — CAPP P390

P410 DERIVE OUTSTANDING PRODUCTION REQUIREMENTS

P430 COMPARE OUTSTANDING REQUIREMENTS WITH ORIGINAL PROGRAMME

P440 OUT OF BALANCE — YES → P450 REPLANNING NECESSARY — YES → P120 / P405
NO ↓ NO ↓

1. Restrictions, Breakdowns etc (cam) — CAM M30
2. Time & Attendance Recording (pers) — PERS
3. Mix Balance & Sequence Rules (cam) — CAM M100

P460 RE-PRIORITISE OUTSTANDING REQUIREMENTS

P470 PRODUCE PRIORITISED CURRENT PRODUCTION REQUIREMENTS → CAPP P470

1. Prioritised Current Production Requirements (capp)

Narrative description of
Computer Aided Production Planning network

Long-term planning sub-system (P10-P110)

This is purely a planning activity carried out to assess the viability and effects of future product strategies and plans. These may involve introducing new products, ceasing production of some products, changes in volumes and mix of products, or any combination of all these.

The results of the long-term planning activity enables product and production strategies to be assessed in terms of their effect on resources - current and planned. It will also enable a financial evaluation to be made in both capital and revenue terms of any proposed strategy. While this evaluation is outside the scope of the current work it is the most vital part of the forward planning exercise undertaken in some form, possibly even subconsciously, by every company.

Since the purpose of long-term planning is to examine probable effects, it may be necessary to carry out the exercise on several strategies to compare the results. Because of this it may include or be linked with modelling or simulation packages.

The definition of 'long-term' in calendar terms varies greatly between different industries. For instance, a manufacturer of simple fashion-influenced products may work only a few months or even weeks ahead, while a manufacturer of one-off heavy engineering products may plan years ahead.

Similar factors will also influence the frequency which such an evaluation is carried out.

P10   The product strategy which is to be evaluated must first be formalised into a coherent set of data with identified quantities, dates, etc.

Since the sources of this data may be diverse (sales, marketing, company directorate, etc), the formation of an agreed proposed strategy may be a substantial task.

P20   The product strategy data compiled in P10 will be in terms of product identities associated with time periods and quantities.

These products must first be broken down (if necessary) into the actual details which will be machined, ie. any assemblies of parts must be 'exploded' into lists of parts which will either be bought outside or manufactured.

This is achieved using a 'Bill of Materials', ie. a list of all parts, sub-assemblies and raw materials that constitute a particular product, showing the quantity of each required to produce one-off of the product.

The Bill of Materials may be completed in the case of current products or provisional in the case of products still being designed. In the case of products not yet at even the outline design stage, it will be necessary to input separately details of the critical parts which conceptually will form part of the product in order to make provision for their resource requirements.

The result of this first breakdown will be a list of parts giving quantities of each required, and broken down into time periods. These parts will be defined as either bought-out finished or manufactured - the manufactured parts being in turn associated with a bought-out requirement for raw material.

For each detail a further breakdown is then required into the separate manufacturing facilities needed to produce that detail, together with the elapsed time required by each facility.

As with the Bill of Materials, this information may be obtained from current manufacturing data, synthetic planning information or directly input estimates - depending upon the stage in the product's life.

Thus at the completion of this activity sets of data will be available - one set for each time period being considered.

Each of these sets of data will consist of a set of detail identities (probably part numbers) with a quantity required during the period being considered.

Where a detail is to be manufactured, not purchased, there will be a further set of data giving each of the production facilities required plus the time required of that facility to complete the quantity specified.

P30 Given the information produced at P20, it is now possible to sum the required capacity of each manufacturing facility during a time period.

Manufacturing facility information is held at the level most suitable within a particular application, which may be, for example, individual machine, machine group, FMS cell, process area - or any combination.

Having determined the capacity required of any particular production facility, this should now be compared with the capacity which it is planned to be available during the time period under consideration. This available capacity should take account of any known variations expected during that period; for example, commissioning of new or enhanced equipment, removal of facilities, planned increases or decreases in working periods.

A comparison of required capacity with available capacity will allow significant differences to be highlighted and reported on.

P40 If the comparison carried out at P30 has revealed a significant discrepancy between the manufacturing facilities available and those required for the proposed product strategy, it is necessary to take decisions (probably management decisions) as to what course of action, if any, is necessary as a result.

If spare capacity is revealed it may be unnecessary to take any action unless this is a continuing pattern, in which case removal of redundant facilities could be considered if appropriate - or an increase in the manufactured volumes if other considerations (eg. sales) allow.

If a shortage in capacity is to be met by increasing the utilisation of facilities (eg. overtime working, extra shifts), then the information on available facilities can be modified and P30 repeated.

If it is decided to cope with a shortfall in capacity by modifying, enhancing or increasing facilities (in conjunction with CAPE and CAM activities) the results of this planned action can also be reflected by a modification of the 'available facilities' information and repeating of P30.

An alternative to any of the above may be to modify the proposed product strategy to take account of the facilities which will be available. In this case a total re-run through the system may be necessary to check the total effects of this revised product strategy.

In practice, any combination of all of the above types of action, plus others, may be taken within P40 - but the end objective is to produce a proposed product strategy which is satisfactory to the manufacturing facilities which will be available at the relevant time.

P50   This activity is, in concept, the same as that at P30 (q.v.). In this case the task is to study the requirements for all items which need to be purchased to support the proposed product strategy. These include bought-out finished parts, raw materials for subsequent manufacturing, tooling, etc.

P60   This activity is, in concept, the same as that at P40 (q.v.), embracing such options as alternative suppliers of materials, alternative parts or tooling and amendment to the proposed strategy.

P70 and P80
        A repeat of the above types of activity with respect to storage facilities and space, plus transport - particularly automatic conveyors, automatic guided vehicle systems, etc.

P90 and P100
        A repeat of the above types of activity with respect to levels of and manpower, taking account of skills and abilities.

P110   When a product strategy has been evolved which is satisfactory, it is published as the basis for future planning in all company activities.

Medium term planning (P120-P390)

This set of activities is concerned with producing manufacturing programmes in response to actual identified requirements. The manufacturing programmes will be achievable, based on the knowledge of production resources, facilities and materials which should be available at the relevant dates. These programmes will be used as a basis for forward planning in such areas as purchasing, maintenance, labour loading, etc., and will also form the basis of shop loading procedures.

The definition of 'Medium Term' in calendar terms will vary greatly between different industries, depending on such things as forward order visibility, and lead times required for purchasing, planning, etc. Similarly the frequency with which such an exercise is carried out and the length and number of time periods covered will vary from application to application.

The general form of the exercise is to evaluate the requirements, in terms of production facilities and resources, for a desired production plan. Where the requirements do not match the available resources within a particular time period, action must be taken to reconcile them at a satisfactory level. The easiest form of action may be to take special action to amend or supplement the basic resources available. Where this still fails to meet the requirement, the desired date for production may be moved to a date at which resources are satisfactory. However, this action carries with it the danger of merely moving the problem (rather than eliminating it) and possibly compounding it at a later date due to 'knock-on' effects.

It is far more satisfactory, with anything other than small temporary imbalances, to produce a new manufacturing plan which is achievable.

P120 To establish the requirements from production activities, it is necessary to compile a desired list of items to be produced, with required dates and quantities.

There are many sources of such a list, such as firm customer orders, anticipated customer orders, statistical patterns of demand expectation, re-order point calculations - or any combination of these and other sources. The end objective, however, is a list of product identities, dates and quantities representing desired production.

It is necessary to take into account at this stage production which has already been achieved or which has already been planned for production in a period prior to the one now being considered.

P130 It may be considered sufficient for this list merely to be passed forward as a production requirement for direct evaluation against available resources and facilities. Alternatively, it may be prudent to carry out an outline checking exercise to test whether the desired programme has any obvious areas of imbalance which would cause rejection when submitted to detailed resource and facility evaluation. Removal of these imbalances from the submitted list will save both time and processing costs.

In many organisations this process is to some extent carried out subconsciously when the P120 requirements are being established - certain products have well-known maximum production rates, and production would not be sought at a rate greater than the maximum - even if potential sales existed.

P140 If an outline capacity check is to be carried out, the first step is to derive a set of production rates which are known to be realistic and generally achievable. A suitable base for this would be those rates derived from the long-term planning exercise, since they have been checked and verified against the facility and resource levels anticipated in the period under consideration.

P150   If additional factors exist which were not taken into account when the plan derived at P140 was set up, these should be used to modify the planned production rates. These may be changes in facility or resource levels which have resulted in a changed production capability, or a change in the product mix within known facility resource levels. Any changes made, however, must be made in the belief that the revised production rates which result are realistic and achievable.

P160   Having on one hand a set of desired production volumes, and on the other a set of achievable rates, the next step is to compare the two. For each time period being considered the desired production volumes are compared with the planning rates, and significant differences noted.

P170   Each of these significant differences is considered and its outline effect (increase or decrease) on the production facilities and resources considered. Small changes may be considered to be worthy of more detailed examination rather than outright rejection, and a change in mix which results in some increases and some decreases may again be considered worth examining further rather than rejecting.

However, a predominant change in one direction (particularly increase) may be considered unworthy of further consideration, as may a change in a particular area which is shown to have a requirement in excess of a known maximum. In such cases the production requirement list can be aborted before going into a lengthy detailed examination phase - thus saving further elapsed time and processing costs.

P180   If a desired production plan has been rejected, it must be redrawn in an attempt to produce a realistic achievable plan. To aid this process, the differences highlighted in P160 and P170 can be used both to show the areas in which revision is necessary, and the size and type of changes which should be attempted. However, the overall responsibility for the new plan must still rest with the sources of the original plan at P120.

P190   When a plan exists which is considered reasonable, this is made available for further detailed examination and evaluation.

P200   The list of production requirements by date from P190 needs to be 'exploded' to produce a list for all parts at all manufacturing levels - ie. a timed requirement for all parts from bought-out finished or raw materials, through any intermediate manufacturing, assembly or sub-assembly levels, to the finished product.

To enable this to be carried out, the system must have available to it the relationships between all parts making up the finished product (a 'Bill of Materials') and a knowledge of the production lead times necessary in the manufacturing/assembly process.

P210   Having achieved such a list of all requirements by time periods, the remaining processes can be considered time period by time period. Accordingly the time period to be considered next is selected.

P220   Each individual item can now be considered. By considering relevant factors for this part - eg. outstanding production plans, outstanding orders on external suppliers, current stocks - a calculation is made of the anticipated stock at the beginning of the time period being considered.

P230 and P240
        This anticipated stock holding should, where appropriate, be compared with stock holding parameters for that part. Any increase necessary to bring the holding of the part to a minimum stock holding level should be added to the requirement. Similarly, where expected stocks exceed the desired stockholding level, the requirement can be reduced by the relevant quantity.

P250   The requirement for material to be supplied from outside the manufacturing environment can now be considered. This may be bought-out finished parts, or raw materials for subsequent manufacturing activities.

The supply requirement is compared with the known capabilities of suppliers, to see whether the calculated requirement will cause supply problems.

P260   If a potential supply problem is highlighted, a decision must be taken as to whether this is a critical problem. There may be exceptional courses of action available (eg. persuading a supplier to increase his rate of supply, finding extra sources of supply, temporarily running stocks below safety levels) which will enable sufficient material to be found to support the proposed production level.

Alternatively, the production requirements will have to be amended in line with the available level of supply. This may be done either by moving the production period to one in which capacity does exist, or, in the case of major imbalance, by preparing a new production plan.

P270   The production requirement for each item is now examined and, using a set of manufacturing times, the production time required from each manufacturing element is derived.

Such manufacturing elements may be individual operations, groups of operations, specific sections of the total manufacturing process, FMS cells, or whatever breakdown of the total manufacturing activity is deemed most appropriate within any particular application.

P280   From P270, the amount of production time required from each manufacturing element has been derived and totalled. It is now necessary to compare this with the actual time available. This is calculated by taking the basic capacity of each element, and amending this in line with any known changes applying during the time period being considered, eg. holidays, scheduled maintenance, planned overtime working, installation of new equipment.

The comparison between the available time and the required time will highlight any areas of significant difference, and the magnitude of the discrepancy.

P290   The areas where such a discrepancy exists and the size of the discrepancies are analysed to see whether the discrepancies can be accommodated, eg. by extra working hours, usage of alternative equipment, outsourcing.

If it is considered impossible to cope with the loading required, then the production requirements will have to be amended in line with the capacities available, by moving the planned production date to one where capacity still exists, or by preparing a new production plan.

P300   A similar procedure to that carried out in P280 is repeated for storage and transportation facilities, eg. storage space, automatic store handling capabilities, AGV capacities, mechanical conveyor capacities, robot handling capabilities.

Similarly, areas of overload (and significant underutilisation) are highlighted for subsequent analysis.

P310   The areas of concern are analysed in a similar way to the analysis carried out at P290. If possible differences in capacity requirements will be accommodated by special action, but failing this, production must either be rescheduled for a period in which capacity exists, or a new production plan prepared.

P320 and P330

The type of exercise performed at P280/P290 and P300/P310 is repeated. This time the labour requirements in terms of numbers, skills and other special requirements are considered.

P340   Using a set of requirements for tooling and fixtures (including tool life) associated with the parts to be manufactured, a requirement for tooling and fixtures can be generated to support the manufacturing programme.

P350 and P360

By considering the requirements calculated in P340, and known requirements for previous periods, stock levels and delivery dates for orders already placed, a calculation is made of the tooling and

fixturing which will be available to support the manufacturing plan for the period in question. The added requirement can thus be calculated, and the viability of acquiring this by the due date considered.

Where a problem exists, decisions on the action to take can be taken in a similar way to the considerations made in P290/P310/P330.

P370   When the plan for the period in question has been totally and successfully evaluated, the results calculated and the initial plan on which those results are based, are recorded.

It is important to appreciate that the evaluation of the plan must be capable of being carried out against any production related resource which may, in a particular environment, be a potential constraint.

P380   If further time periods exist in the submitted plan for which evaluations and checks have still to be made, the next time period is now selected.

P390   If all time periods have been evaluated and checked, the results are made available to all activities requiring this information.

Short term planning (P400-470)

This set of activities is concerned with the actual sequencing and part-by-part planning of the production process and its associated facilities and functions.

The forward horizon is limited to that which is the necessary minimum required to organise the supply of available resources to support the production process. At this stage it is necessary to consider what it is desired to produce and match this with the resources and facilities which are actually available. In this, it is accepted that despite the medium term plan's formulation of an achievable manufacturing programme (and the necessary materials ordering, maintenance plans, etc., based on this), in the real short term unforeseen problems will occur, eg. plant breakdowns, non-delivery of materials, urgent changes of requirement. These must be taken into account by the short term planning function, in order to allow production to continue in the most efficient way of those products which have the highest priority, and which are capable of being manufactured.

This production should, in most cases, accord to the medium term plan, but must have the capability of reacting to exceptional states and emergencies as they occur.

The frequency with which this exercise is repeated will vary according to the type and nature of the manufacturing facility being controlled, from continual to any time period considered appropriate.

P400　　Since the medium term plan will in most cases cover several time periods, it is first necessary to select that part of the plan which is relevant to the current period.

P405　　At this point any immediate short term amendments can be incorporated. While it is accepted that in many production environments such amendments are inevitable and must be acknowledged, it must also be borne in mind by anyone authorising such amendments that a degree of disruption to other planned production will probably result. There may also be conflict with action taken previously, eg. the ordering of special materials or tooling whose lead times require action based on the medium term manufacturing programme.

P410　　In order to produce a current list of requirements it is necessary to remove from the programme those items already completed or in progress.

P420　　Information must be derived identifying those items complete or in progress. This will probably be direct from the manufacturing control system, but elements may also be derived from storage or quality systems - depending upon local requirements and configurations.

P430 to P450

An optional stage, which will be of value in some circumstances, is to check the continuing validity of the original manufacturing programme. If the balance and mix of the original programme is checked against that of the achievement so far, a level of compliance can be measured.

If the balance is significantly different (measured by locally applied parameters) from that conceived in the original programme, a warning should be generated to warn that a potential problem situation exists.

If, after examination, it is considered that the situation is still acceptable, the process can continue. If, however, it is considered that action is required then this action can be instigated, eg. the generation of a new manufacturing programme which takes account of the factors which have caused the detected imbalance.

P460   Having obtained a list of outstanding manufacturing requirements, the next step is to allocate a latest priority to each of them. The rules which govern the setting of relative priorities may be many, depending upon local circumstances. However, the system must allow these rules to be weighted and considered in the optimum way to generate a prioritised requirement list which best fits the local conditions at the time at which it is generated. Equally, the mechanisms must exist which allow the relevant selection criteria to be continuously updated in order to reflect the current manufacturing and decision-making status, at any particular time.

P470   Having set the latest priorities on all items remaining on the current manufacturing programme, a list of items in priority order is made available to the actual machine loading activity.

The frequency with which a list is issued by carrying out the short term scheduling activity will depend upon the type of business and local circumstances (eg. frequency of change of selection criteria). At the extreme it may be a nearly continuous activity, or possibly only once per shift in a stable environment.

The frequency of issue and the length of the list may also be influenced by needs for support activities to anticipate manufacturing patterns, eg. tool setting, fixturing, material movement. Issues may be made on a single item basis.

Computer Aided Production Planning (CAPP) sub-systems

The Flowchart for the Production Planning activities in the ESPRIT Pilot Study has been broken down into sub-systems:

1   Long term planning
       (Nodes P10 to P110)

2   Order assimilation
       (Nodes P120 to P190)

3   Medium term planning
       (Nodes P200 to P390)

4   Short term planning
       (Nodes P400 to P470)

Long term planning sub-system (P1)

P1.1 Sub-system description

This is purely a planning activity carried out to assess the viability and effects of future product strategies and plans. These may involve introducing new products, ceasing production of some products, changes in volumes and mix of products, or any combination of all these.

The results of the long-term planning activity will enable product and production strategies to be assessed in terms of their effect on resources - current and planned. It will also enable a financial evaluation to be made in both capital and revenue terms of any proposed strategy. While this evaluation is outside the scope of the current work it is the most vital part of the forward planning exercise undertaken in some form, possibly even subconsciously, by every company.

Since the purpose of long-term planning is to examine probable effects, it may be necessary to carry out the exercise on several strategies to compare the results. Because of this it may include or be linked with modelling or simulation packages.

The definition of 'long-term' in calendar terms will vary greatly between different industries. For instance, a manufacturer of simple fashion-influenced products may work only a few months or even weeks ahead, while a manufacturer of one-off heavy engineering products may plan years ahead.

Similar factors will also influence the frequency with which such an evaluation is carried out.

Long term planning sub-system

P1.2 Flowchart

Computer Aided Production Planning

P1.2. SUB SYSTEM FLOWCHART Sheet 1 of 1

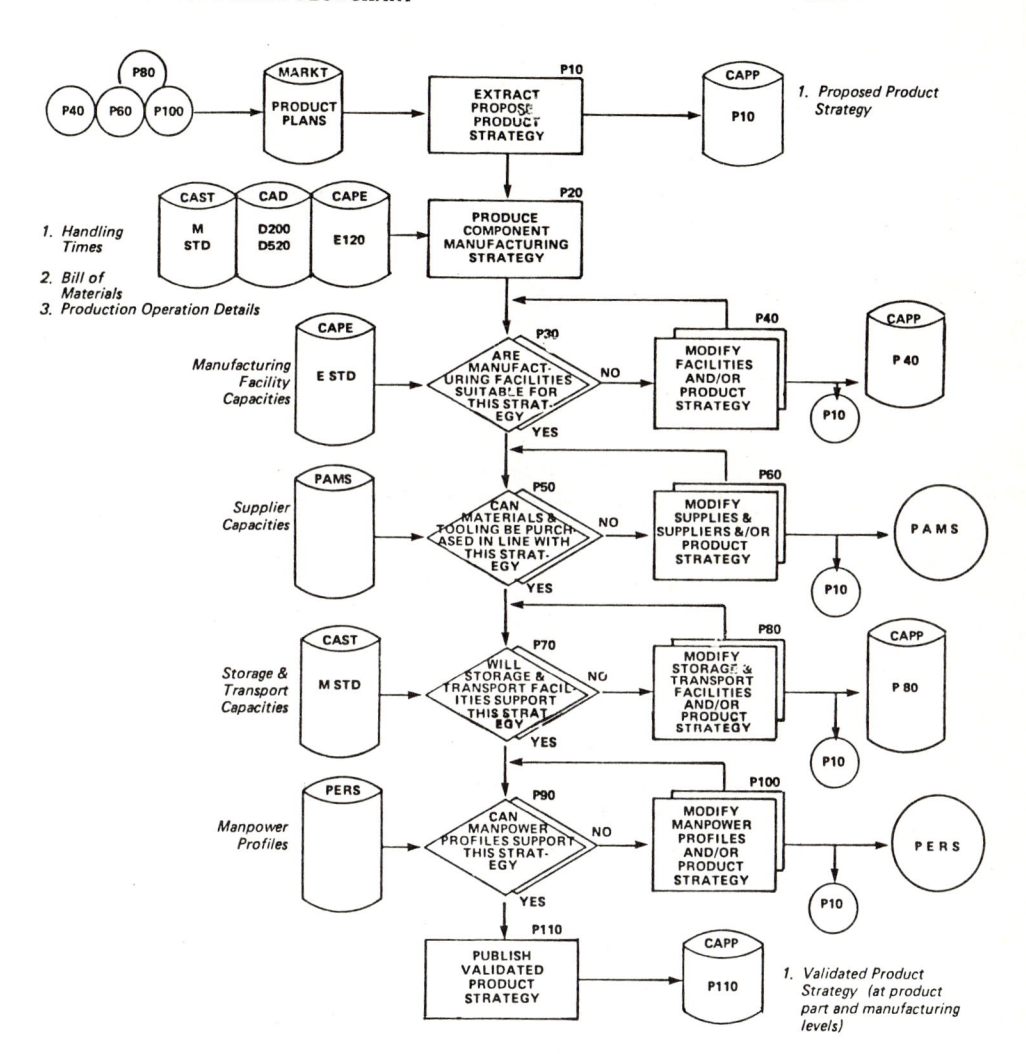

P1.3 **Design rules - Long term planning**

P1.3.1 **Input data**

The sub-system must be capable of accepting the minimum data items listed below:

Data is required relating to:
- Product Strategy (Section P1.3.1.1)
- Bill of Materials, ie. Structure of Product (Section P1.3.1.2)
- Capacity Requirements (Section P1.3.1.3)
- Available Capacity (Section P1.3.1.4)

P1.3.1.1 **Information relating to product strategy**
- **Product Identity**

    A unique identification of each manufactured product which forms part of the proposed product strategy.

- **Quantity**

    The quantity of each product which the strategy proposes to produce within each time period covered by the strategy - each time period will therefore have associated with it a unique set of product identities and quantities

- **Time Period**

    Each time period covered by the strategy must be uniquely defined, and allow for definition by any combination of:

    Identity (1986; Period 6; 3rd Quarter, etc.)

    Length (Year; month; 30 days, etc.)

    Date (April-June 1985; 4-18 May 1986, etc.)

    The precise length of time period and method of definition will be determined by local requirements.

P1.3.1.2 **Information relating to bill of materials**
(Structure of Product)
- **Product Identity**

    The unique identity of each manufactured product being considered.

- **Constituent Parts**

    The unique identities of all the details required to construct the manufactured product.

- **Quantity Required**

    For each detail within a product, the quantity of that detail required to manufacture one manufactured product

    eg. No. off, length, weight.

- **Source**

    For each detail, a definition of whether it is purchased in its completed form, or alternatively whether production operations are carried out on it.

    The Bill of Materials may be a direct one-level list of parts, or a multi-level structured list identifying intermediate assemblies - depending on other local requirements.

It must be possible to interrogate an existing Bill of Materials if this exists for current products, and merge this with other inputs for details relating to planned products.

P1.3.1.3 **Information relating to capacity requirements**

For each detail requiring production operations to be carried out on it, the following information must be capable of being assimilated.

- **Detail Identity**

    A unique identity for each detail (eg. part number).

- **Facility Identity**

    A unique identity for any facility used in the manufacture of the identified detail for which capacity checking is required (eg. machine tool number, cutting tool number, labour skill grade, pallet type).

- **Facility Type**

    An identification of the type of facility according to local grouping (eg. machine tool, cutting tool, storage unit).

- **Volume or Quantity**

    The amount of use of each identified facility required to produce one detail (eg. 20 minutes, 1 storage rack).

    Thus each detail identity will be associated with a number of facility identity/type/volume records, sufficient to cover all of the production resources required by that detail, and for which capacity requirement checks are required.

    It must be possible to build-up this information from a multiple number of sources, including both existing records and files for known parts and synthetic values or estimates input to cover planned parts.

P1.3.1.4 **Information relating to available capacity**

For each facility for which capacities have been identified the following information must be capable of being assimilated.

- **Facility Identity**

    A unique identity for any facility being considered. The identity must be the same as that used for capacity requirement purposes.

- **Facility Type**

    Again, as in capacity requirement.

- **Facility Capacity**

    The normally available capacity for each identified facility. The units of measurement used must be the same as those used for capacity requirement purposes.

    It must be possible to build-up this information from a number of different sources or files, including different sets of information existing for current facilities, and estimated capacities specifically entered for planned facilities.

P1.3.2 **Input reformatting**

The sub-system must be capable of reformatting, as required, input data into the form required by the sub-system. It is assumed that the input data will be structured in a disciplined and identified manner.

P1.3.3 Time periods

The time periods used for long term planning purposes may be of any length, but each should be capable of identification as outlined in Input Section P1.3.1.1. The various time periods involved in any one planning exercise may themselves be of differing lengths.

P1.3.4 Capacity types

Any factor which is production related and quantifiable by detail should be capable of use in long term planning. The quantity used should be directly related to the type of capacity being considered, eg. physical quantity, time, space.

P1.3.5 Reporting by exception

Reporting from the system should be possible in total or by exception, eg. reporting only on those items where a mis-match in capacity exists in excess of a user defined level.

P1.3.6 Synthetic figures or estimates

Where synthetic figures or estimates have been used for products or details not currently being manufactured, this must be identifiable on any output if required.

P1.3.7 Modification to capacity level

The capacity available overall or from any individual facility; type of facility or facility group should be capable of modification by a fixed amount or percentage, in order to assess the effect of potential facility changes, shift pattern variations, etc.

P1.3.8 Output data

The sub-system must be capable of producing data to the minimum levels specified in this section.

It must be in a form which will facilitate reformatting by a subsequent module, ie. all items must be structured in a disciplined and identified manner.

Data is required relating to:

- Facility Information (Section P1.3.8.1)
- The details to be produced (Section P1.3.8.2)

P1.3.8.1 Information relating to facilities

- **Facility Type**

    An identification of the type of facility according to local grouping (eg. machine tool, cutting tool, labour).

- **Facility Identity**

    A unique identity of any facility used in manufacturing for which a capacity check has been carried out (eg. machine tool number, cutting tool number, labour skill type, pallet type).

- **Time Period**

    The identity of each particular time period for which the long term planning exercise has been carried out for a particular facility.

- **Detail Identity**

    The identity of any detail using a specific facility during a specific time period.

- **Quantity Required**

  The quantity, volume, time, etc., of any particular facility required by the detail in the time period.

- **Total Quantity Required**

  The total of all the above quantities summed for a particular facility within a particular time period.

- **Total Capacity**

  The total available capacity of a particular facility in a particular time period, expressed in the same units as the quantity required, above.

- **Capacity Surplus or Shortfall**

  The difference between the total capacity and the total quantity required.

P1.3.8.2 **Information relating to detail**

- **Detail Identity**

  The unique identity of the detail being considered.

- **Time Period**

  as in P1.3.1.1

- **Quantity**

  The production volume of the detail proposed for the time period.

- **Facility Identity**

  The identity of any facility used in the manufacture of the detail under consideration.

- **Volume or Quantity**

  The volume, quantity, time, etc., of the specified facility required to produce the specified quantity of the identified detail during the time period under consideration.

Order assimilation sub-system (P2)

P2.1 Sub-system description

This set of activities is required to collect orders together prior to manufacturing so that the forward manufacturing requirement can be dealt with in an ordered and disciplined manner.

The word "ORDER" is used to describe a production requirement, ie. a commitment to procure raw materials, and use resources to carry out manufacturing processes in order to produce end products. The requirement for those end products may be to satisfy the known need of a customer, or it may be an internally generated speculative requirement in order to have finished products available to meet customer requirements which are forecast for the future. The important factor is that the order is an authority to produce a specified volume of a specified product at a specified date.

The facility to carry out, if required, an outline feasibility checking of the size and balance of the bank of orders created is also included.

Order assimilation sub-system

P2.2 Flowcharts

P2.2. SUB SYSTEM FLOWCHART

Sheet 1 of 1

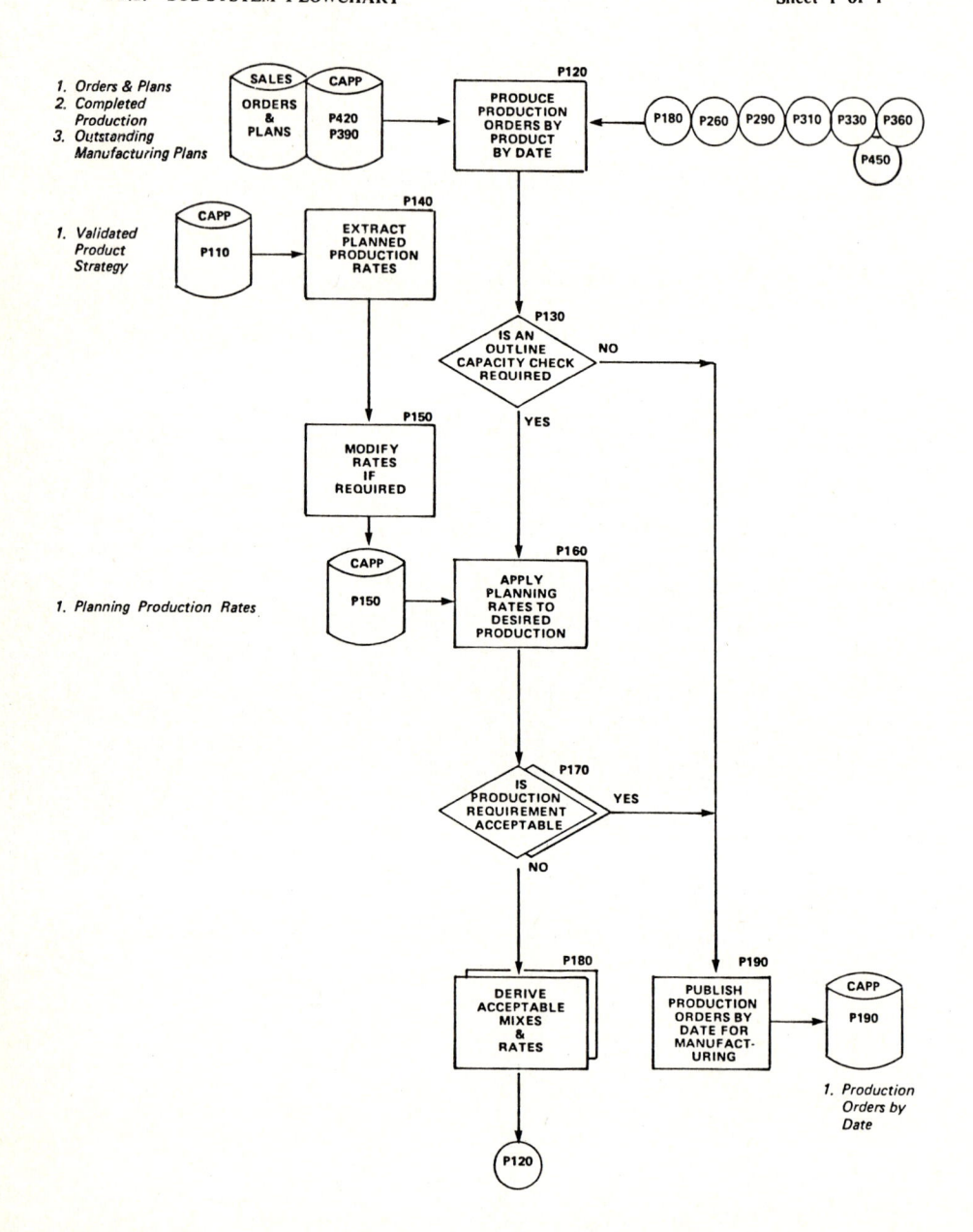

P2.3 Design rules - order assimilation

P2.3.1 Input data

The sub-system must be capable of accepting the minimum data items listed below:

P2.3.1.1 Information relating to desired production

- **Order Identity**

    A unique identity by which a specific authorised production requirement can be referenced.

- **Product Identity**

    A unique identity by which any product included in any order can be referenced.

- **Date**

    The date at which the completed product is required. It may be in the form of a calendar date, a time period, (eg. a week), or a precise time within a calendar date - depending on the nature of the manufacturing being carried out.

- **Quantity**

    The quantity or volume of production of the identified product required at the specific date on a specific order.

- **Priority**

    An additional indication of specific priority attached to an order, to be read independently of, or in conjunction with, the required date.

    The use of such an indication (if used) will depend on the nature of the manufacturing operation, but may be used, for example, to show absolute priority irrespective of date; or to denote a specific condition, eg. "for exhibition" which has particular implications.

P2.3.1.2 Information relating to production rates

- **Product Identity**

    A unique identity of a product as in 2.3.1.1.

- **Quantity**

    A quantity or volume of production of the identified product which can be produced in a given time period under normal circumstances.

- **Time Period**

    The time period on which the above quantity is based (eg. one week, one month).

P2.3.1.3 Information relating to time periods

ie. The periods of time over which production is to be planned and executed.

- **Period Identity**

    A unique identity of each time period to be used, eg. Period 1, 1986; week 43, 1985; 27 July 1989.

- **Period Duration**

    The start and end points of each period identified, defining a duration.

P2.3.2 **Input reformatting**

The sub-system must be capable of reformatting, as required, input data into the form required by the sub-system. It is assumed that the input data will be structured in a disciplined and identified manner.

P2.3.3 **Date and time period**

The sub-system must be capable of relating any given time and date with the time period to which it corresponds.

P2.3.4 **Unacceptable loads**

If an unacceptable load condition is detected, the system must be capable of producing a list (prioritised if required) of desired production versus capability for subsequent evaluation and action.

P2.3.5 **Production rates**

The planning production rates must be capable of being derived from the long-term production plan or created independently. In either case they should be capable of modification to take account of changes in production capability.

P2.3.6 **Order input**

The sub-system must be capable of receiving information on new order requirements as a batch submission or in an "as required" flow, building up this information until further action is required.

P2.3.7 **Output data**

The sub system must be capable of producing data to the minimum levels specified in this section. It must be in a form which will facilitate reformatting by a subsequent module, ie. all items must be structured in a disciplined and identified manner. Output will normally be required in a consolidated batch form for subsequent input to a production planning sub system.

- **Order Identity**

    A unique identity by which each specific authorised production requirement can be referenced.

- **Product Identity**

    A unique identity by which any product included in any order can be referenced.

- **Quantity**

    The quantity or volume of production of the identified product related to a specific order.

- **Time Period**

    The planned time period during which a specific product/order is required to be completed.

- **Date**

    The specific date or time within a time period by which completion of a specified product/order is required.

- **Priority**

    Any additional prioritisation information as outlined in 2.3.1.1.

Medium term planning sub-system (P3)

P3.1   **Sub system description**

This set of activities is concerned with producing a set of manufacturing programmes in response to actual identified requirements. The manufacturing programmes will be achievable, based on the knowledge of production resources, facilities and materials which should be available at the relevant dates. These programmes will be used as a basis for forward planning in such areas as purchasing, maintenance, labour loading, etc., and will also form the basis of shop loading procedures.

The definition of 'Medium Term' in calendar terms will vary greatly between different industries, depending on such things as forward order visibility, and lead times required for purchasing, planning, etc. Similarly the frequency with which such an exercise is carried out and the length and number of time periods covered will vary from application to application.

The general form of the exercise is to evaluate the requirements, in terms of production facilities and resources, for a desired production plan. Where the requirements do not match the available resources within a particular time period, action must be taken to reconcile them at a satisfactory level. The easiest form of action may be to take special action to amend or supplement the basic resources available. Where this still fails to meet the requirement, the desired date for production may be moved to a date at which resources are satisfactory. However, this action carries with it the danger of merely moving the problem (rather than eliminating it) and possibly compounding it at a later date due to 'knock-on' effects.

It is far more satisfactory, with anything other than small temporary imbalances, to produce a new manufacturing plan which is achievable.

Medium term planning sub system

P3.2 Flowchart

P3.2. SUB SYSTEM FLOWCHART

Sheet 1 of 2

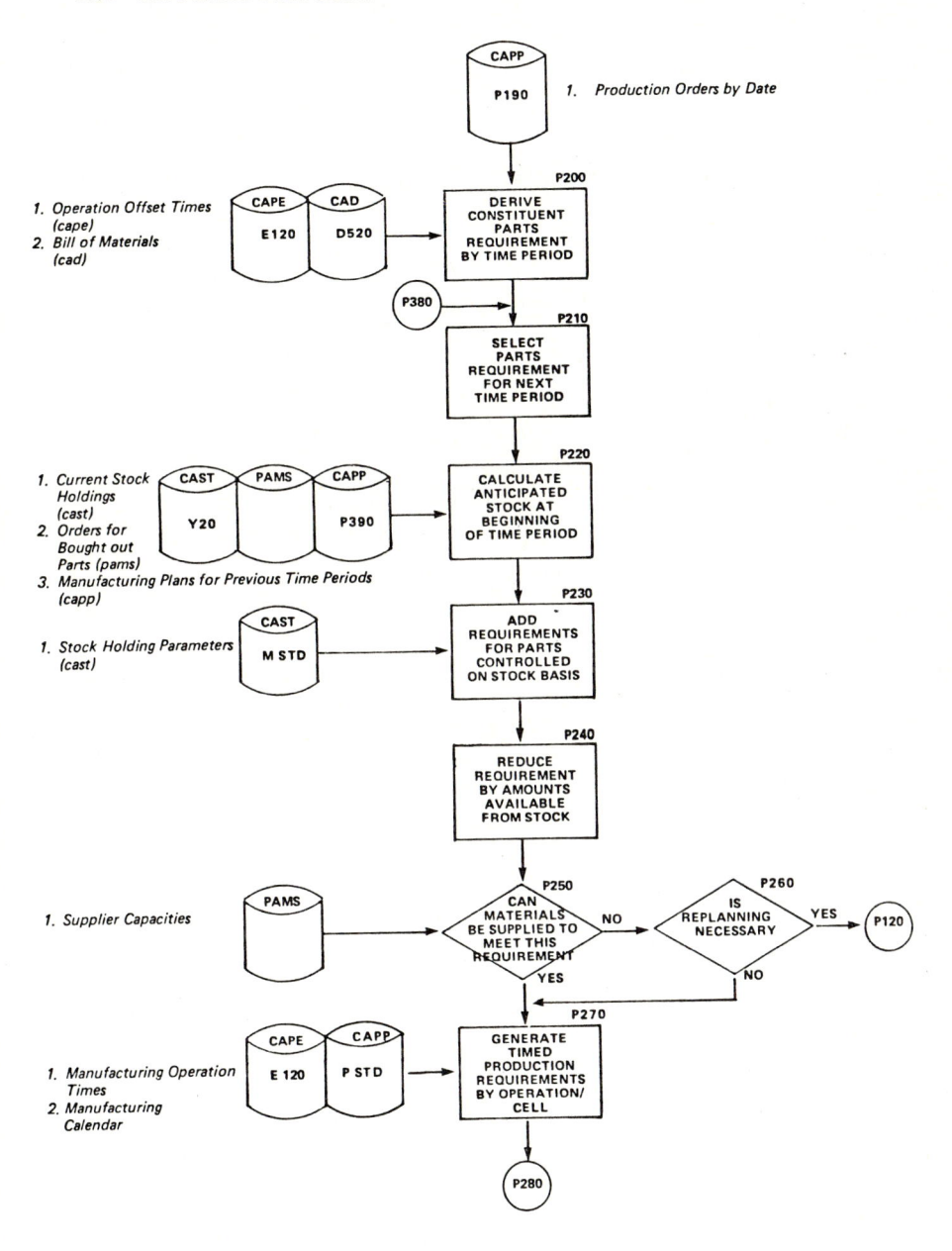

P190 (CAPP) — 1. Production Orders by Date

P200 DERIVE CONSTITUENT PARTS REQUIREMENT BY TIME PERIOD
- CAPE E120 — 1. Operation Offset Times (cape)
- CAD D520 — 2. Bill of Materials (cad)

P380

P210 SELECT PARTS REQUIREMENT FOR NEXT TIME PERIOD

P220 CALCULATE ANTICIPATED STOCK AT BEGINNING OF TIME PERIOD
- CAST Y20 — 1. Current Stock Holdings (cast)
- PAMS — 2. Orders for Bought out Parts (pams)
- CAPP P390 — 3. Manufacturing Plans for Previous Time Periods (capp)

P230 ADD REQUIREMENTS FOR PARTS CONTROLLED ON STOCK BASIS
- CAST M STD — 1. Stock Holding Parameters (cast)

P240 REDUCE REQUIREMENT BY AMOUNTS AVAILABLE FROM STOCK

P250 CAN MATERIALS BE SUPPLIED TO MEET THIS REQUIREMENT
- PAMS — 1. Supplier Capacities

NO → **P260** IS REPLANNING NECESSARY — YES → P120
NO ↓
YES ↓

P270 GENERATE TIMED PRODUCTION REQUIREMENTS BY OPERATION/CELL
- CAPE E 120 — 1. Manufacturing Operation Times
- CAPP P STD — 2. Manufacturing Calendar

P280

P3.2 SUB SYSTEM FLOWCHART

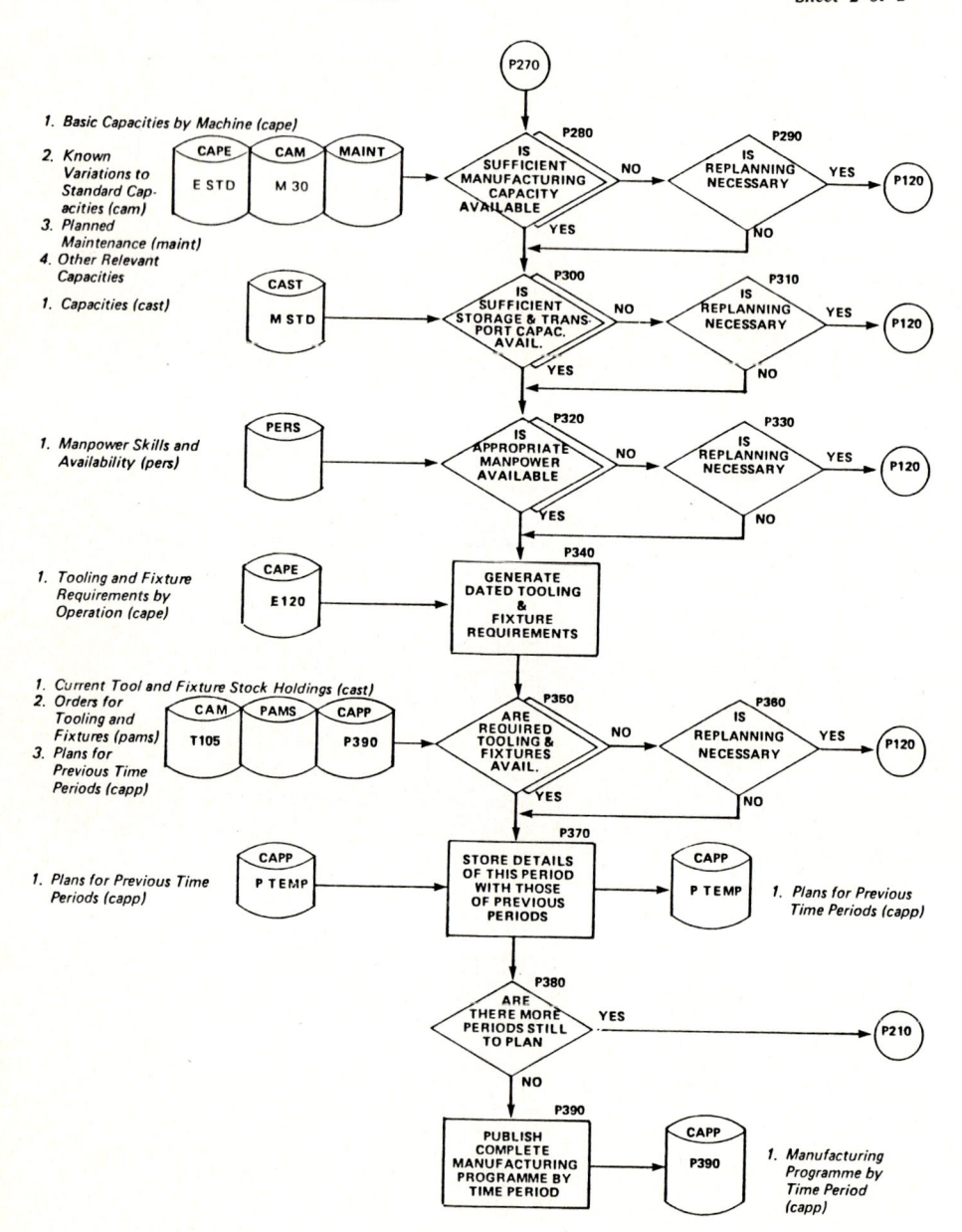

P3.3	**Design rules - medium term planning**
P3.3.1	**Input data**

The sub-system must be capable of accepting the minimum data items listed below; it is anticipated that a complete batched input will be submitted.

- **Order Identity**

    A unique identity by which each specific authorised production requirement can be referenced.

- **Product Identity**

    A unique identity by which any product included in any order can be referenced.

- **Quantity**

    The quantity, or volume of production, of the identified product related to a specific order.

- **Time Period**

    The planned time period during which a specific product/order is required to be completed.

- **Date**

    The specific date or time within a time period by which completion of a specified product/order is required.

- **Priority**

    Any additional prioritisation information attached to an order, to be read independently of, or in conjunction with, the required date.

P3.3.2	**Input re-formatting**

The sub-system must be capable of re-formatting, as required, input data into the form required by the sub-system. It is assumed that the input data will be structured in a disciplined and identified manner.

P3.3.3	**Priority**

The facility must exist to hold an indication of relative priority against every order input to the sub-system. This may be a specific priority indicator, a special status indicator, a required date, or any combination.

It must be possible to subsequently amend these priorities.

If calculation of priority is based on various data using an algorithm to derive the priority, automatic recalculation must be possible if any of the base data items change.

P3.3.4	**Other facilities and resources**

It must be possible to add to the facilities and resources specifically mentioned any additional production-related facilities and resources required by a particular user of the sub-system. In such a case it must be possible to provide a relationship between level of production and the additional resource, such that the amount of resource required to support any given level and mix of production can be calculated.

P3.3.5	**Reporting of resource mismatches**

Where a significant mismatch occurs in required facilities/resources versus available facilities/resources, reporting must be available at a detailed level in order to allow reasoned decisions and amendments to be made.

For example, in the case of a manufacturing capacity shortage at a particular machine, a breakdown of all parts requiring that machine during the time period in question with volumes, times, orders, relative priorities, etc.

The level of significance at which mismatches are reported should be determined by the user of the sub-system

P3.3.6 **Automatic rescheduling**

If the sub-system moves production of any part to a time period other than that resulting naturally from lead time considerations (eg. balancing of capacities) the capability must exist to report on all such re-scheduling actions.

P3.3.7 **Temporary capacity changes**

The system must allow temporary changes to be made to standard facility and resource levels, to an identified new level and for an identified period, eg. to cover holidays, planned maintenance, scheduled overtime working.

Permanent changes are regarded as part of the standard maintenance routines associated with external files of information accessed by the sub-system.

P3.3.8 **Reporting levels**

When a successful plan has been calculated and completed, it must be possible to report on any aspect of the plan in a flexible manner to meet the needs of any individual user of the sub-system.

eg.
- Reporting should be possible at any level, ie. by resource, product identity, order identity, etc.
- Reports should be capable of being at a summarised level where such a summary level can be logically identified.

The structure, aspects reported on, content and format of reports must be capable of change in a rapid and inexpensive manner, in order to meet changing user requirements.

P3.3.9 **Retention of input data**

The initial data input to the sub-system must be retained until a successful plan has been completed totally. It must be retained in a form which will allow modification of the input (as opposed to complete re-entry) in the event of a significant problem forcing a change in the data input.

P3.3.10 **Replanning**

It must be possible to subsequently replan a section of the overall plan for which a plan has already been produced. When this happens, and requirements differ from those already published, the differences must be available for all normal reporting levels.

P3.3.11 **Manufacturing order identity**

It must be possible to generate a specific manufacturing order identity for each detail manufacturing requirement within a time period. This may be the same as (if appropriate) or at least referenced back to, the order identity by which the final product is referenced.

P3.3.12 **Output data**

The sub-system must be capable of producing data to the minimum levels specified in this section.

It must be in a form which will facilitate reformatting by a subsequent module, ie. all items must be structured in a disciplined and identified manner.

A set of output data must be provided for each identified time period.

- **Manufacturing Order Identity**

    The unique identity by which a specific authorised production requirement of a detail has been referenced.

- **Detail Part Identity**

    A unique identity which references each detail requiring manufacturing/procurement, etc. to achieve a product/order manufacturing requirement. Where, due to the manufacturing process, the nature and therefore the identity of the detail changes, both initial and final identities must be given.

- **Quantity**

    The quantity, or volume of production, of each identified product or detail required in the identified time period.

- **Date**

    The specific date or time by which each completed product or detail is required.

- **Priority**

    Any additional prioritisation information attached to a detail/product/order, to be read independently of, or in conjunction with, the required date. It may also indicate any special status or type of order category defined by the user.

- **Resource/Facility**

    A unique identity by which each resource or facility which is required in the production of each detail within the specified time period can be identified.

- **Resource Measurement**

    The unit of measurement used to quantify the particular resource (hours, minutes, litres, metres, numeric quantity, kilos, kilowatts, etc., etc.).

- **Resource Requirement**

    The amount of each specific resource required for each detail/product/order within the specified time period.

Output will normally be required in a consolidated batch form for subsequent input to a variety of systems (eg. Short Term Planning, Material Procurement, Tooling Planning, Labour Requirement Planning, Energy Planning). Each of these systems will have differing requirements in terms of the specific data items required, and ease of structuring and producing such varied outputs is therefore a requirement of this system.

Short term planning sub-system (P4)

P4.1 Sub-system description

This set of activities is concerned with the actual sequencing and part-by-part planning of the production process and its associated facilities and functions. It achieves this by releasing orders for production in required quantities and frequencies.

The forward horizon is limited to that which is the necessary minimum required to organise the supply of available resources to support the production process. At this stage it is necessary to consider what it is desired to produce and match this with the resources and facilities which are actually available. In this, it is accepted that despite the medium term plan's formulation of an achievable manufacturing programme (and the necessary materials ordering, maintenance plans, etc., based on this), in the real short term unforeseen problems will occur, eg. plant breakdowns, non-delivery of materials, urgent changes of requirement. These must be taken into account by the short term planning function, in order to allow production to continue in the most efficient way of those products which have the highest priority, and which are capable of being manufactured.

This production should, in most cases, accord to the medium term plan, but must have the capability of reacting to exceptional states and emergencies as they occur.

The frequency with which this exercise is repeated will vary according to the type and nature of the manufacturing facility being controlled, from continual to any time period considered appropriate. In the same way the number of new parts released will vary according to local circumstances.

Short term planning sub-system

P4.2 Flowchart

P4.2. SUB SYSTEM FLOWCHART　　　　　　　　　　　　　　　　　　Sheet 1 of 1

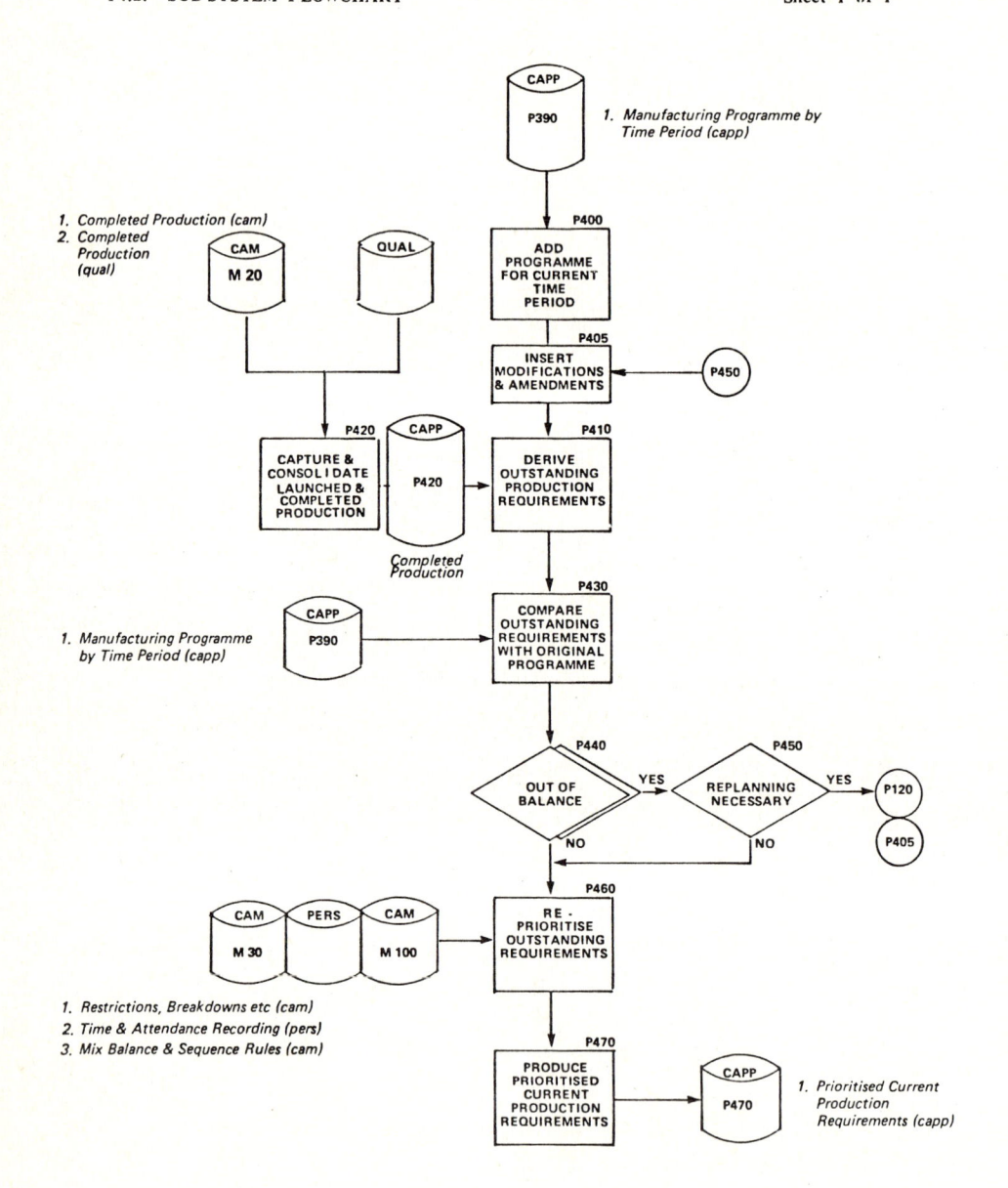

Computer Aided Production Planning 145

P4.3   **Design rules - short term planning**

P4.3.1   **Input data**

The sub-system must be capable of accepting the minimum data items listed below for an identified time period:

- **Manufacturing Order Identity**

  The unique identity by which a specific authorised production requirement has been referenced.

- **Detail Part Identity**

  A unique identity which references each detail requiring manufacture. Where, due to the manufacturing process, the nature of the detail changes sufficiently to require different identities before and after the process, both identities must be given and identified.

- **Quantity**

  The quantity, or volume of production, of the identified detail(s) required.

- **Date**

  The specific date or time by which the detail is required.

- **Priority**

  Any additional prioritisation information attached to a detail/order, to be read independently of, or in conjunction with, the required date. It may also indicate any special status or type of order category defined by the user.

It is anticipated that data will be accepted in batch mode.

P4.3.2   **Modifications**

The sub-system must be capable of accepting modifications to the data held at any time. Such modifications may include adding or deleting manufacturing orders, or amending information on an order. Any modifications must be capable of being protected to ensure that they are only carried out by appropriate authorities.

P4.3.3   **Released orders**

Orders, once released for production, must remain released even if not executed within the normal time horizon - unless specifically restored or removed. If they are removed or restored as unreleased outstanding requirements, this action must be communicable to any dependent functions, eg. tooling, fixturing.

P4.3.4   **Out of balance parameters**

The parameters used to define potential out of balance situations must be capable of being defined locally, and updated as required.

P4.3.5   **Capture of completion information**

The sub-system must be capable of capturing information on details launched onto production and (if applicable) completed, from any combination of locally defined sources.

P4.3.6   **Prioritisation rules**

The sub-system must allow as many locally defined prioritisation rules and algorithms to be incorporated as required. All rules and algorithms must be capable of modification with minimal effort and delay as and when this is required. Any such modifications must be protected to ensure that they are not made from unauthorised sources.

P4.3.7 **Output data**

The system must be capable of producing data to the minimum levels specified in this section. It must be in a form which will facilitate reformatting by a subsequent module, ie. all items must be structured in a disciplined and identified manner.

Output may be interactive, or occur in a succession of small batches.

- **Manufacturing Order Identity**

   The unique identity by which a specific authorised production requirement has been referenced.

- **Detail Part Identity**

   A unique identity which references each detail requiring manufacture.

- **Rough Detail Identity**

   Where the identity of the detail changes due to the manufacturing process, the identity of the rough or raw material input to the manufacturing process.

- **Quantity**

   The quantity, or volume of production, of each identified detail which is required for this manufacturing order.

- **Date**

   The specific date or time by which the detail is required.

- **Priority**

   Any additional prioritisation information attached to a detail/order, to be read independently of, or in conjunction with, the required date.

   It may also indicate any special status or type of order category defined by the user.

Chapter 6

Computer Aided Manufacture
Computer Aided Storage and Transportation

Introduction

The investigation into these two topics highlighted a potential problem with regard to the basic objectives of the project. Whilst the project is intended to demonstrate and define the close interrelationships between the topics it is an essential part of the definition process that each topic can be considered separately. In this case it became very difficult to define Computer Aided Manufacture properly without introducing activities that were defined as part of Computer Aided Storage and Transportation. Therefore it has been decided that these two topics will be considered as one entity. In order to demonstrate this problem and justify our decision there follow two separate discourses, one for Computer Aided Manufacture and one for Computer Aided Storage and Transportation. The activity definitions (ie. the flowcharts and narrative descriptions) have been created without separation into the two major topics.

In creating the activity definitions it rapidly became apparent that for reasons of clarity and understanding it was necessary to introduce an extra step in the definition process. This step is represented by the diagram opposite this page. The main message that this chart is intended to convey is that there are many activities that will take place simultaneously within the overall CAM/CAST process. Basic flowcharts did not adequately demonstrate this overlap because of their sequential flow characteristic.

Another peculiarity of the activities within the CAM/CAST areas is that multiple occurrences of the same activity may exist within one CIM system and these separate activities occur simultaneously and independently. For example, in a CIM system that incorporates an overhead monorail delivery system and an AGV system, two distinct and independent delivery management systems would normally be preferable to a combined module. Also, in a large system, it may be preferable to implement resource scheduling recursively. For example, a main resource scheduler may control the movements of tools and work pieces between multiple machine cells (a turning cell, a milling cell and a grinding cell say), and lower level resource schedulers could control the movement of tools and workpieces to and from local buffers and the individual machines within the cell. For these reasons also basic flowcharts cannot adequately represent the CAM/CAST activities.

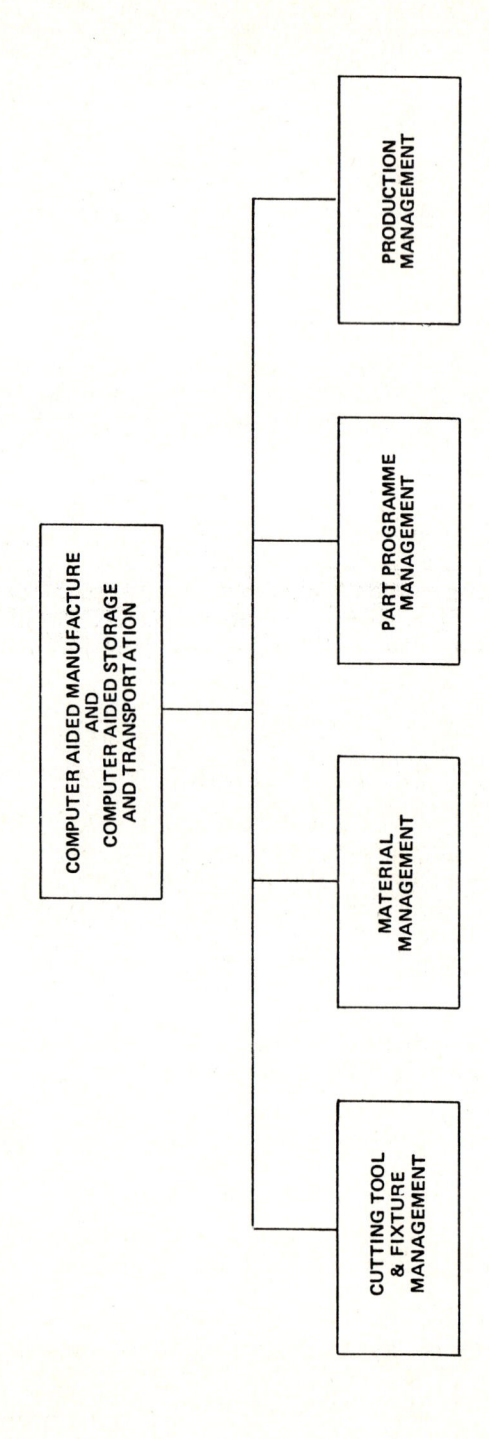

```
COMPUTER AIDED MANUFACTURE
         AND
COMPUTER AIDED STORAGE
    AND TRANSPORTATION
         |
    ┌────┬────┬────┬────┐
    │    │    │    │
CUTTING  MATERIAL  PART      PRODUCTION
TOOL     MANAGEMENT PROGRAMME MANAGEMENT
& FIXTURE           MANAGEMENT
MANAGEMENT
```

Computer Aided Manufacture

Introductory discourse

Overview

Machine tool technology and manufacturing systems have gone through dramatic changes in the past few years due to the advent of the micro-processor and have made the unmanned factory a viable proposition.

It is now foreseen that the management of the factory as well as individual machine control will be totally computer controlled.

The area of Computer Aided Manufacture will include machine tool control systems, component loading from the transport system, tool management, component inspection and support services such as coolant control/swarf disposal and factory maintenance.

1 **Machine control systems**

The level of capability of machine control systems for Numerical Control (NC) systems in the typical manufacturing facility can vary greatly depending upon the age of equipment and the market for which the control system is intended. In order to provide for the range of equipment already installed without requiring machine replacement at an unacceptable capital cost, it is desirable that the design rules include a provision for older generation control systems. Among considerations to be addressed will be :

a) **Hard-wired NC control systems**

Where the functions of the NC controller are derived from hard-wired electronic logic, part-programmes are traditionally read into the machine tool control character by character from punched paper tape, as the component is machined. There is no facility for part-programme retention in the machine control; therefore, additional storage and communications capability will be required for incorporation of this type of machine into an integrated system.

b) **CNC machine control with small capacity single programme storage**

The use of Computer Numerical Control (CNC) to replace hard-wired logic gives additional power to retain part-programmes at the machine. Although, traditionally, these part-programmes have been input from punched paper tape, many machines are now being fitted with simple serial interfaces which allow data input. Design rules in this area will include requirements for conversion interfaces to provide a suitable secure communications protocol for message checking, since the simple serial interface rarely includes error checking more comprehensive than character parity.

Programme changes for this type of control will require transmission of a replacement programme from the central storage.

c) **CNC machine control with multi-programme capability**

Where there is sufficient part-programme storage memory in the machine control together with part-programme labelling facilities, it may be possible for all the programmes normally used to be kept at the machining centre. In this case, the design rules will provide for selection between these programmes by the central supervisory system, in addition to the facility of down-loading programmes for new components.

d) **DNC communications interfaces**

One of the main areas of design rule specification will be the communications interface requirements for machine control systems. A minimum requirement will be the reception of part-programmes from the supervisory computer through simple unidirectional input. With more sophisticated control systems, including CRT display of programmes and diagnostic messages, the requirements for communications protocol will include provision for bidirectional communication of management information such as production quantities and condition monitoring of the machine and tooling. Provision will also be made for tooling requirements, in addition to the machine compliment of duplicate tooling, to be requested from the supervisory control when tool condition monitoring has indicated that tool life is exceeded and no further duplicate tools are available at the machine.

2 **Material handling and fixturing**

Material handling and fixturing can be a very high proportion of the investment of Flexible Manufacturing Systems (FMS).

Use of modular fixture design must be encouraged to reduce investment, storage and control. It is very important that fixture designers are aware of CAD/CAM requirements to enable efficient operation of the system.

Computer Aided Storage and Transportation

Introductory discourse

Overview

This subject involves the application of a wide range of techniques to an infinite variety of production needs, with a growing requirement that the adopted approach must provide increased productivity, flexibility and re-usability. As such, it is necessary to consider the topics of unit storage, unit transportation, unit identification and control of unit sequencing. It is desirable to integrate all transportation systems within a site, so goods reception and mass transportation (and ultimately traffic control) are also topics for consideration.

Storage techniques

Units may be stored in order to achieve space efficiency, rapid access to items in random order, or to provide simple buffering of unit flows with or without re-sequencing capability. The units may be raw material stocks, work-in-progress or finished items, or fixtures, pallets or tools used during the process.

The techniques available include high bay warehousing using fully automated placement and extraction, multi-lane FIFO storage systems with re-sequencing capability and single lane FIFO storage with simple input/output gating. In each case it is necessary to balance the conflicting requirements of efficient storage, rapid unit access and continuous throughput capability. These compromises are often complicated by the need to provide graceful degradation and by factors such as storage units being a multiple of the items to be forwarded, or alternatively, the case where a forward requirement is a multiple of the stored units. Indeed often these two factors exist within one storage facility as a "picking" facility.

Unit transportation

The requirement to transport a unit from one workplace to another has always been an area of considerable interest to production engineers. The number of techniques in use is considerable and can be illustrated as ranging from the man carrying a unit, through powered carriers and conveyor systems to the passing of liquids, for example, paint by pipe work. Many new techniques are under development including wire guided robot-carriers, free ranging automatic trucks and conveyorised, individually routing, carriers.

Unit identification

An area of specific concern lies in the rapidly evolving area of unit identification. Techniques available include - sensing of the unit by weight, outline, colour or material identification, remote or touch reading of a coded identifier attached to the unit, or coded identification of the unit carrier within the transportation system.

The recognition of coded identifiers includes the use of infra-red or laser readers and bar coded identifiers, erected peg identifiers for mechanical or electrical control of conveyor switching, solid state externally energised code units and magnetic wave guide transporter units activating strategically placed transmitter/receivers.

Early indications exist for possible use of magnetically encoding and reading the unit itself with the advantage of building in extra identification as the unit progresses through the production process.

Unit sequencing control

One of the major advantages in automating the storage and transportation methods lies in the information such systems provide. The collection of data regarding work-in-progress enables control to be exercised over the sequencing of units to forward production work-stations in line with production plans whilst taking into account unit availability and facility breakdowns. Such control systems can be designed to optimise sequencing decisions in order to prioritise production volume, profit levels, order ageing or stock investment in line with operating policies.

Mass transportation

The arrival of materials at a site, and their passage between entrance gate, reception office and unloading deck, normally involves treating many storage units as a single lot. Several of the techniques and technologies relevant to unit transportation can also be applied to the control of mass transportation.

Goods reception

The original validation of incoming material is important not only for the identification of the goods, but also for the recognition of any batch numbers and modification levels associated with specific delivery lots; which may in turn be necessary for categorising and bonding individual storage units.

Computer Aided Manufacture
Computer Aided Storage and Transportation

Summary

The above discourses can now be referenced to give an example of the close interlinking of the two areas.

Material Management (CAM) must make available a certain component from a specific stores location.

Stores Control (CAST) is however responsible for accessing the component and delivery out of the storage facility.

Cutting Tool and Fixture Management (CAM) must make available the fixture to the load operation and must make available the tools needed for the next (machining) operation.

In any significant machining process the tools and fixtures will be stored in a storage facility (CAST). In addition if the cutting tool store is not immediately adjacent to the work centre then a delivery system (CAST) must be instructed by the resource scheduler (CAM).

This brief, simplistic example demonstrates the considerable interlinking of activities that can only be correctly assigned to either CAM or CAST.

Therefore it has been decided to combine CAM and CAST into a single topic for the purpose of flowcharting.

Four sub-topics have been extracted from the combined topic, and flowcharts and design rules have been generated for the sub-topics. The sub-topics are:

a)   Cutting Tool and Fixture Management
b)   Material Management
c)   NC Part-Programme Management
d)   Production Management

A brief description of the four sub-topics follows:

Cutting tool and fixture management

This activity is concerned with providing to the work centres all the cutting tools and/or work holding devices that have been declared as needed for specific operations. Due to the problems of wear and tear it is also important to consider the refurbishment of such equipment. The need to carry out cutting tool refurbishment and replacement is readily accepted but often the other aspects of refurbishment (such as checking that fixtures have not worn to the detriment of accurate alignment of components) are overlooked.

Material management

In order to be able to produce machined components it is essential to have a steady supply of correct 'rough' material. Rough material being used here to imply only that further operations are to be carried out. To be able to supply this material it is necessary to know the condition and location of such material so that it can be supplied to the machining processes in a timely and orderly fashion.

NC Part-programme management

The levels of mechanisation addressed by this pilot project covered all levels between 'unlinked NC machine operation' and 'Flexible Manufacturing Systems'. All of these imply the creation of part-programmes that define the machining operation and the maintenance of such programmes in a readily usable form. Many differing techniques are available varying from racks of paper tapes to fully integrated DNC linked libraries of part-programmes

Production management

This sub-topic encompasses the management of the delivery and presentation of the resources prepared and made available by the first three sub-topics (ie. tools and fixtures, workpieces, part-programmes), and control of their use at the fixed facilities within the CIM system (ie. the machine tools, robots, storage buffers, fixturing stations, etc.). Tasks are despatched in an appropriate sequence to satisfy the prioritised production plan created by the CAPP activity, and on-going tasks are monitored so that contingency action may be initiated in the event of any task failures.

Chapter 7

Computer Aided Manufacture
Computer Aided Storage and Transportation
Sub-topics

Cutting tool and fixture management

Introduction

It is quite often found that in manufacturing areas the control of cutting tools and associated durable items appertaining to both tool and workholding are sadly neglected.

This section covers aspects relating to the control of all items used in a machining area which, due to wear mechanisms are subject to replacement either on a regular or occasional basis.

Items are categorised into two basic groups, namely:

a) **Consumable**; relating to the **cutting tool** the part which actually forms or removes component material.

b) **Durable**; identified as any item appertaining to toolholding rearwards from the cutting edge, and is applicable to all forms of tool set-ups.

   Items on workholding fixtures subject to regular replacement are also listed under this heading.

Computer Aided Manufacturing/Computer Aided Storage and Transportation

Cutting tool and fixture management

Flowcharts

CAM / CAST
CUTTING TOOL & FIXTURE MANAGEMENT

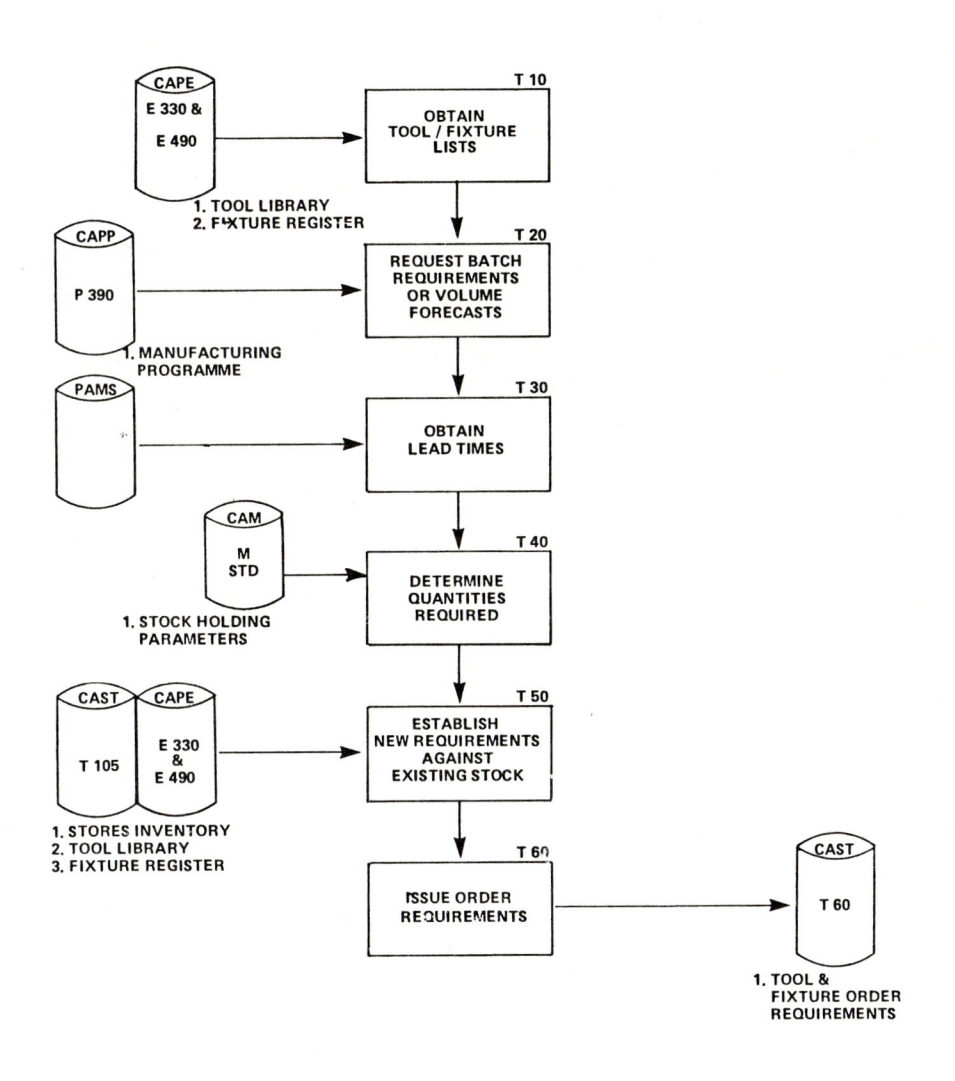

Sheet 1 of 4

CAPE E 330 & E 490 → **T 10** OBTAIN TOOL / FIXTURE LISTS
1. TOOL LIBRARY
2. FIXTURE REGISTER

CAPP P 390 → **T 20** REQUEST BATCH REQUIREMENTS OR VOLUME FORECASTS
1. MANUFACTURING PROGRAMME

PAMS → **T 30** OBTAIN LEAD TIMES

CAM M STD → **T 40** DETERMINE QUANTITIES REQUIRED
1. STOCK HOLDING PARAMETERS

CAST T 105, **CAPE** E 330 & E 490 → **T 50** ESTABLISH NEW REQUIREMENTS AGAINST EXISTING STOCK
1. STORES INVENTORY
2. TOOL LIBRARY
3. FIXTURE REGISTER

T 60 ISSUE ORDER REQUIREMENTS → **CAST** T 60
1. TOOL & FIXTURE ORDER REQUIREMENTS

CAM / CAST
CUTTING TOOL & FIXTURE MANAGEMENT

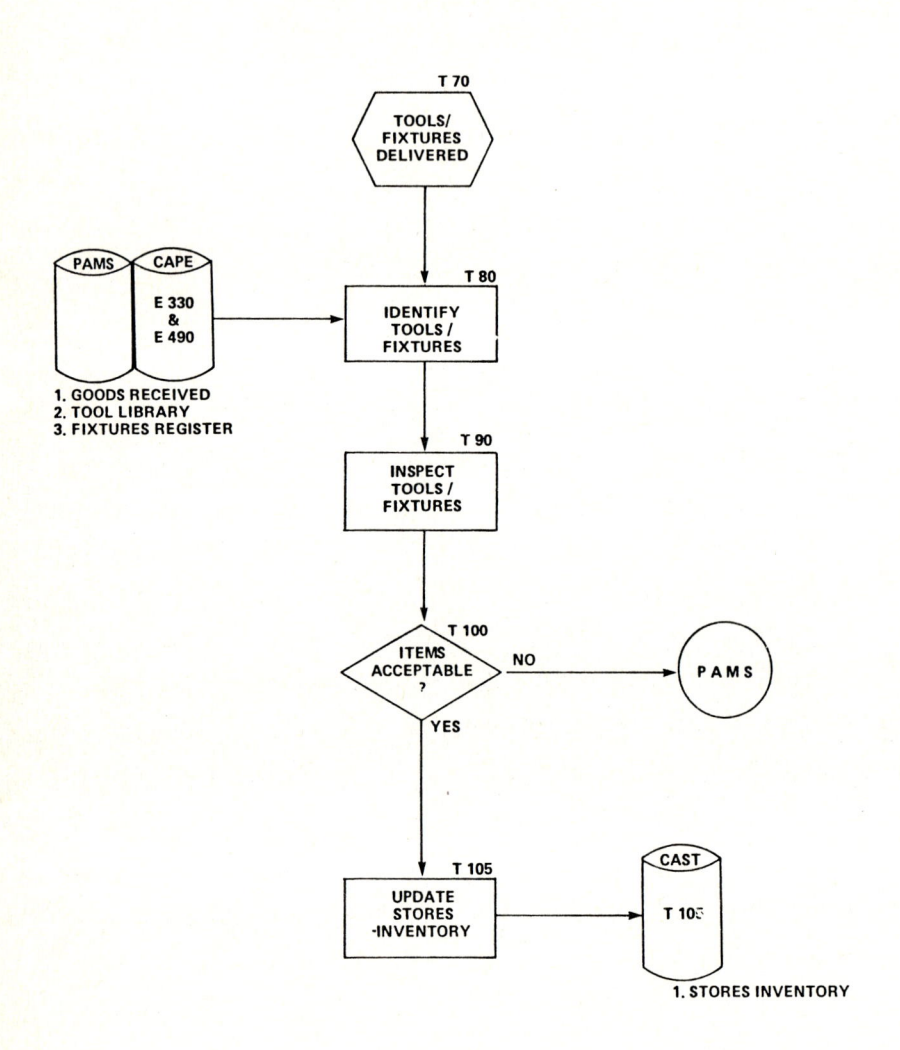

Sheet 2 of 4

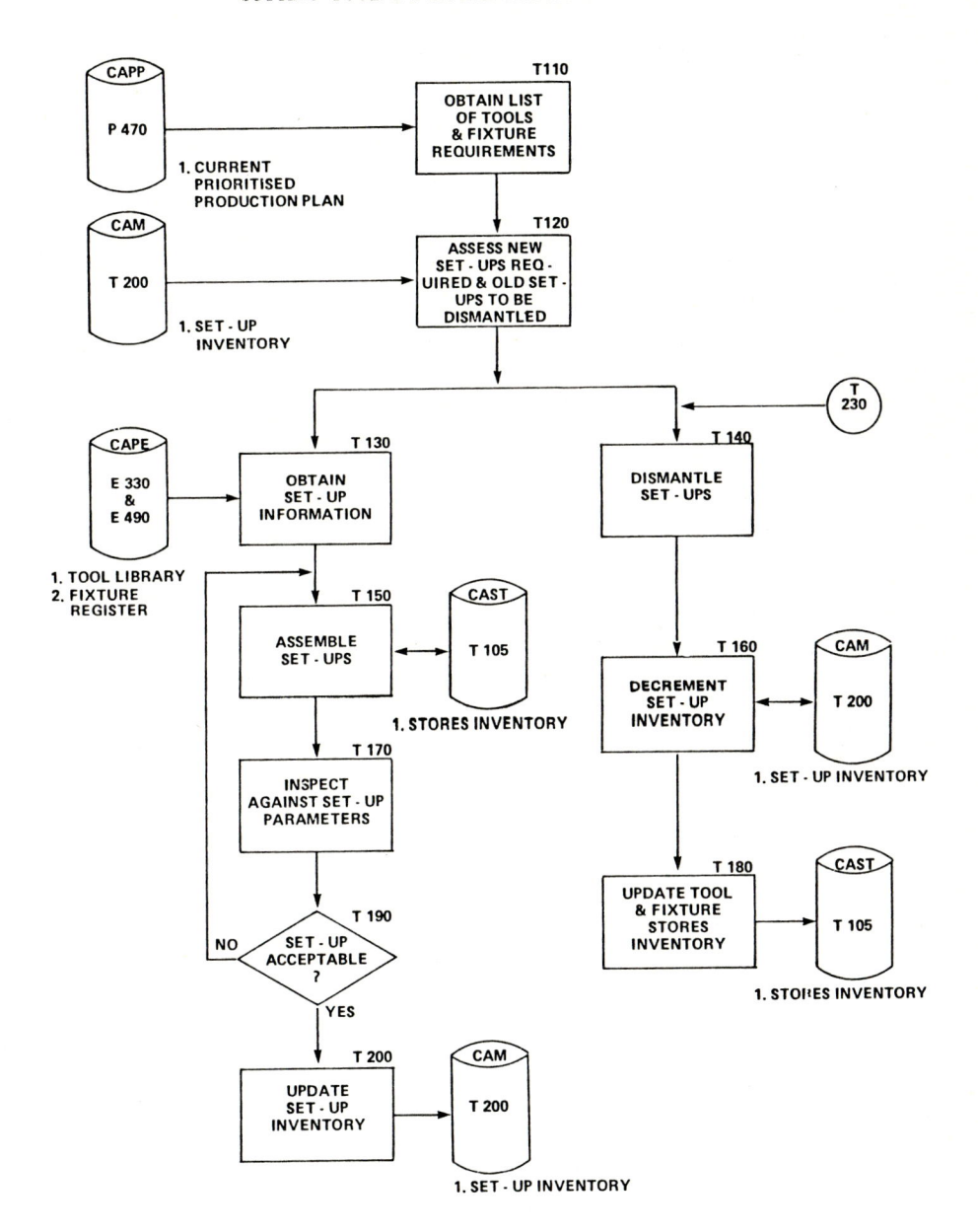

CAM / CAST Sheet 3 of 4
CUTTING TOOL & FIXTURE MANAGEMENT

CAM / CAST
CUTTING TOOL & FIXTURE MANAGEMENT

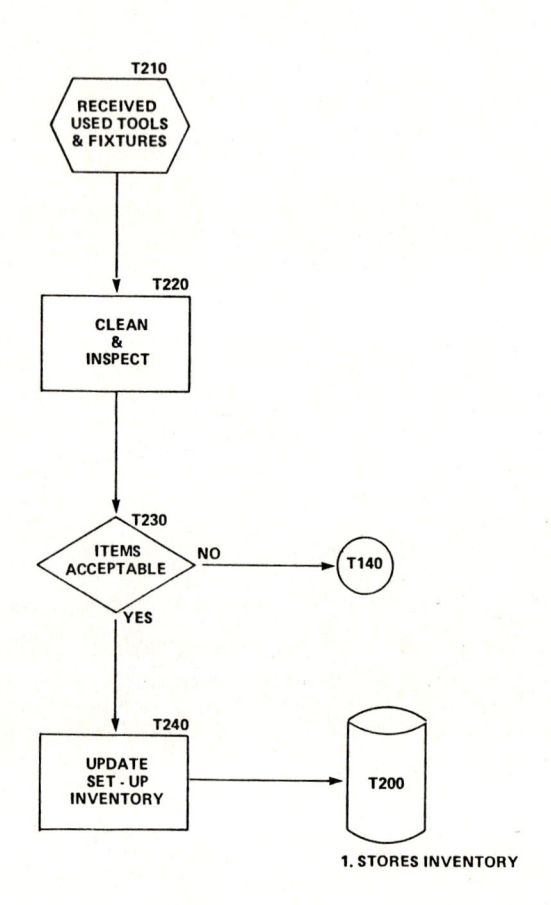

Sheet 4 of 4

```
                    T210
                  ┌───────┐
                 ╱ RECEIVED ╲
                ⟨ USED TOOLS ⟩
                 ╲ & FIXTURES╱
                  └───┬────┘
                      │
                      ▼
                    T220
                  ┌─────────┐
                  │  CLEAN  │
                  │    &    │
                  │ INSPECT │
                  └────┬────┘
                       │
                       ▼
                     T230
                   ╱ ITEMS ╲    NO
                  ⟨ACCEPTABLE⟩──────▶ ( T140 )
                   ╲       ╱
                    ╲ YES ╱
                      │
                      ▼
                     T240
                  ┌─────────┐
                  │ UPDATE  │          ┌──────┐
                  │ SET-UP  │─────────▶│ T200 │
                  │INVENTORY│          └──────┘
                  └─────────┘      1. STORES INVENTORY
```

Narrative description of
cutting tool and fixture management

T10   From information generated in nodes E330 and E490, both cutting tool and fixture data can be extracted to establish a list of total process requirements for a new or existing component.

T20   In addition, it is necessary to obtain from data generated in CAPP node P390, planned production schedules either in the form of batch requirements as applied to a short order situation, or predicted volume forecasts in the case of repetitive long-term components.

T30   Allied to the process requirements and planned production schedules is the need for projected lead times. This information will be obtainable from Purchasing in a generic form for both consumable and durable items.

T40   It will be necessary at this stage to calculate the total quantity requirement of each individual item, appertaining to usage during the machining process.

The requirements will be calculated from the following information:

- Cutting time per piece/number of hits.
- Pieces per hour/shift.
- Total piece requirement/batch run.
- Item lead time.
- Required minimum stockholding level.*

    Note*   The requirement of stock levels is purely dependent upon management philosophy, ie. three months usage or just-in-time.

T50   The resultant quantities will now have to be segregated to establish new requirements from existing stock. This will be achieved by obtaining data from the stores inventory control, node T105 and nodes E330 and E490, whereby items listed as existing will be identified together with a code number and verification of existing stock levels.

T60   Once the total quantities of both consumable and durable items have been compiled and identified as new parts required or existing coded stock, they are then stored on an information file. Purchasing will assess this information to raise orders for tools and fixtures.

T70   Whenever new tools/fixtures, or replacements, arrive then they must be handled.

T80   Information on expected tool deliveries must be accessed to enable the tools and fixture delivered to be properly identified and coded.

T90   Items delivered will be inspected for full quality and specification assessment. The checking data is obtained from nodes E330 and E490, the tool library and fixture register.

T100 Items inspected may not be acceptable, in which case Purchasing will be notified to instigate return to the supplier.

T105 Items deemed acceptable will be accepted and the tool and fixtures stores inventory updated to indicate that these items are now available for production use.

T110 It is necessary to obtain total quantity requirements of tools and fixtures needed to support the current production plan.

T120	The inventory of existing set-ups is accessed and compared with the production requirement. As a result the new set-ups required can be established and any existing set-ups that are not required in the current production plan can be identified and dismantled to make tool holders available.
T130	Tool library and fixture register information is accessed to determine the precise equipment needed to create new set-ups.
T140	Existing set-ups that are deemed to be not required are dismantled into component parts.
T150	New set-ups are assembled using component parts from the stores inventory.
T160	Dismantled set-ups are removed from the list of available set-ups.
T170	New set-ups must be checked against parameters. A set-up must conform to the specification. A tolerance may be given in which case the precise setting must be recorded for this particular set-up.
T180	Component parts of dismantled set-ups are incorporated into the tool and fixture stores inventory.
T190	If a particular set-up does not conform with the specification, it must be altered to achieve conformity.
T200	Acceptable set-ups are added to the current set-up inventory available for production. Precise setting data must be included in this information.
T210	Used tool and fixture set-ups will be returned by production.
T220	The set-ups will need cleansing prior to inspection. On completion of the washing process all set-ups will be visually inspected to determine suitability for re-use.
T230	Set-ups deemed to be worn beyond use will be dismantled for re-use of good component parts and the refurbishment or replacement of worn components. Set-ups that pass inspection will be kept for subsequent re-use.
T240	Set-ups that are suitable for re-use are re-identified to the set-up inventory list, indicating that they are still available.

Computer Aided Manufacturing/
Computer Aided Storage and Transportation
Cutting tool and fixture management sub-systems

The flowchart for the Cutting Tool and Fixture Management activities within the ESPRIT pilot study has been broken down into sub-systems:

1 Evaluation of 'the total quantity requirements'
   (Nodes T10 to T60)

2 Stores inventory maintenance
   (Nodes T70 to T105)

3 Production set up control
   (Nodes T110 to T240)

Total quantity requirement sub-system(T1)

T1.1 **Sub-system description**

This sub-system is concerned with the task of generating the total quantity requirements of all singular items which, due to wear mechanisms are subject to replacement either on a regular or occasional basis.

The first task is to obtain information regarding total process requirements, planned production schedules and projected lead times. The sub-system then examines all the available information for each individual item concerned and calculates the total quantities needed to sustain continuous production for the duration of the process.

Finally, the resultant quantities will be segregated to establish new requirements from existing stock and stored on an information file.

Quantity requirement sub-system

T1.2 Flowchart

T1.2   SUB SYSTEM FLOWCHART

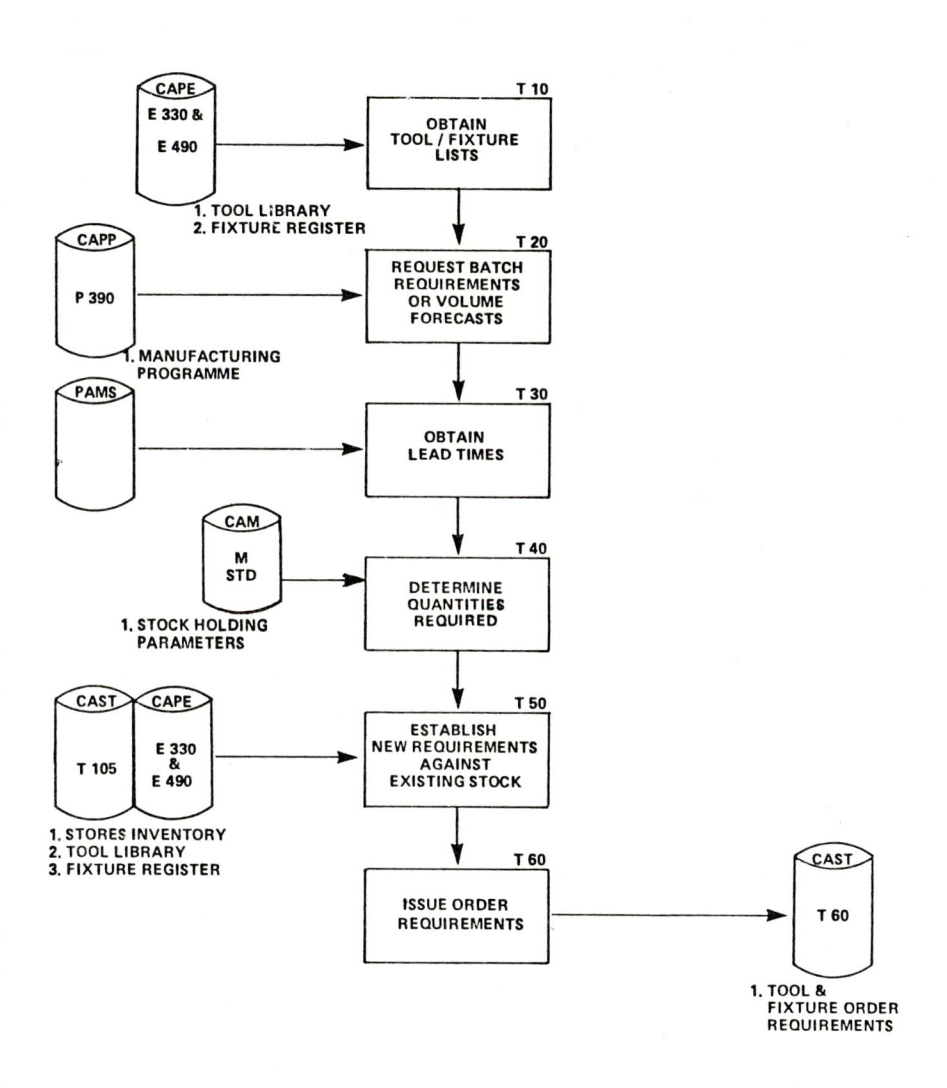

```
CAPE
E 330 &
E 490          ──▶  T 10
                    OBTAIN
                    TOOL / FIXTURE
                    LISTS
1. TOOL LIBRARY
2. FIXTURE REGISTER
                          │
CAPP                      ▼
P 390          ──▶  T 20
                    REQUEST BATCH
                    REQUIREMENTS
                    OR VOLUME
                    FORECASTS
1. MANUFACTURING
   PROGRAMME
                          │
PAMS                      ▼
               ──▶  T 30
                    OBTAIN
                    LEAD TIMES
                          │
CAM                       ▼
M STD          ──▶  T 40
                    DETERMINE
                    QUANTITIES
                    REQUIRED
1. STOCK HOLDING
   PARAMETERS
                          │
CAST    CAPE              ▼
T 105   E 330  ──▶  T 50
        &           ESTABLISH
        E 490       NEW REQUIREMENTS
                    AGAINST
                    EXISTING STOCK
1. STORES INVENTORY
2. TOOL LIBRARY
3. FIXTURE REGISTER
                          │
                          ▼
                    T 60
                    ISSUE ORDER   ──▶   CAST
                    REQUIREMENTS         T 60

                                    1. TOOL &
                                       FIXTURE ORDER
                                       REQUIREMENTS
```

T1.3 Design rules - total quantity requirement

T1.3.1 Input data

The system must be capable of accepting the minimum data requirement listed below, for any consumable or durable item.

- **Identity**

  An unambiguous reference by which the item can be selected - a code number is a common example.

- **Dimensioning**

  The fully dimensioned specification of any item in terms of basic dimensions, tolerances and machining requirements.

- **Material**

  The type of material from which the item is to be made. Where appropriate this should include the grade or composition of the material, eg. "high speed steel grade M2" rather than "high speed steel".

- **Planned Production Volume**

  The estimate of the volume or rate of production allied to the item. This is required to establish order quantity and stock levels.

- **Lead Time**

  The estimated time in generic form that an item may take from the stage of order to delivery.

- **Existing Coded Stock**

  The quantity of items bearing this identity currently available from stores.

T1.3.2 Input re-formatting

Data will be received in a batch format. The sub-system must be capable of re-formatting input data into the form required by the sub-system. It is assumed that the input data will be structured in a disciplined and identified manner.

T1.3.3 Scheduling scope

When a total process list is received, the sub-system must have the capability of re-scheduling to both individual item requirement and operation order.

T1.3.4 Item categorisation

The sub-system must be capable of categorising singular items into their relevant groups.

   ie: Consumable; relating to the cutting tool, the part which actually forms or removes component material.

   Durable; identified as any item appertaining to toolholding rearwards from the cutting edge, and is applicable to all forms of tool set-ups. Items on workholding fixtures subject to regular replacement are also listed under this heading.

T1.3.5 Output data

The system must carry out (with manual assistance if necessary) the compilation of total quantity requirements of each individual item recognised as either a consumable or durable part.

This must be structured in a disciplined and identified manner.

T1.3.6 **Output data format**

The system must be capable of outputting total quantity requirements for a complete process. Each item will be identified by part or code number, followed by a full description of the part and finally total order requirements in terms of initial order quantity and stockholding levels.

Stores inventory maintenance sub-system (T2)

T2.1 **Sub-system description**

This sub-system is concerned with the task of coding and holding stock of all consumable and durable items, the total quantities of which have been correlated in sub-system 1.

After receipt and inspection of the order requirement and the coding of all relevant items is complete, the total consignment will be held in bonded store awaiting production use.

Stores inventory maintenance sub-system

T2.2 Flowchart

T2.2   SUB SYSTEM FLOWCHART

Sheet 1 of 1

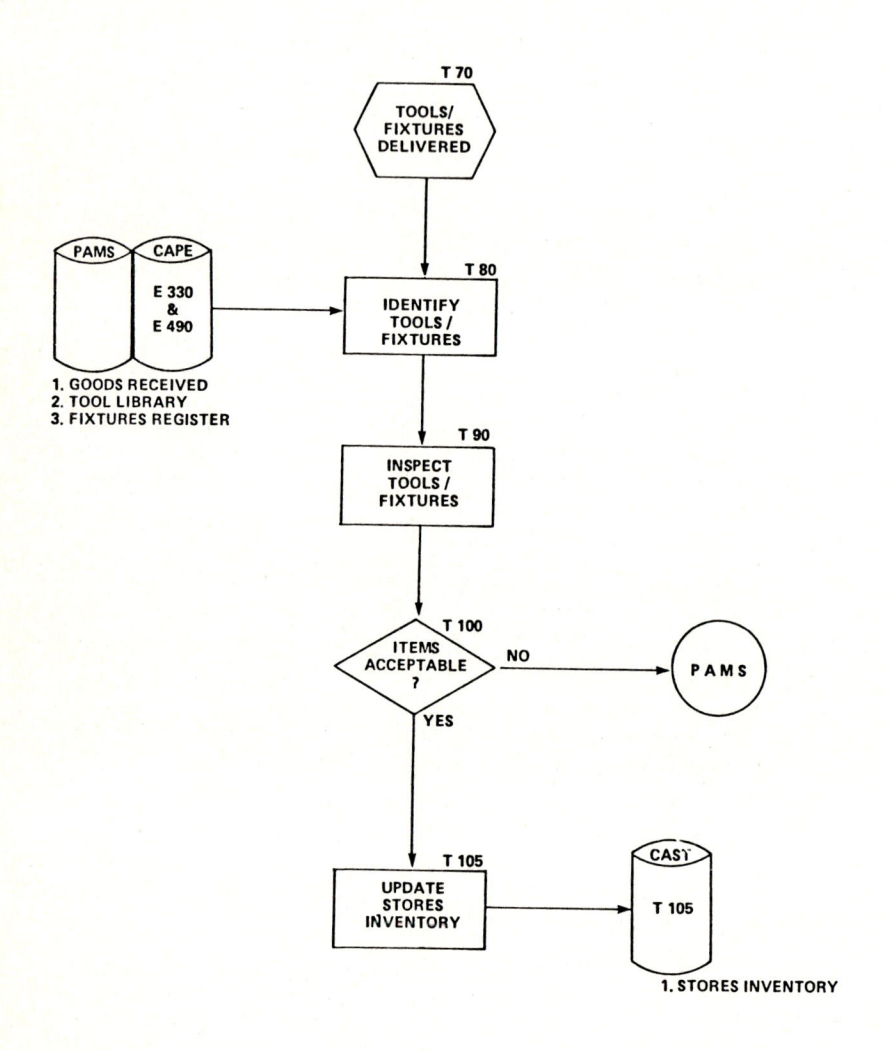

CAM CAST sub-topics

T2.3   **Design Rules - stores inventory maintenance**

T2.3.1   **Input Data**

The system must be capable of accepting the minimum data requirement listed below, for any consumable or durable item.

- **Identity**

  An unambiguous by which the item can be selected - a code number is a common example.

- **Dimensioning**

  The fully dimensioned specification of any item in terms of basic dimensions, tolerances and machining requirements.

- **Material**

  The type of material from which the item is to be made, where appropriate this should include the grade or composition of the material, eg. "high speed steel grade M2" rather than "high speed steel".

- **Requirement list**

  The total requirements applicable to a complete process must be given in a logical order. Each item will be identified by part or code number, followed by a full description of the part and finally total order requirements in terms of initial order quantity and stockholding levels.

T2.3.2   **Input Re-formatting**

Data will be received in a batch format. The sub-system must be capable of re-formatting input data into the form required by the sub-system. It is assumed that the input data will be structured in a disciplined and identified manner.

T2.3.3   **Scheduling scope**

When a total process list is received the sub-system must have the capability of re-scheduling to both individual time requirement and operation order.

T2.3.4   **Item categorisation**

The sub-system must be capable of categorising singular items into their relevant groups.

     ie.   Consumable; relating to the cutting tool, the part which actually forms or removes component material.

           Durable; identified as any item appertaining to toolholding rearwards from the cutting edge, and is applicable to all forms of tool set-ups. Items on workholding fixtures subject to regular replacement are also listed inder this heading.

T2.3.5   **Inventory coding**

From the total durable and consumable item requirement for a complete process it is expected that certain items will be categorised as a new addition. The sub-system must be capable of entering the new parts (manually if necessary) onto file allocating new code numbers designated to specific item areas and entering stores stockholding quantities.

It is also necessary that the sub-system is capable of (manually if necessary) adjusting stock levels of existing coded items accordingly.

T2.3.6 **Output data**

The system must be capable of outputting total process data for any individually coded item to any relevant manufacturing area, as and when required.

Production set-up control sub-system

T3.1 **Sub-system description**

This sub-system is concerned with the task of assembling individual durable and consumable items into completed set-ups, applicable to a process requirement and to maintain production continuity throughout the duration of that process.

The first task is to access the total process requirements and obtain the necessary individual items needed to build into the relevant set-up condition.

Secondly, when production condition is go, to despatch the completed set-ups to the relevant machining area for loading onto workcentres.

Finally, the sub-system will be required to maintain all fixture and tool set-ups in a usable condition until the process is terminated.

Production set-up control sub-system

T3.2 Flowchart

CAM CAST sub-topics

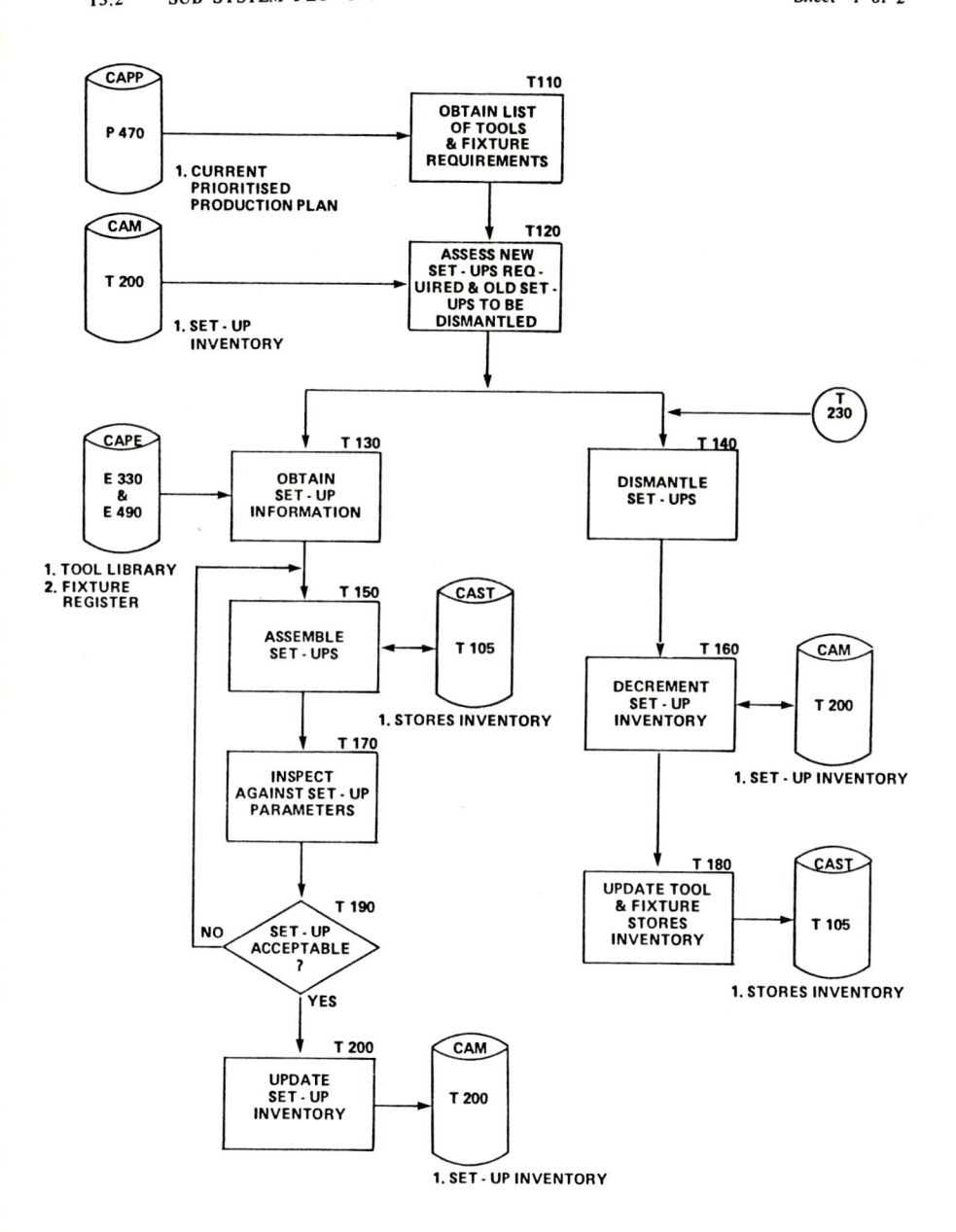

T3.2   SUB SYSTEM FLOWCHART   Sheet 1 of 2

```
  CAPP                          T110
  P 470  ───────────────→  OBTAIN LIST
                            OF TOOLS
      1. CURRENT            & FIXTURE
      PRIORITISED          REQUIREMENTS
      PRODUCTION PLAN
                                 │
   CAM                           ▼
                              T120
  T 200  ───────────────→  ASSESS NEW
                           SET-UPS REQ-
      1. SET-UP            UIRED & OLD SET-
      INVENTORY            UPS TO BE
                           DISMANTLED
                                 │
        ┌────────────────────────┴────────────────────┐
        │                                              │       T 230
        ▼                                              ▼     ◄──
   CAPE          T 130                            T 140
  E 330  ──→  OBTAIN                           DISMANTLE
    &         SET-UP                            SET-UPS
  E 490       INFORMATION
                  ▲                                  │
  1. TOOL LIBRARY │                                  │
  2. FIXTURE     │                                  │
     REGISTER    │   T 150        CAST             ▼
                 │  ASSEMBLE                    T 160            CAM
                 │  SET-UPS  ◄──→  T 105       DECREMENT
                 │                              SET-UP   ◄──→   T 200
                 │        1. STORES INVENTORY   INVENTORY
                 │      T 170                                1. SET-UP INVENTORY
                 │  INSPECT
                 │  AGAINST SET-UP                 │
                 │  PARAMETERS                     ▼
                 │                              T 180           CAST
                 │      T 190                UPDATE TOOL
                 │  NO  SET-UP                & FIXTURE   ──→  T 105
                 └──◄  ACCEPTABLE              STORES
                        ?                     INVENTORY
                        │YES                             1. STORES INVENTORY
                        ▼
                     T 200        CAM
                   UPDATE
                   SET-UP    ──→  T 200
                   INVENTORY
                                1. SET-UP INVENTORY
```

T3.2   SUB SYSTEM FLOWCHART

Sheet 2 of 2

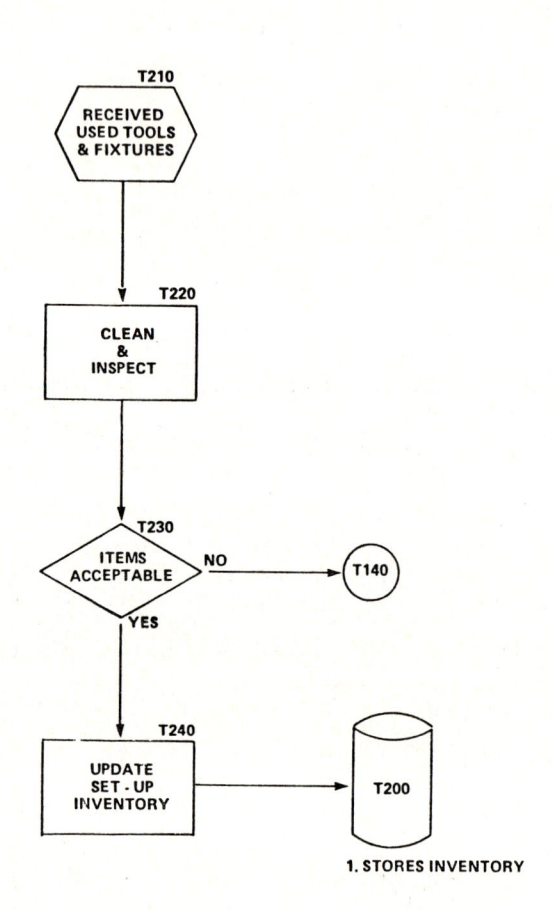

T210
RECEIVED USED TOOLS & FIXTURES

T220
CLEAN & INSPECT

T230
ITEMS ACCEPTABLE — NO → T140
YES ↓

T240
UPDATE SET-UP INVENTORY → T200

1. STORES INVENTORY

T3.3	**Design Rules - stores inventory maintenance**
T3.3.1	**Input Data**

The system must be capable of accepting the minimum data requirement listed below, for any consumable or durable item.

- **Identity**

    An unambiguous by which the item can be selected - a code number is a common example.

- **Dimensioning**

    The fully dimensioned specification of any item in terms of basic dimensions, tolerances and machining requirements.

- **Material**

    The type of material from which the item is to be made, where appropriate this should include the grade or composition of the material, eg. "high speed steel grade M2" rather than "high speed steel".

- **Requirement list**

    The total requirements applicable to a complete process must be given in a logical order. Each item will be identified by part or code number, followed by a full description of the part and finally total order requirements in terms of initial order quantity and stockholding levels.

T3.3.2	**Input Re-formatting**

Data will be received in a batch format. The sub-system must be capable of re-formatting input data into the form required by the sub-system. It is assumed that the input data will be structured in a disciplined and identified manner.

T3.3.3	**Process condition**

The system must be able to determine at any time for an ongoing process, the predicted usage of each consumable or durable item related to that process.

T3.3.4	**Item assembly format**

The sub-system must be capable of producing setting data applicable to both tool set-up assemblies and fixture assemblies, in terms of dimensions and tolerancing.

T3.3.5	**Item sub-assembly**

The sub-system must be capable of categorising singular items into their applicable assembled groups, thus producing sub-assemblies of both tool and fixture set-ups.

These sub-assemblies will consist of items described as both consumable and durable.

- ie. Durable; identified as any item appertaining to toolholding rearwards from the cutting edge, and is applicable to all forms of tool set-ups. Items on workholding fixtures subject to regular replacement are also listed inder this heading.

    Consumable; relating to the cutting tool, the part which actually forms or removes component material.

T2.3.6 **Output data**

The sub-system must be capable of outputting information relating to process condition regarding tool and fixture refurbishment. This must include warning of impending tool change and or process termination.

Material Management

Introduction

Material Management comprises the procedure needed to maintain control over material to be used in the production process. It covers both precursive action and responsive activity needed to aid the operation of the production process.

The broad areas of material management include the following:

a)   Forward Provisioning
b)   Material Receipt
c)   Material Issue
d)   Storage Facility Instruction

Forward Provisioning

- This is a housekeeping activity with two main objectives. The first of these is to take account of expected deliveries in order to be ready to receive new deliveries. The second objective is to be prepared to issue material to production just in time to meet production needs.

Material Receipt

- This is the activity concerned with acceptance of material into the domain of the material management area. Forward provisioning will have taken account of peak delivery problems, but this activity must be aware of constraints such as decanting workload, storage locations available to achieve optimum use of the facility.

Material Issue

- The essential aspect of this activity is the issue to production of the discrete material required to maintain production against plan. This must however be balanced against the material receipts in order to obtain optimum usage of the storage facility. For example, it is typical of automated high-bay warehouse facilities that material issues and material receipts are 'paired' whenever possible to reduce stacker crane movements that do not carry material to a minimum.

Storage Facility Instruction

- In order to receive or issue material it is essential that material management is able to instruct the storage facility as to which tasks it must carry out. It is equally important that material management is aware of facility breakdowns so that it can issue instructions that are able to be performed.

Computer Aided Manufacturing/Computer Aided Storage and Transportation

Material Management

Flowcharts

CAM / CAST
MATERIAL MANAGEMENT

Sheet 1 of 3

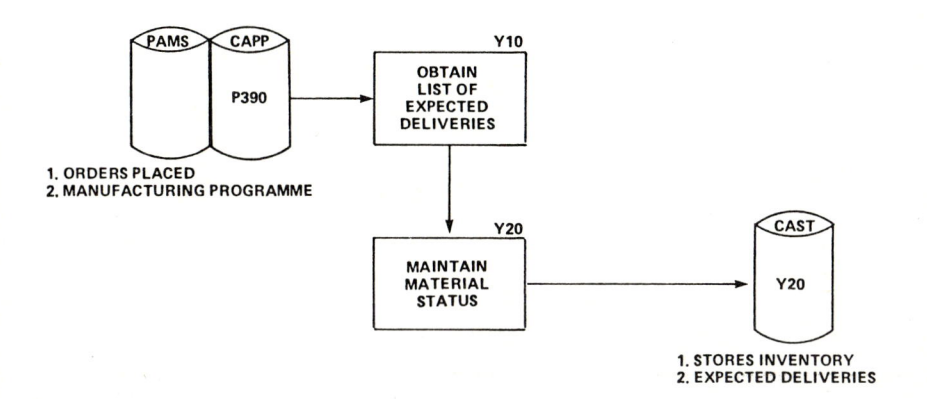

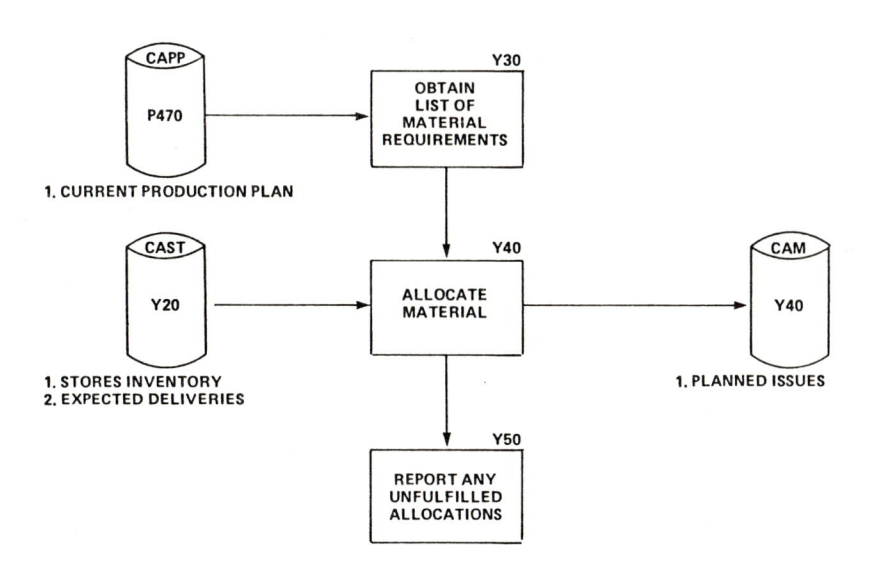

PAMS / **CAPP** P390
1. ORDERS PLACED
2. MANUFACTURING PROGRAMME

Y10 OBTAIN LIST OF EXPECTED DELIVERIES

Y20 MAINTAIN MATERIAL STATUS

CAST Y20
1. STORES INVENTORY
2. EXPECTED DELIVERIES

CAPP P470
1. CURRENT PRODUCTION PLAN

Y30 OBTAIN LIST OF MATERIAL REQUIREMENTS

CAST Y20
1. STORES INVENTORY
2. EXPECTED DELIVERIES

Y40 ALLOCATE MATERIAL

CAM Y40
1. PLANNED ISSUES

Y50 REPORT ANY UNFULFILLED ALLOCATIONS

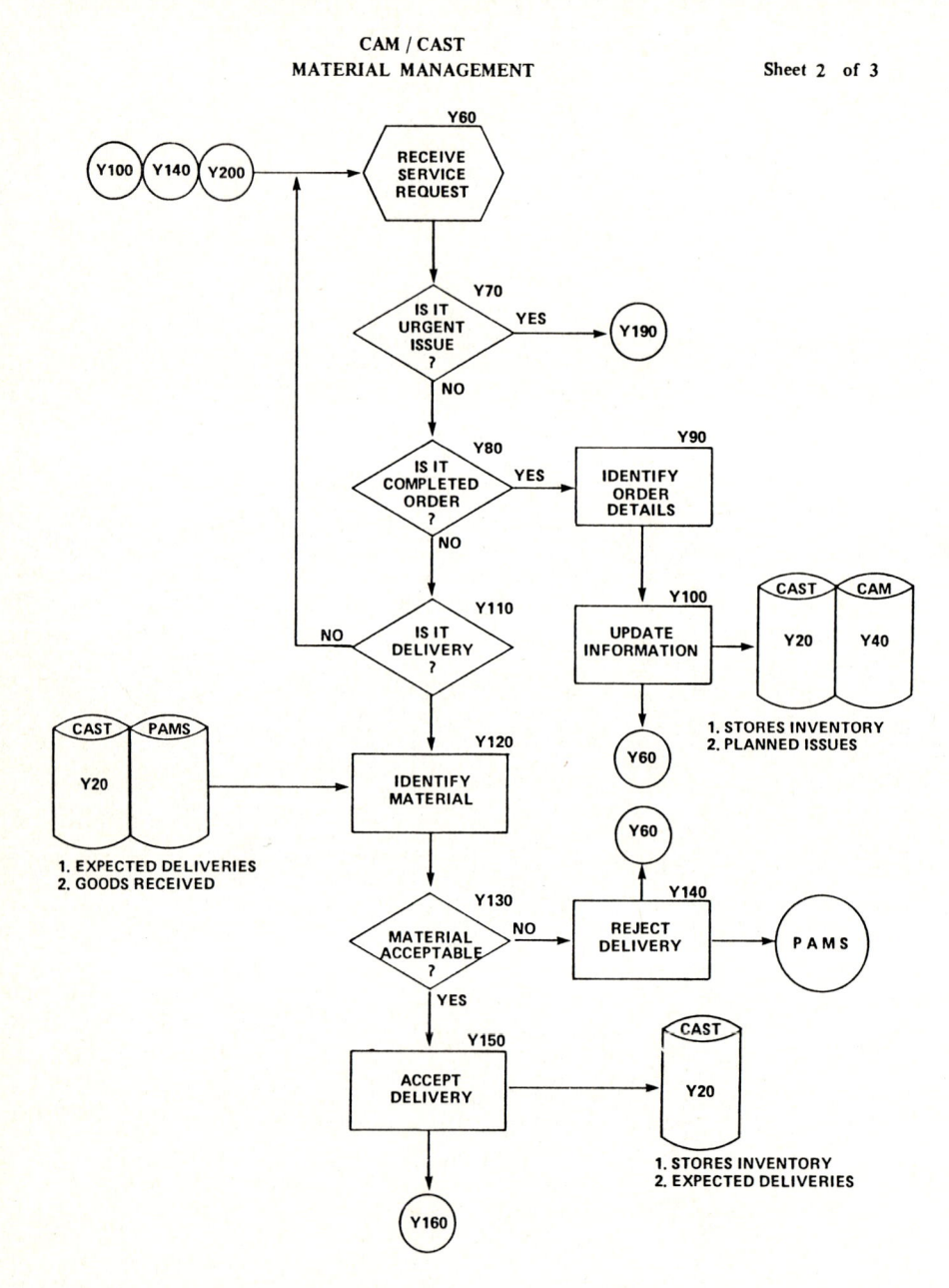

CAM / CAST
MATERIAL MANAGEMENT

Sheet 2 of 3

```
                    Y60
  Y100  Y140  Y200  RECEIVE
                    SERVICE
                    REQUEST
                       │
                       ▼
                     Y70
                   IS IT          YES
                   URGENT  ─────────►  Y190
                   ISSUE?
                      │ NO
                      ▼
                     Y80                    Y90
                   IS IT          YES    IDENTIFY
                   COMPLETED  ─────────►   ORDER
                   ORDER?                  DETAILS
                      │ NO                    │
                      ▼                       ▼
                     Y110                   Y100          CAST   CAM
              NO   IS IT                   UPDATE          Y20    Y40
         ◄───────  DELIVERY?             INFORMATION
                      │                       │        1. STORES INVENTORY
                      │                       ▼        2. PLANNED ISSUES
  CAST  PAMS          │                      Y60
  Y20                 ▼
                    Y120
  1. EXPECTED      IDENTIFY
     DELIVERIES    MATERIAL
  2. GOODS                                   Y60
     RECEIVED        │                        ▲
                     ▼                        │
                    Y130          NO        Y140
                   MATERIAL   ─────────►   REJECT   ─────►  PAMS
                   ACCEPTABLE?             DELIVERY
                     │ YES
                     ▼
                    Y150                    CAST
                   ACCEPT    ─────────►      Y20
                   DELIVERY
                     │                   1. STORES INVENTORY
                     ▼                   2. EXPECTED DELIVERIES
                    Y160
```

**CAM / CAST
MATERIAL MANAGEMENT**

Sheet 3 of 3

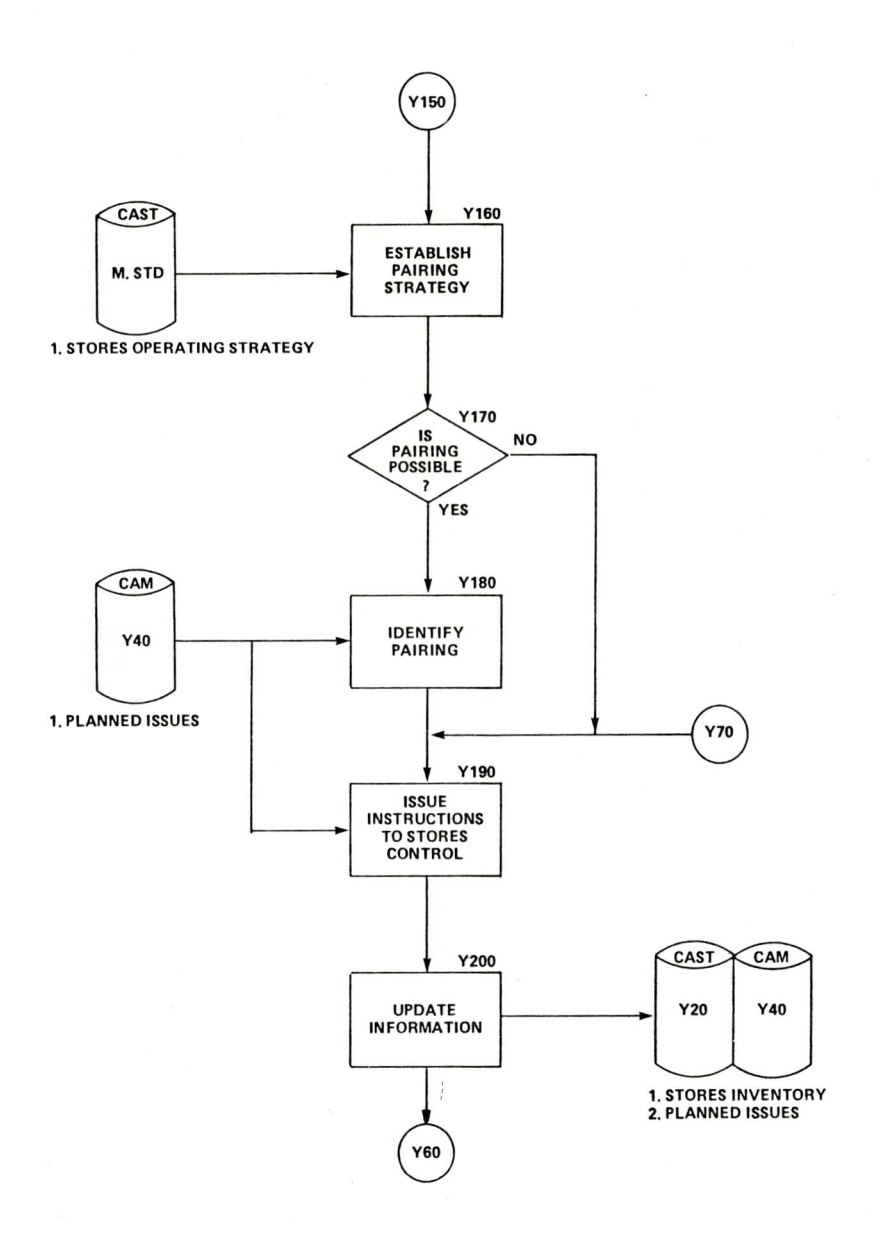

- Y150
- Y160 **ESTABLISH PAIRING STRATEGY**
  - CAST: M. STD
    - 1. STORES OPERATING STRATEGY
- Y170 **IS PAIRING POSSIBLE ?** — NO → Y70
  - YES
- Y180 **IDENTIFY PAIRING**
  - CAM: Y40
    - 1. PLANNED ISSUES
- Y190 **ISSUE INSTRUCTIONS TO STORES CONTROL**
- Y200 **UPDATE INFORMATION** → CAST Y20, CAM Y40
  - 1. STORES INVENTORY
  - 2. PLANNED ISSUES
- Y60

Narrative description of Material Management

Y10 A part of the precursive action is to prepare for the receipt of material required as soon as such information is available from the Production Planning process. In addition if any items in the plan are 'bought-out' then it will be necessary to obtain the latest delivery information from Purchasing.

Y20 Having obtained the information on expected deliveries, it is necessary to maintain up-to-date information on the planned material receipts.

Y30 At a later point in time Production Planning will have created a firm manufacturing plan for the latest time period which must be obtained by Material Management.

Y40 Using the firm manufacturing plan information together with knowledge of current material holding and planned receipts it is necessary to carry out a material allocation process. This process will include not only the checking that suitable material is currently available but will include identification of actual stock location. For example, in an automated high bay stores facility the stock holding will be distributed throughout the store facility in order to maintain stock availability in the event of partial facility failure. The allocation process must therefore assess the planned material requirement and identify the individual stock holdings that must be extracted to meet the requirement in total. At the end of the allocation process all stock extractions must have been pre-determined so that a prioritised list of stores extractions exists.

Y50 Ideally there should be no problems in making the allocations. However, due to last minute changes in the stock status (such as a delivered batch being declared as unusable) or breakdown of the stores facility, it may be that certain allocations cannot be fulfilled. It is necessary to identify these shortfall conditions. These unallocated requirements must be reported to both the Resource Scheduler so that the affected orders are not sequenced and the Production Planning process so that the delay can be considered against the wider implications than the directly affected manufacturing process.

Y60 The nature of Material Management activities in this part of its process are basically responsive activities dependent on requests received. It is therefore necessary to receive the request and identify the source and nature of the request in order to decide on the action to take place. Three basic forms of request for service have been identified as follows.

- Request to receive a delivery of material from an external source. This may be either goods receipts into the company or stock transfer within the company. Material Management activity does not differ in its response to such a request.
- Request to receive material from the associated manufacturing process.
- Request to issue material to the associated manufacturing process. This request is only needed when Material Management has failed to issue as a natural follow-on activity following action dependent on one of the other type of service requests. (see Y180).

Y70 If the request is for an urgent issue of material to the manufacturing process then a high priority issue will be enacted.

CAM CAST sub-topics

Y80   If the request is for receipt of components that have been worked on by the associated manufacturing process, then the correct response will take place.

Y90   If the request is identified as a completed batch (or part batch) returned by the associated manufacturing process then it is necessary to identify the material correctly. This must include not only the component identity but also the material status and its machining history. The material status can be fully finished and inspected, or part-finished, or needing re-work and further inspection.

Y100   The details on the completed order are used to update information on material in stock and planned issues, since the completion of the order may result in allocated stock being freed for other use and planned issues no longer required.

Y110   If the service request is for receipt of an external delivery then the relevant activity will ensue.

Y120   Once it has been established that an external delivery is to be received it is necessary to check the identity of the delivered material and relate this to an expected delivery as identified in Y20.

Y130   A decision must be made as to whether it is possible, or even expedient, to accept the delivery. Delivery will not be accepted without previous information on such an event since this may cause congestion in the stores facility. It is also prudent to refuse delivery if the current status of the storage facility prevents successful holding of the material. Factors to be considered include current stock holding of component, current storage capacities and where necessary current decanting facilities.

Y140   If the delivery is rejected then it is necessary to issue a return instruction to the carrier as well as informing the sending function that the material is to be returned.

Y150   If the delivery is accepted then a series of material movements is created. This will include the off-loading, decanting if necessary, and placement into the storage facility.

Y160   All accepted requests for delivery of material will have generated stores movement transactions. For efficient stores operation it is normal to carry out a pairing operation. This is a procedure that is designed to reduce unnecessary movements to a minimum. For example, if a set of components are to be placed in the store at position X, then it is prudent to identify a need to extract another set of components from a position very close to X (or at least on the route back from X to the storage input point). Various strategies can be applied and may vary both with the type of facility and the components in consideration. In addition the pairing strategies can vary in accordance with dynamic needs that depend on both management operating decisions and current congestion/breakdowns within the facility. It is thus necessary to continually review the situation to ascertain the pairing strategy to be used at this time.

Y170   It is necessary to determine whether a pairing should be made. If not then the relevant stores control transaction is issued directly.

Y180   If a pairing should be made then the second part of the pairing must be identified taking account of the known first part. This first part is a movement of material of a distinct type to a unique empty stores location. The second part must therefore be an extraction from the stores. This extraction will normally be one of the prioritised list of materials wanted during the current time period by the associated manufacturing process as created by Y40. The pairing decision takes account of the movements identified and the status of the stores facility.

Y190   Instructions to the stores control process must include the precise movements including the selected stores input point and the selected stores output point.

Y200   All the while that stores instructions are being generated it is necessary to continually review and update the status of material so that decisions taken are based on the most accurate information.

Computer Aided Manufacturing/
Computer Aided Storage and Transportation

Material Management sub-systems

The flowchart for the Material Management activities has been broken down into sub-systems:

1   Assessment of forward deliveries
         (Nodes Y10 to Y20)

2   Material requirement allocation
         (Nodes Y30 to Y50)

3   Material co-ordination
         (Nodes Y60 to Y200)

Assessment of forward deliveries sub-system (Y1)

Y1.1 Sub-system description

This sub-system is concerned with the task of assessing the effect of planned deliveries of material upon the stores reception area and the storage capacity needed.

The first task is to obtain latest information on planned deliveries. This information must include the date and time of delivery, the material expected, its quantity, its packaging as delivered and as stored. The sub-system then examines the effect of these deliveries upon the stores reception area, the decanting workload where needed, and the storage capacity needed to hold the material.

Finally the sub-system updates the material status data so that subsequent sub-systems have available the information on expected deliveries.

Assessment of forward deliveries sub-system

Y1.2 Flowchart

Y1.2. SUB SYSTEM FLOWCHART

Sheet 1 of 1

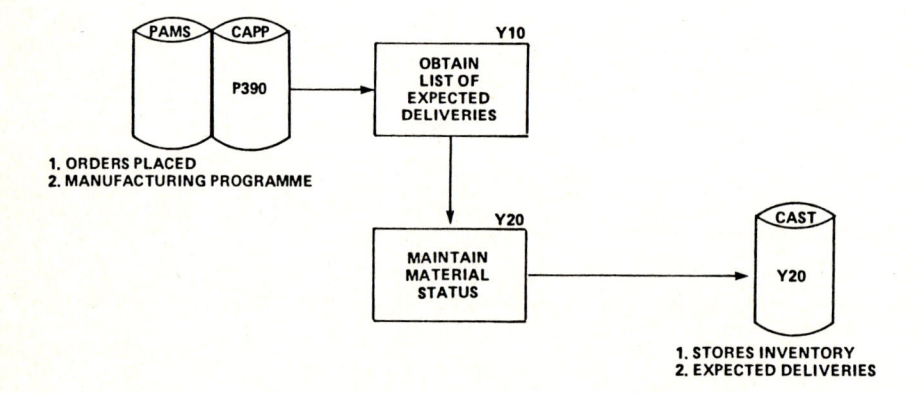

```
   PAMS   CAPP              Y10
   ┌──────────┐        ┌──────────┐
   │   P390   │───────▶│ OBTAIN   │
   │          │        │ LIST OF  │
   └──────────┘        │ EXPECTED │
                       │ DELIVERIES│
1. ORDERS PLACED       └─────┬────┘
2. MANUFACTURING PROGRAMME   │
                             ▼
                          Y20                    CAST
                       ┌──────────┐          ┌──────────┐
                       │ MAINTAIN │─────────▶│   Y20    │
                       │ MATERIAL │          │          │
                       │  STATUS  │          └──────────┘
                       └──────────┘          1. STORES INVENTORY
                                             2. EXPECTED DELIVERIES
```

Y1.3	**Design rules - assessment of forward deliveries**
Y1.3.1	**Input data**

The sub-system must be capable of accepting the minimum data items listed below:
Data is required relating to:
- The expected deliveries (Section Y1.3.1.1)
- The storage criteria (Section Y1.3.1.2)
- The storage capacity available (Section Y1.3.1.3)

Y1.3.1.1 **Description of data items**
Data relating to expected deliveries

- **Component Identity**
  The identity must be an unambiguous reference by which the delivered material will be recognised. Typically this would be a component part number.

- **Batch Identity**
  Each separate batch of material to be delivered and then considered as an entity must be given an unambiguous reference.

- **Batch Quantity**
  The total quantity that is to be considered as the separate batch.

- **Usage Detail**
  The batch will usually be identifiable as required to fulfil a specific manufacturing order. Such usage information may be repeated if the batch is for more than one manufacturing order.

- **Part Lot Identity**
  A total batch may be delivered in separate lots. Each part lot must be unambiguously identified.

- **Delivery Detail**
  This information must include details of date and time of delivery of the part lot.

- **Packaging**
  The type of container in which the components in the part lot will be delivered, together with the quantity contained. This information may be repeated if the part lot will comprise more than one container.

Y1.3.1.2 **Information relating to storage criteria**

- **Component Identity**
  Each separate component to be stored must have an unambiguous reference, typically a part number, so that storage criteria can be matched to expected deliveries.

- **Storage Packaging**
  The type of standard container in which components are to be held, together with the quantity that can be contained.

Y1.3.1.3 **Information relating to storage capacity**
The sub-system must be capable of considering the current and planned utilisation of the storage facility. To enable decisions to be made on the storage of expected deliveries it is necessary to pre-allocate space and workload for expected components to avoid congestion and overloading within the material management domain. Therefore the following information is needed.

- **Storage Capacity**
  The total quantity of storage locations that are empty and have not been pre-allocated must be available.
- **Storage Constraints**
  The restrictions on use of storage locations must be given. These can be standard restrictions caused by the storage facility design (such as certain locations are able to hold specific sizes of container only) or usage restrictions caused by the current occupancy of adjacent locations. In addition, certain locations may be restricted currently due to known facility failures.

Y1.3.2 **Input re-formatting**
Data will be received in a batch format (see Section 7). The sub-system must be capable of re-formatting, as required, input data into the form required by the sub-system. It is assumed that the input data will be structured in a disciplined and identified manner.

Y1.3.3 **Time period identification**
The sub-system must be able to accept a variable time period over which expected deliveries are to be assessed. This facility is needed when the time period for assessment differs from that covered by the planning process. In addition, it will be necessary to accept a variable time bucket for work content.

Y1.3.4 **Delivery assessment**
The sub-system must examine each delivery in order to establish the activities that may be required during reception, the activities that may be needed to decant the delivery into suitable storage containers and then pre-allocate storage space.

Y1.3.5 **Reception assessment**
All deliveries that are expected to be received during the selected time bucket must be assessed in terms of time taken to clear the delivery so that the next can be accepted. This activity can be relatively complex if the delivery capacity allows for multiple deliveries to occur simultaneously.

Y1.3.6 **Decanting assessment**
Where the delivery packaging is unsuited to immediate storage, the sub-system must be able to calculate, within a specified time bucket, the amount of decanting activity that must be catered for and calculate the number of storable containers that the delivery will require.

Y1.3.7 **Storage assessment**
Once the final storage packaging is defined it is necessary to identify which storage locations may be used to hold the delivered material. This process must take account of the order of delivery as it pre-allocates storage locations. The pre-allocation does not however signify that the material when actually delivered will automatically occupy the pre-defined locations, but does allow for over-capacity loadings to be identified in time for action to be taken.

Y1.3.8 **Storage characteristics**
During the pre-allocation of storage, it is necessary that account is taken of storage constraints. This will allow the potential warnings on over-capacity utilisation to be as representative as possible of the actual conditions that will occur.

CAM CAST sub-topics

Y1.3.9 **Recursive processing**
In cases where decanting into suitable containers has been identified, there is likely to be a choice of suitable container sizes. If the selected container size then causes an overload of storage capacity it may be necessary to reconsider the decanting choice and then re-try for storage assessment.

Y1.3.10 **Problem reporting**
Throughout the process of the sub-system there are numerous occasions where the process will be unable to continue due to lack of information. For example, a particular component delivery may need decanting but information is not available regarding suitable storage container types. In all such cases, the sub-system must report the problem but continue processing all other deliveries.

Y1.3.11 **Overload reporting**
The sub-system must be capable of comparing its assessment of reception workload, decanting workload, and storage capacity required against known maxima considered acceptable and then reporting any overloads so that action to prevent the overload or make extra resources available can be instigated.

Y1.3.12 **Output data**
The sub-system must be capable of producing data to the minimum levels shown below. It must be in a form which will facilitate re-formatting by a subsequent module, ie. all items must be structured in a disciplined and identified manner.

Y1.3.12.1 **Description of data items**
Data relating to expected deliveries and storage activity.

- **Component Identity**
  The identity must be an unambiguous reference by which the delivered material will be recognised.

- **Batch Identity**
  Each separate batch of material to be delivered must be given an unambiguous reference.

- **Batch Quantity**
  The total quantity that is to be delivered as a particular batch must be given.

- **Usage Detail**
  The planned usage of the material comprising the batch must be unambiguously identified.

- **Part Lot Identity**
  Each delivery of part of a batch must be identified separately.

- **Delivery Date**
  The date and time of the delivery must be given.

- **Packaging**
  The type of container and number of containers together with individual container load quantities must also be known for each delivery.

- **Storage Packaging**
  The type of container and number of containers into which a delivery will be decanted must be produced.

- **Special Constraints**
  Any special constraints to be borne in mind during delivery acceptance must be noted. For example, if the component to be stored is not to be placed adjacent to certain other types of component then this information must be available for later modules.

Material requirement allocation sub-system (Y2)

Y2.1   **Sub-system description**

This sub-system is dedicated to the task of producing a prioritised list of material issue transactions needed to serve the production process.

The first task is to obtain a current list of production material requirements. Once this prioritised list is available it is necessary to access the material status data in order to allocate specific stocks to meet the production plan. In carrying out this procedure it is necessary to allocate stock by taking into account stock status, age and position in the facility. This balancing task takes account of stock rotation and stores facility operating constraints. For example, if three separate batches of stock exist spread across several discrete storage bays then the allocation must be from the oldest batch taking care to use such stock from different storage bays.

In carrying out the stock allocation process it will sometimes be necessary to take account of expected deliveries where existing stock does not fulfil the demand. In such cases the sub-system must make such tentative allocations obvious to other systems that may rely on the allocation.

Even so, there will be occasions where the demand cannot be fulfilled. In such cases the sub-system must report the shortfall in allocation to enable other systems to modify their actions.

Material requirement allocation sub-system

Y2.2 Flowchart

Y 2.2. SUB SYSTEM FLOWCHART

Sheet 1 of 1

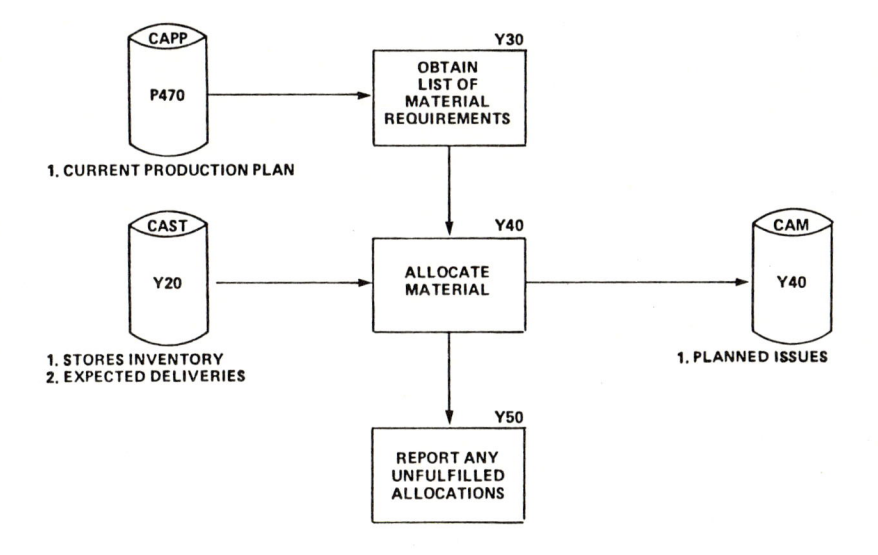

CAPP / P470
1. CURRENT PRODUCTION PLAN

Y30 — OBTAIN LIST OF MATERIAL REQUIREMENTS

CAST / Y20
1. STORES INVENTORY
2. EXPECTED DELIVERIES

Y40 — ALLOCATE MATERIAL

CAM / Y40
1. PLANNED ISSUES

Y50 — REPORT ANY UNFULFILLED ALLOCATIONS

Y2.3 Design rules - material requirement allocation

Y2.3.1 Input data

The sub-system must be capable of accepting the minimum data items listed below:

Data is required relating to:

- Material required by production (Section Y2.3.1.1).
- Current stockholding and expected stocks (Section Y2.3.1.2).

Y2.3.1.1 Description of data items

Data relating to material required.

- **Order Identity**
  The identifier for the manufacturing order.
- **Order Quantity**
  The quantity of components to be produced.
- **Finished Component Identifier**
  The part number of the finished component associated with the order.
- **Rough Component Identifier**
  The part number of the material as supplied to the manufacturing process.
- **Scaling Factor**
  The indicator to show how many finished components are produced from a single rough component. The reverse situation is equally important in general terms although less likely in machining processes.
- **Order Priority**
  An indicator giving the relative ranking of the order compared to others in the same time period. This may be a priority code number or a date required by the manufacturing process.

Y2.3.1.2 Information relating to stockholding

- **Rough Component Identity**
  The identifier for the material as held on stock.
- **Batch Identity**
  Material must be identified as belonging to a specific batch.
- **Batch Quantity Detail**
  For a given batch of components the quantity in stock must be available together with individual stock locations and quantities.
- **Usage Detail**
  The planned usage of the material as previously identified must be available. This can be either the identity of finished components or the combined identity of finished components within a specific manufacturing order depending on the restrictions on usage that are to be applied.
- **Expected Deliveries Detail**
  For each expected delivery that adds stock to a specific batch, or creates a new batch, that is not currently received, there must be information on the identity of the part lot, its delivery date and time, its packaging and planned requirements.

Y2.3.2 Input re-formatting

Data will be received in batch format   The sub-system must be capable of re-formatting, as required, input data into the form required by the sub-system. It is assumed that the input data will be structured in a disciplined and identified manner.

Y2.3.3 **Processing scope**
The sub-system must be capable of choosing material from current and expected stock so that it can produce a prioritised list of stores extractions that will give an orderly and timely flow of material into the production process under the management of the Material Co-ordination sub-system.

Y2.3.4 **Time period identification**
The sub-system must be able to perform material allocation for time periods defined by the requirements of the business overall, eg. 8-hour shifts, 24-hour days, 5-day weeks. All time-based planning processes will be similarly versatile. However, the sub-system must be able to consider both multiple planning time periods and also sub-sets of planning time periods. For example, if the production process utilises a minimally manned night shift then the material allocation sub-system must consider the requirements for two shifts whilst creating a stores extraction list to be carried out in the day shift only. Also the sub-system must be able to carry out a revised allocation process for, say, 2 hours only when necessary to allow for significant urgent change in production plan. Therefore the sub-system must be capable of accepting instructions regarding the time period for which allocations are to be made and also the time span in which the requirements must be encapsulated.

Y2.3.5 **Stock requirement calculation**
For each manufacturing order the sub-system must calculate the number of components to be issued and assign a unique priority to that requirement. The priority is based on the planned production priority.

Y2.3.6 **Stock Allocation**
The sub-system must identify which extraction of material or set of extractions from the storage facility will fulfil the calculated requirement. In so doing account must be taken of information relating to stock usage. For example, if a batch of components are identified as pertaining to a particular manufacturing order that is different from the order giving rise to the requirement then that stock cannot be allocated. If the requirement is to use stock to produce component A but component A is not a defined usage of the stock in question then that stock cannot be allocated. When stock is allocated to a specific order and therefore dedicated to production of a specific batch of components then the stock so allocated becomes unavailable for further allocation.

Y2.3.7 **Delivery allocation**
If the requirement for a particular component cannot be completely fulfilled from current stock then the sub-system must examine details of expected deliveries in an attempt to complete fulfilment of the order. In so doing account must be taken of delivery date and time as well as abiding by any stock usage information. If such a process is used then the potential stores extractions must be marked to indicate their nature. This is necessary to give the Material Co-ordination sub-system information to avoid moving such material into the storage facility when its issue will be immediately required and to give warning that the production process has become reliant on particular stock deliveries.

Y2.3.8 **Stores extraction identification**
Once all material requirements that can be fulfilled have been addressed then a list of stores movements (issues) must be drawn-up taking account of the requirement priorities and timings and known facility constraints.

Y2.3.9   **Shortfall reporting**
          The planning processes will normally cater for all known problems so that the material requirement allocation sub-system should be able to fulfil all requirements. However, late deliveries and recent facility failures within the storage area may mean that all requirements cannot be fulfilled. In this event it will be necessary for the sub-system to report all such shortfalls to both the production process itself so that machines are not allocated to orders that cannot be carried out due to material shortage and also to the planning process so that contingent action may be instigated.

Y2.3.10   **Output data**
          The sub-system must be capable of producing data to the minimum levels shown below. It must be in a form which will facilitate re-formatting by a subsequent module, ie. all items must be structured in a disciplined and identified manner.

Y2.3.10.1 **Description of data items**
          Data relating to Stores Extractions:

- **Order Identity**
  The identifier of the order for which material is required.

- **Finished Component Identity**
  The identifier of the component which will be produced from the material.

- **Rough Component Identity**
  The identifier of the material to be extracted.

- **Batch Identity**
  The identifier of the specific batch of material to be used.

- **Sequence Number**
  The relative priority of the extraction. Where more than one extraction is needed to fulfil the order and all such extractions needed have identical internal storage priority, then a common sequence number may be used, but an arbitrary internal storage priority is preferred.

- **Stock Movement Detail**
  The address of the storage location to be accessed to carry out the extraction and the address to which it is to be delivered. If it is only a potential extraction dependent on an expected delivery then the stock location address will be insignificant.

- **Potential Extraction Marker**
  For extractions planned on the basis of expected deliveries, then a marker must be given. This should indicate the expected date and time of the stock being available. In addition, the quantity required must be given based on standard container loads.

Material co-ordination sub-system (Y3)

Y3.1 **Sub-system description**

This sub-system is concerned with those tasks incorporated in the receiving of material, the issuing of material and the instructing of the storage facility to carry out the physical input and output of material.

When requested to receive material the sub-system must be given the details of the material being delivered. This will include the material identity, the quantity, its packaging and where possible its specific usage. This information will be compared with expected deliveries and associated storage detail such as palletisation required. This information together with knowledge of current stores reception workload is used to make the decision as to whether to accept the delivery or not. In some cases the request will be caused by the return from the production process of finished components.

When new material is accepted for storage it must be given a specific storage location taking account of current stores operating strategies. In so doing the stores input transaction is matched with a stores extraction request from the list of movements determined by the Material Requirement Allocation System. The paired transaction is then issued as an instruction to the stores facility.

Occasionally an urgent request will be made by the production process for material which has not been made available by the above process. In this case stores facility instructions for the issue will be made as a priority.

Material co-ordination sub-system

Y3.2 Flowchart

Y 3.2. SUB SYSTEM FLOWCHART

Sheet 1 of 2

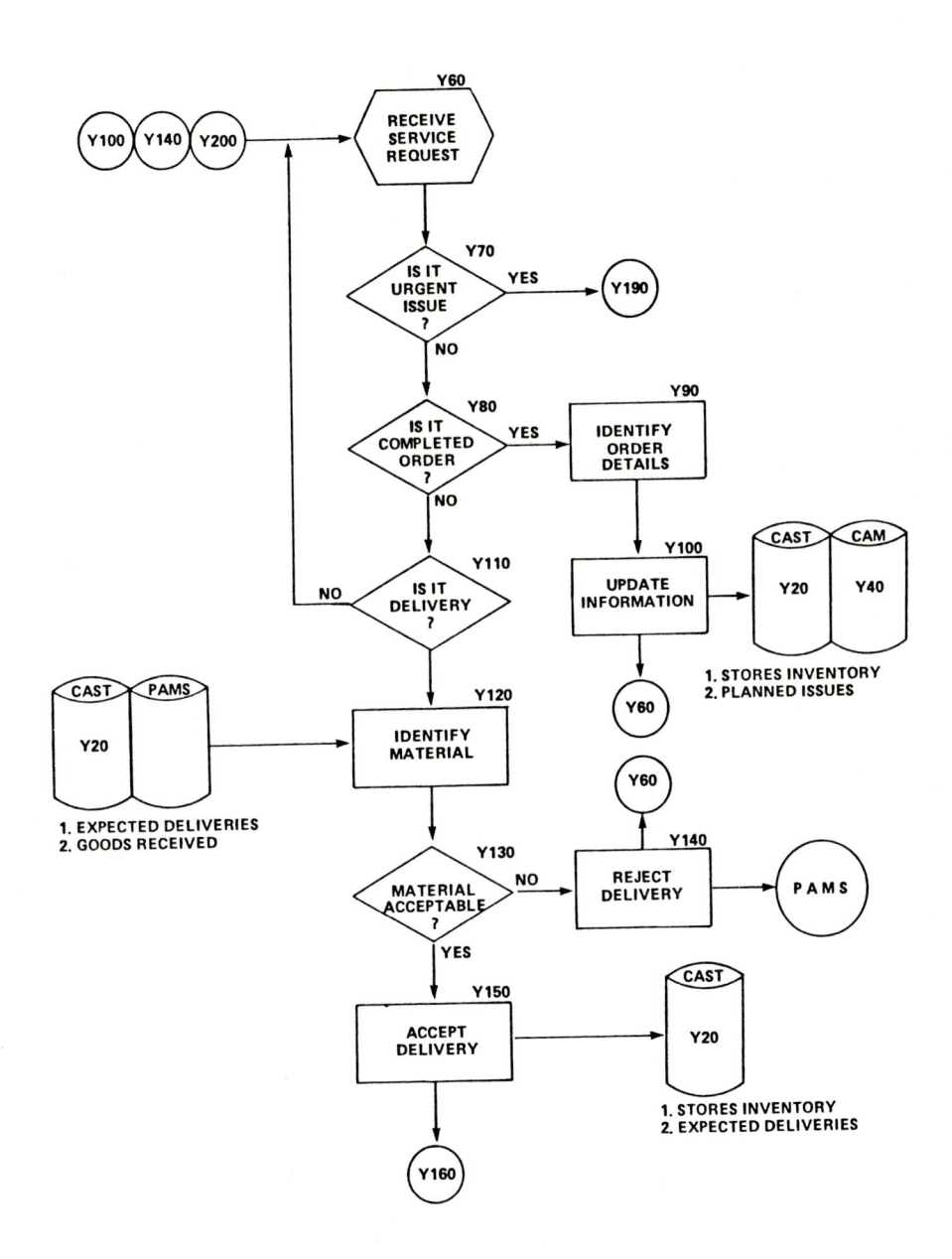

Y 3.2   SUB SYSTEM FLOWCHART

CAM CAST sub-topics

Y150 → Y160 ESTABLISH PAIRING STRATEGY

CAST / M. STD → Y160
1. STORES OPERATING STRATEGY

Y170 IS PAIRING POSSIBLE? — NO → Y70
YES ↓

CAM / Y40 → Y180 IDENTIFY PAIRING
1. PLANNED ISSUES

Y190 ISSUE INSTRUCTIONS TO STORES CONTROL

Y200 UPDATE INFORMATION → CAST Y20 / CAM Y40
1. STORES INVENTORY
2. PLANNED ISSUES

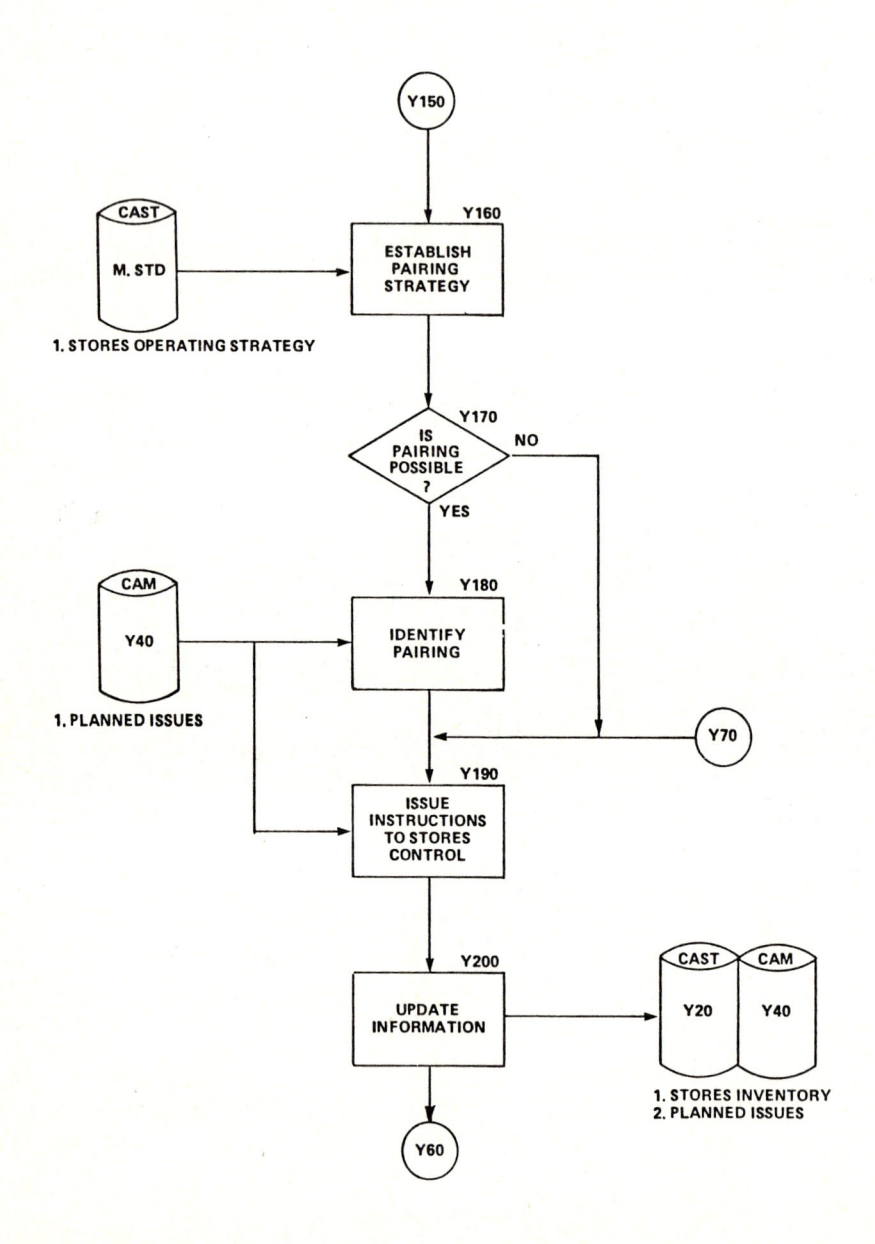

Y60

Y3.3	**Design rules - material co-ordination**
Y3.3.1	**Input data**

The sub-system must be capable of accepting the minimum data items listed below: Data is required relating to:

- Stores extractions to meet production requirements (Section Y3.3.1.1).
- Material deliveries (Section Y3.3.1.2).
- Urgent production requests (Section Y3.3.1.3.).
- Returns from production (Section Y3.3.1.4).

Y3.3.1.1 **Description of data items**

Data relating to stores extractions:

- **Sequence Number**
  The relative priority of the stores extraction.
- **Order Identity**
  The identifier of the manufacturing order for which the stores extraction is required.
- **Finished Component Identity**
  The identifier of the finished component or components which are the subject of the manufacturing order.
- **Rough Component Identity**
  The identifier of the material to be extracted.
- **Batch Identity**
  The identifier of the specific batch of material to be used.
- **Stock Movement Detail**
  The address of the storage location to be accessed and the address to which the extracted material is to be delivered.
- **Potential Extraction Marker**
  A marker to indicate that an extraction is required once the stock has been delivered. The marker must include the date and time that delivery will be available and the quantity to be then extracted, based on standard container loads.

Y3.3.1.2 **Information on delivery of material**

- **Component Identity**
  The unique identifier for the material to be delivered.
- **Batch Identity**
  The unique identifier of the batch of material
- **Part Lot Identity**
  The identifier of the particular delivery comprising part of a batch of material.
- **Packaging**
  The type of container and number of containers together with individual container load quantities.

Y3.3.1.3 **Information on urgent production requests**

The Material Allocation sub-system will have taken account of all production requirements and it is the objective of the Material Co-ordination sub-system to issue such requirements in an orderly and timely manner. However, to cater for urgent requests to issue material to production the sub-system must be able to accept as input the manufacturing order number.

Y3.3.1.4 Information on production returns

The sub-system must be able to accept information on order completion from production. Each completion must indicate:

- **Order Identity**
  The unique reference of the manufacturing order.

- **Finished Component Status**
  The quantity completed and quality standard must be accepted.

Y3.3.2 Input re-formatting

Data will be received in batch format for the stores extraction data whereas all other data will be received interactively. The sub-system must be capable of re-formatting, as required, input data into the form required by the sub-system. It is assumed that the input data will be structured in a disciplined and identified manner.

Y3.3.3 Delivery acceptance

The sub-system must be able to accept requests for reception of material. In so doing it must check that the delivery is expected by matching delivery data with that held for the expected deliveries. The sub-system must carry out the reception process and if necessary the decanting process. The sub-system must be able to warn a designated authority when it cannot match actual delivery details with expected delivery details.

Y3.3.4 Delivery rejection

In cases where the delivery cannot be matched with an expected delivery detail then the system will reject the delivery. It may be necessary at this time to prompt for a response (either human or from a goods reception system) in order to allow for the rejection to be overridden. Such occasions may occur where a delivery is significantly early, probably several days early, but rejection may lead to future material shortage.

Y3.3.5 Order completion process

The sub-system must be capable of accepting notification of completed orders from production. This feed-back allows the sub-system to decide on further action. If the order is signalled as completed then all outstanding stores extractions can be cancelled. This apparent illogicality can occur if the planning process strategies specify extra material to the process due to a less than 100% effective conversion ratio, ie. the production process inherently produces a significant number of scrap components. It may also occur due to last-minute reductions in order quantities inside the material management reaction time.

When the completion signal indicates correct quantity and quality then the sub-system can archive its records for that order. If the quality is suspect then the records will be kept extant for subsequent analysis to deduce the source of the quality problem. This may be due to material deficiencies in which case all material forming a batch will require analysis.

Y3.3.6 Urgent production requests

The sub-system must cater for urgent requests from the production process. In so doing it must recognise the order number and raise the priority of the allocated extractions to meet the demand. This will only occur when the storage facility is unable to handle the planned load due to breakdowns or other unforeseen circumstances.

CAM CAST sub-topics

Y3.3.7   **Stores input allocation**
When a delivery is accepted the sub-system must identify the number of locations and their addresses that will be used to hold the delivered stock. In so doing the sub-system must take account of the type of container and/or component to be stored and use the stores operating strategy in current use. For example, in a high-bay warehouse store facility it is important to spread the stockholding across as many storage aisles as possible in order to maintain availability when cranes breakdown and to balance floor loadings.

Y3.3.8   **Pairing identification**
As the need for stores input movements are generated the sub-system must check its list of stores extractions required and where possible pair a high priority extraction with the next input movement. In all cases the sub-system must balance the conflicting requirements such that congestion is avoided in the stores reception area, production requirements are fulfilled in an orderly fashion and stores facility utilisation is maximised. The system must be able to be tuned to vary this balance as experience in operation provides further statistics on performance. There may be an opportunity in this area for an 'expert' systems approach so that the function is self-tuning.

Y3.3.9   **Stores instruction**
Once any possible pairing decision has been made, or the need for an unpaired stores movement has been established then the precise instruction to the storage facility must be issued.

Y3.3.10   **Output data**
The sub-system must be capable of producing data interactively
   to the minimum levels shown below. It must be in a form which will facilitate re-formatting, ie. all items must be structured in a disciplined and identified manner.

Y3.3.10.1   **Description of data items**
Data relating to Stores Instruction:

- **Input Address**
  The unique reference of the stores location into which material is to be placed.
- **Input Identity**
  The unique identifier for the material to be stored.
- **Batch Identifier**
  The batch number pertaining to the stored material.
- **Input Quantity**
  The quantity of the components to be stored.
- **Input Source**
  A unique reference for the point where the material to be stored is to be found.
- **Output Address**
  The unique reference of the stores location to be accessed to extract material.
- **Output Identity**
  The identifier for the stored material.
- **Batch Identifier**
  The batch number pertaining to the stored material.

- **Output Quantity**
  The quantity to be extracted.
- **Output Description**
  An unambiguous reference to the point to which the stored material is to be delivered.

If the stores instruction is for a single movement then either the input set or output set of data items will be insignificant.

Part-programme management

Introduction

It is essential in an integrated manufacturing system to be able to prepare part-programmes in accordance with the requirements of the planned production schedule. This will ensure that verified and proven part-programmes are available in the library when they are required. Such a degree of control will also tend to ensure the load on the part-programming department is reasonably evenly distributed.

A particular approach has been adopted with regard to the forms in which part-programmes are held in the library. Part-programmes are first held in CLDATA form where they remain until the particular machine tools which will use these programmes have been selected. They are then post-processed as required.

It is believed that in the foreseeable future certain elements in a part-programme can only be proved on the machine tool concerned. However, it is also acknowledged that as CADCAM systems develop the amount of machine tool involvement in prove-out should significantly diminish. It is for these reasons that this system embraces both areas of part-programme prove-out.

CAM CAST sub-topics

**Computer Aided Manufacturing/Computer Aided Storage and Transportation
Part-programme management**

Flowcharts

CAM CAST sub-topics

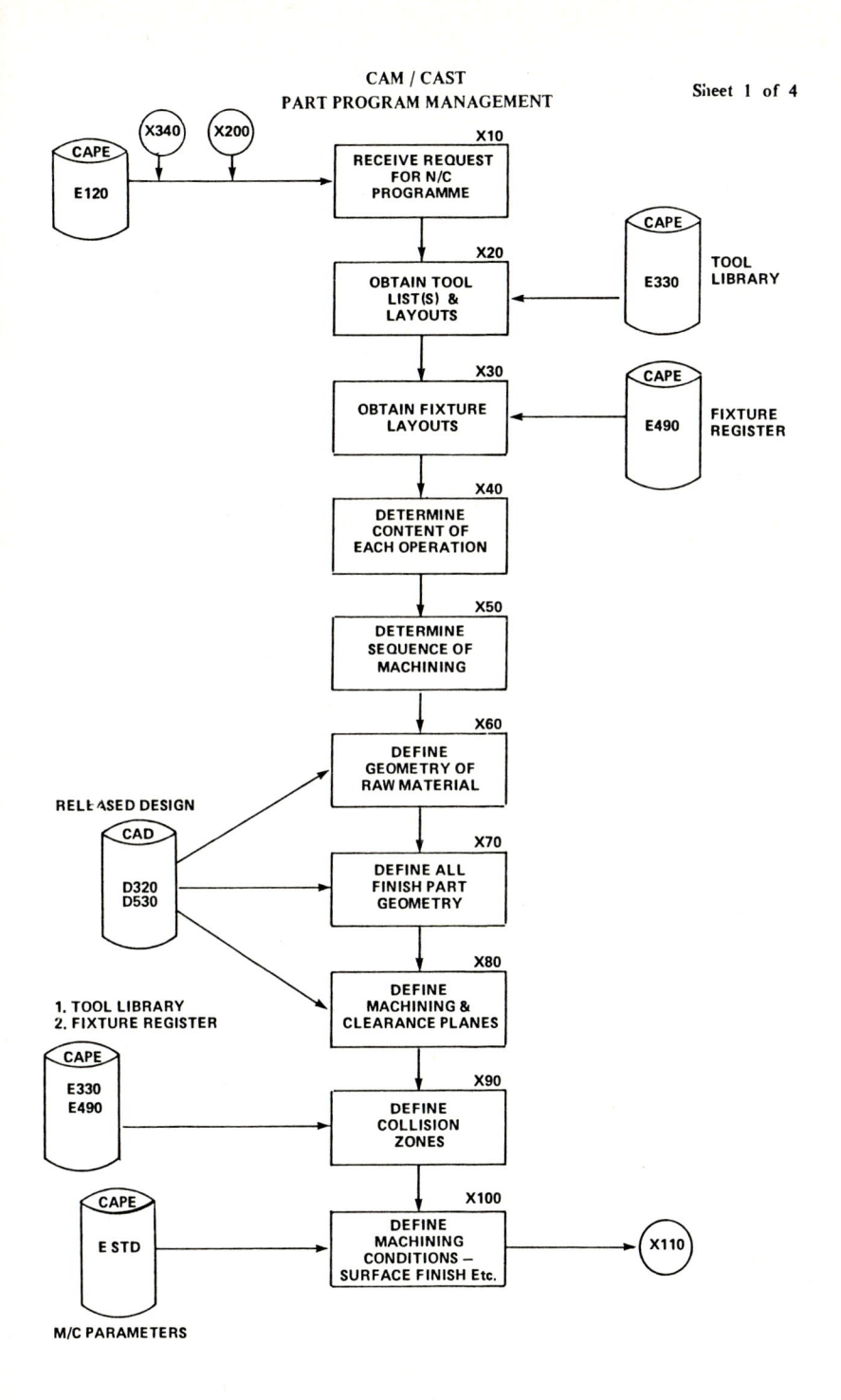

CAM / CAST
PART PROGRAM MANAGEMENT

Sheet 1 of 4

- CAPE E120
- X340
- X200

X10 RECEIVE REQUEST FOR N/C PROGRAMME

X20 OBTAIN TOOL LIST(S) & LAYOUTS ← CAPE E330 TOOL LIBRARY

X30 OBTAIN FIXTURE LAYOUTS ← CAPE E490 FIXTURE REGISTER

X40 DETERMINE CONTENT OF EACH OPERATION

X50 DETERMINE SEQUENCE OF MACHINING

X60 DEFINE GEOMETRY OF RAW MATERIAL

RELEASED DESIGN
CAD D320 D530

X70 DEFINE ALL FINISH PART GEOMETRY

1. TOOL LIBRARY
2. FIXTURE REGISTER

X80 DEFINE MACHINING & CLEARANCE PLANES

CAPE E330 E490

X90 DEFINE COLLISION ZONES

CAPE E STD

X100 DEFINE MACHINING CONDITIONS – SURFACE FINISH Etc. → X110

M/C PARAMETERS

**CAM / CAST
PART PROGRAM MANAGEMENT**

Sheet 2 of 4

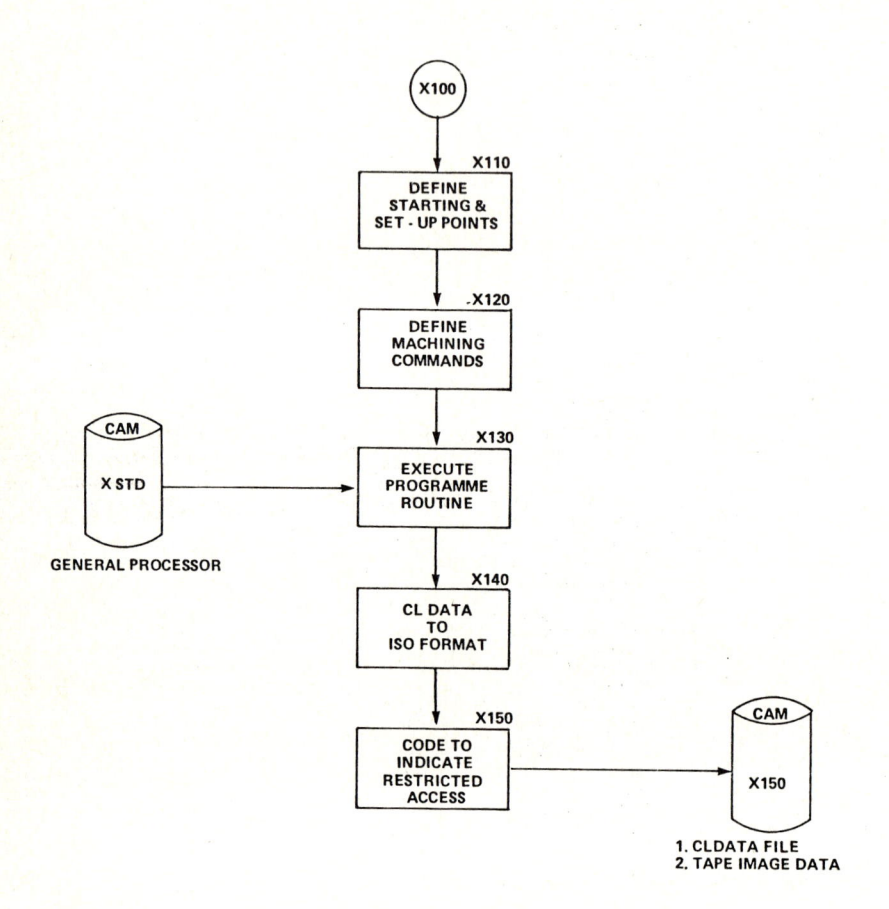

```
                    (X100)
                      │
                      ▼  X110
              ┌─────────────────┐
              │     DEFINE      │
              │   STARTING &    │
              │ SET - UP POINTS │
              └─────────────────┘
                      │
                      ▼  X120
              ┌─────────────────┐
              │     DEFINE      │
              │    MACHINING    │
              │    COMMANDS     │
              └─────────────────┘
                      │
  ╭─────╮             ▼  X130
  │ CAM │     ┌─────────────────┐
  │XSTD │────▶│     EXECUTE     │
  ╰─────╯     │    PROGRAMME    │
              │     ROUTINE     │
GENERAL              └─────────────────┘
PROCESSOR             │
                      ▼  X140
              ┌─────────────────┐
              │     CL DATA     │
              │       TO        │
              │   ISO FORMAT    │
              └─────────────────┘
                      │
                      ▼  X150         ╭─────╮
              ┌─────────────────┐     │ CAM │
              │    CODE TO      │────▶│X150 │
              │   INDICATE      │     ╰─────╯
              │  RESTRICTED     │
              │    ACCESS       │   1. CLDATA FILE
              └─────────────────┘   2. TAPE IMAGE DATA
```

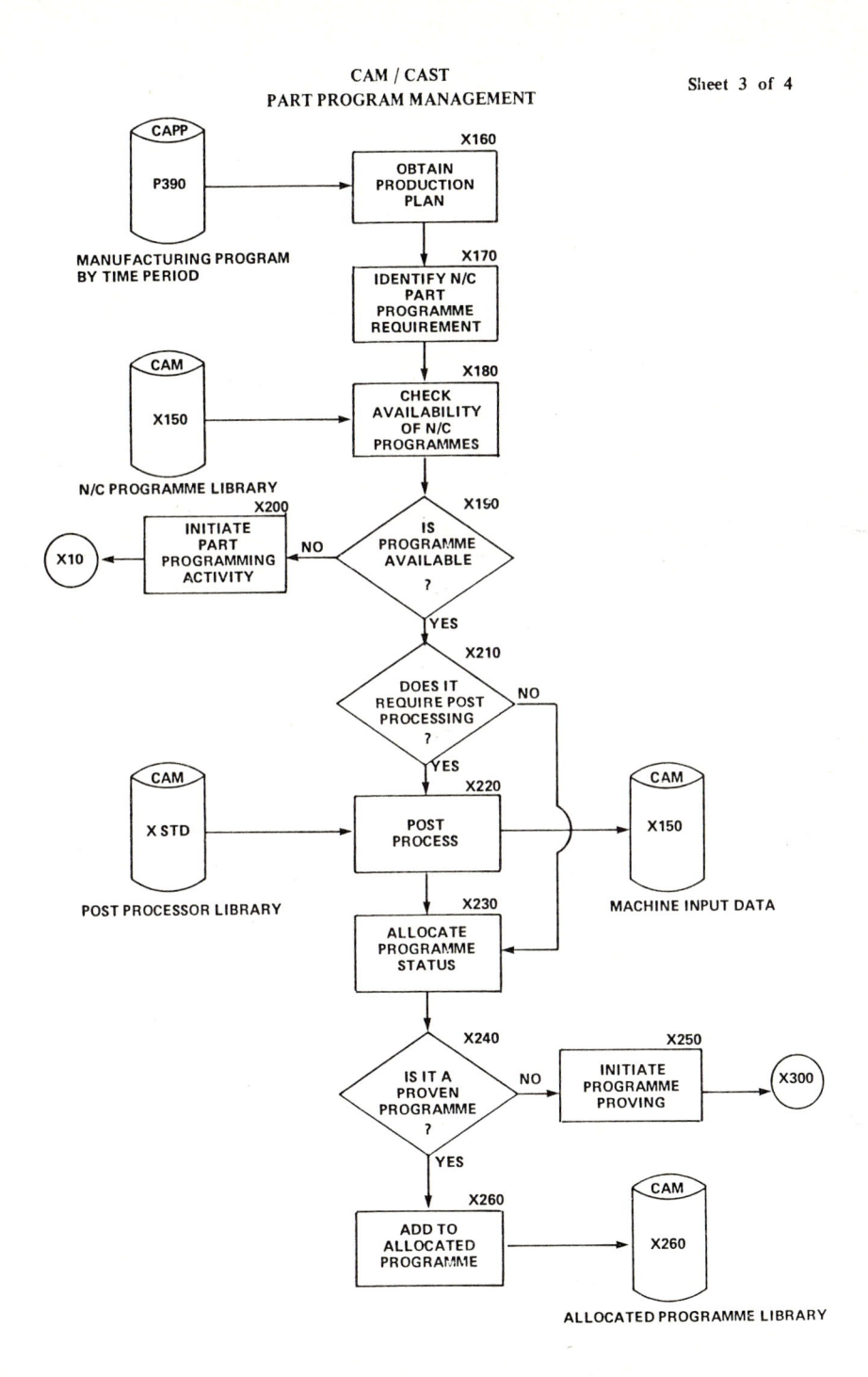

CAM / CAST
PART PROGRAM MANAGEMENT

Sheet 3 of 4

CAM CAST sub-topics

CAM / CAST
PART PROGRAM MANAGEMENT

Sheet 4 of 4

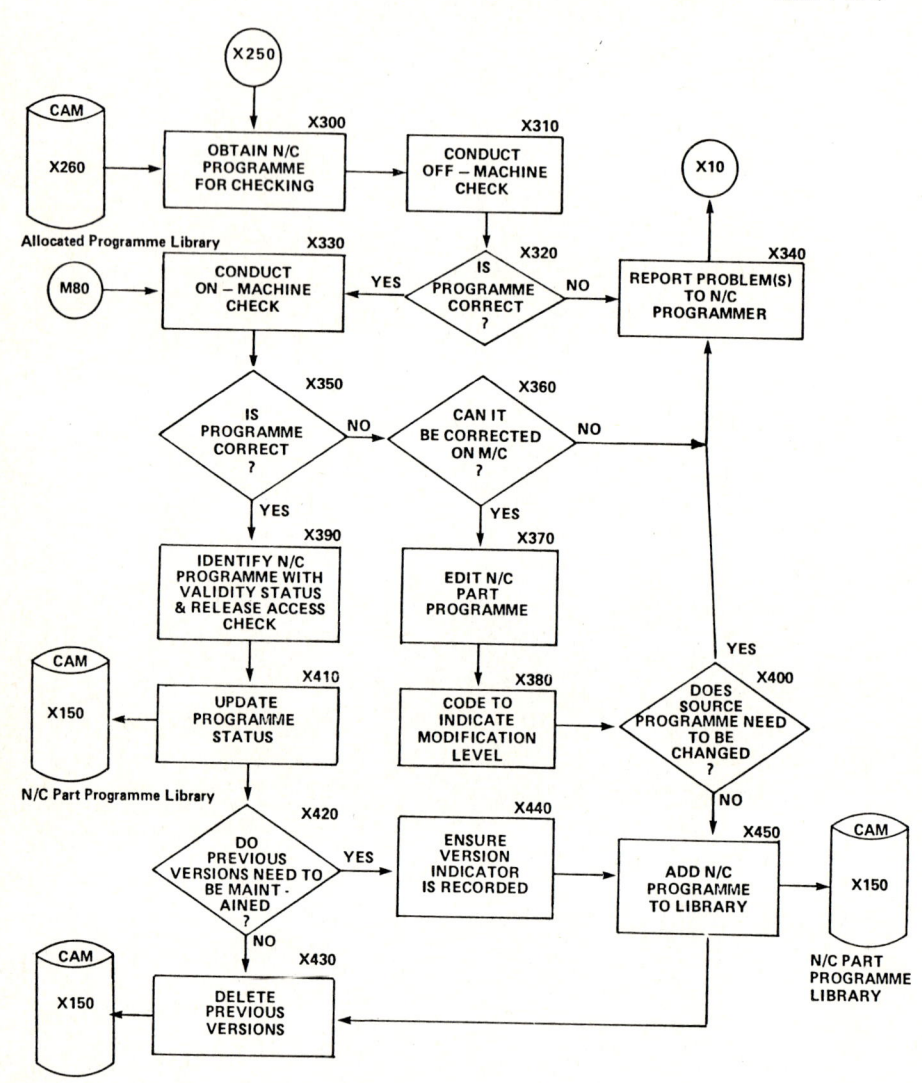

(Flowchart)

- X250 → X300 OBTAIN N/C PROGRAMME FOR CHECKING
- CAM X260 (Allocated Programme Library) → X300
- X300 → X310 CONDUCT OFF – MACHINE CHECK
- X310 → X320 IS PROGRAMME CORRECT?
  - YES → X330 CONDUCT ON – MACHINE CHECK
  - NO → X340 REPORT PROBLEM(S) TO N/C PROGRAMMER → X10
- M80 → X330
- X330 → X350 IS PROGRAMME CORRECT?
  - YES → X390 IDENTIFY N/C PROGRAMME WITH VALIDITY STATUS & RELEASE ACCESS CHECK
  - NO → X360 CAN IT BE CORRECTED ON M/C?
    - YES → X370 EDIT N/C PART PROGRAMME → X380 CODE TO INDICATE MODIFICATION LEVEL
    - NO → X340
- X390 → X410 UPDATE PROGRAMME STATUS
- CAM X150 (N/C Part Programme Library) ← X410
- X380 → X400 DOES SOURCE PROGRAMME NEED TO BE CHANGED?
  - YES → X340
  - NO → X450 ADD N/C PROGRAMME TO LIBRARY → CAM X150 (N/C PART PROGRAMME LIBRARY)
- X410 → X420 DO PREVIOUS VERSIONS NEED TO BE MAINTAINED?
  - YES → X440 ENSURE VERSION INDICATOR IS RECORDED → X450
  - NO → X430 DELETE PREVIOUS VERSIONS → CAM X150 (N/C Part Programme Library)

Narrative description of
N/C Part-programming

X10   When the selection of machining process has been completed a request is received for an N/C part-programme to be prepared.

X20   In order to conduct the programming process it is necessary to acquire the complete description and content of the tool list which has been compiled at nodes E280-E330 of CAPE.

X30   The fixture layouts must be obtained from the fixture library - E490.

X40   The actual content of each operation must be defined in order that the N/C programme may be structured. This is more necessary on multi-process machines such as machining centres.

X50   Having determined the content it is now necessary to define the sequence of machining within each individual N/C operation.

X60   It is necessary to define the geometry of the raw material be it casting, forged billet or bar stock. This will allow the optimum number of cuts to be determined.

X70   Geometric definition of all internal and external features of the component is necessary to provide the data for structure of the machining process, ie. profile sequence, etc.

X80   Location of machining and clearance planes must be defined relative to the reference point of each face.

X90   Collision zones must be defined to avoid any possibility of collision between any part of the machine tool or its cutting tools with the component and its fixtures.

X100   Definition of machining conditions involves definition of all cutting features, ie. cutter diameter, speeds, feeds, rapid moves, surface finish, tolerancing and coolant requirements as appropriate.

X110   Starting and set-up points are required to define the location of each face of the component relative to a common location such as centre of rotation. Definition is required for X, Y, Z and angular displacement, as appropriate.

X120   The programme may now be structured by defining the machining commands according to the sequence of machining as specified at node X50. This activity includes structuring of the profile of any surfaces using a sequence of geometric references which were defined at node X70.

X130   Execution of the programme routine involves processing of the programme by computer using a general-purpose N/C processor such as APT or its derivatives such as EXAPT or other N/C processors which are capable of processing the source data to provide the cutter location data (CLDATA) without reference to the machine tool on which the component will be produced.

X140   As the part-programme is required to be proved before it can be used for scheduled production, it is necessary to include an identification code to restrict access to those who are authorised to conduct prove-out.

X150   Output from the general processing stage is the CLDATA which should be compiled to ISO standard format.

X160   A copy of the desired production plan is obtained from the Production Scheduler - CAPP P390.

X170	Those details within the production plan for which part-programmes are required are established.
X180	By reference to the library of N/C part-programmes (X150) determine availability of the required part-programmes are determined.
X190 and X200	If the part-programme is not available a request must be made to the programme preparation area X10 for the programme to be written.
X210	As the programme is available its status must be determined to see if it has been post-processed or is still in CLDATA. This will enable a decision to be made as to whether it is necessary to conduct the post-processing stage.
X220	The N/C post processor is used to convert the CLDATA format of the part-programme into machine readable format appropriate to the machine on which the detail is to be produced. The post processor is a computer programme which is peculiar to a machine type and therefore a different post processor is required to be used if the detail is to be produced on any other machine type from that which has been previously used. Output of the post processing stage is recorded within the N/C library maintained on X150 in machine readable format.
X230	If the programme is required to be proven its status must be identified to restrict its use to proving runs only.
X240	It is necessary to assess the validity of the N/C part-programme to determine if it is necessary to prove the programme either on or off the machine.
X250	If the programme has not yet been proven it is necessary to initiate the proving process.
X260	The N/C part-programmes are then included within the library of programmes which are allocated for completion of the production plan.
X300	Those N/C part-programmes which are required to be proved are advised by the Production Scheduler (P120) and collected from the N/C part-programme library (X150).
X310	In order to minimise the machine time required for proving of the part-programme it is desirable to employ all available means to pre-check the machine input data. This could involve the use of graphics techniques if they are available.
X320	The result of the off-machine check at node X310 will determine the accuracy of the programme and allow this decision to be made.
X330	If the programme is correct it can now be checked on the machine. The most time-efficient method should be employed but at all times care should be taken to ensure that the machines/fixtures and details are protected from damage. Prove-out could involve cutting "in-air", use of polystyrene models or step-by-step progress through the programme on an actual detail.
X340	If the programme is not correct from either the "on" or "off" machine checks the problem must be reported back to the programme preparation management in order that corrective action can be taken.
X350	When the "on-machine" prove-out is performed the validity of the programme will be determined and the answer to this question will be known.
X360	If a problem exists it may be possible to correct it on the machine. If it is not possible the problem should be reported as described in node X340.
X370	If the programme can be corrected on the machine the programme should be edited to allow the batch of parts to be produced if required.

CAM CAST sub-topics

X380 If the programme has been corrected it is necessary to identify the change by revision of the modification level.

X390 If the programme is correct is must be identified with a validity status to release the access restriction applied at node X230.

X400 If the programme has been modified (edited) on the machine, management must be consulted to determine if it is necessary to modify the source statements, ie. prior to general processing at node X130.

X410 As the programme is now considered to be correct the status of the programme must be incorporated within the N/C programmer library.

X420 Management must be consulted to determine if it is necessary to maintain previous versions of this programme.

X430 If previous versions of the programme are not required to be maintained they should be deleted.

X440 If previous versions are to be maintained it should be ensured that clear modification and version codes have been applied and are recorded.

X450 If the edited part-programme can be used then it must be incorporated within the N/C programme library in its revised form.

CAM CAST sub-topics

Computer Aided Manufacturing/Computer Aided Storage and Transportation

Part-programming management sub-systems

The flowchart for the Part-Programme Management activities has been broken down into the following sub-systems:

1 N/C Part-Programming
    (Nodes X10 to X150)

2 N/C Part-Programming Management
    (Nodes X160 to X260)

3 N/C Part-Programming Prove-Out
    (Nodes X300 to X450)

N/C Part-programming sub-system (X1)

X1.1　　　**Sub-system description**

This sub-system is concerned with the task of creating the machining data instructions for N/C machine tools for those details which have been selected as being suitable for processing within an N/C machine.

The process involves obtaining details of tooling and fixtures designed within CAPE and establishing the sequence of operations within the N/C machining cycle. A major part of this activity involves definition of part geometry, machining technology - speeds and feeds, etc., and collision zones, etc. A set of source statements is then produced to describe the machining sequence in the format required by the Processor which is to be employed. A number of general processors are available which will process the prepared data to provide the Cutter Location Data (CLDATA) to ISO format. The information is then stored in this form until the decision has been made as to which specific machine tool, will be used for producing the detail.

N/C Part-programming sub-system

X1.2 Flowchart

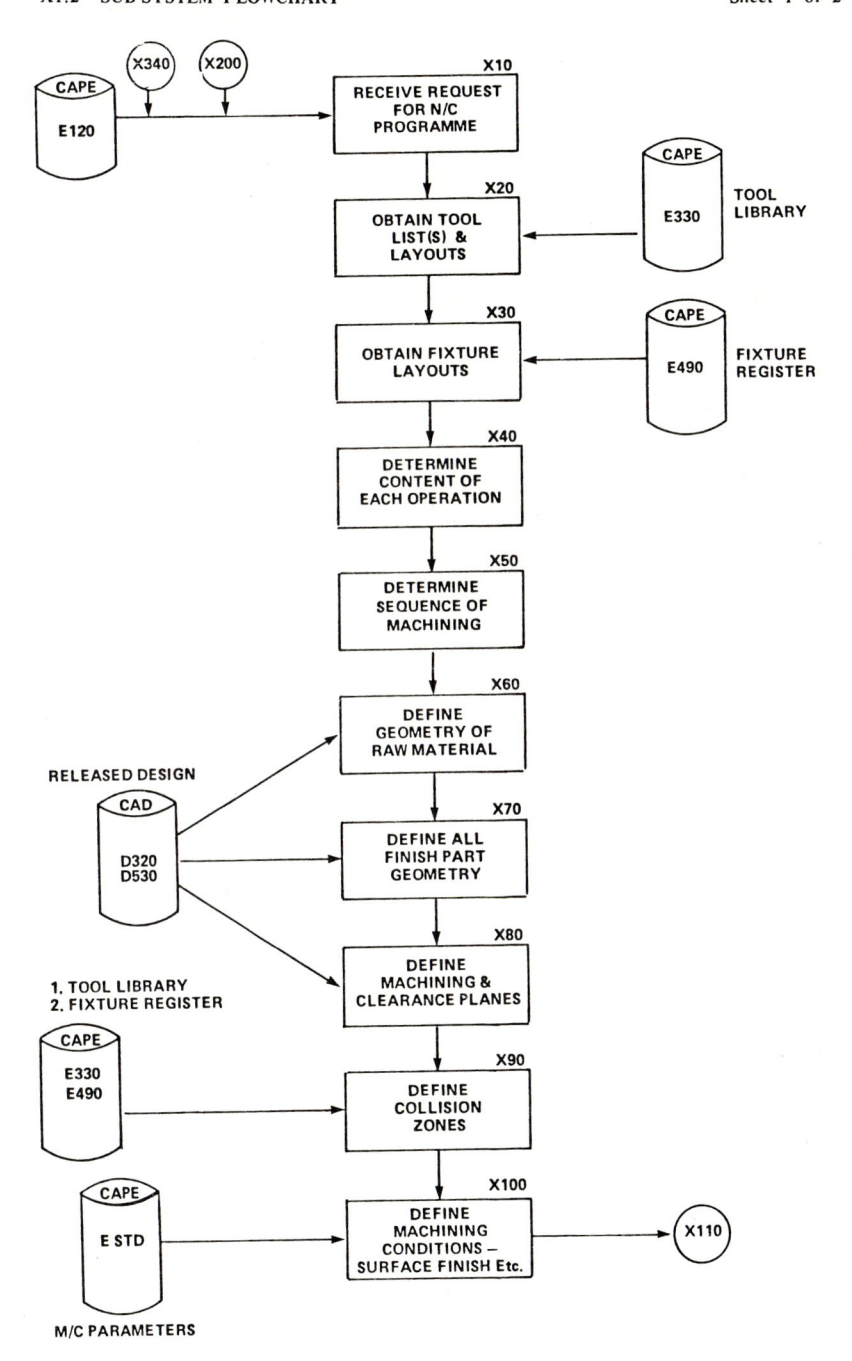

X1.2   SUB SYSTEM FLOWCHART   Sheet 1 of 2

- X340, X200 → **X10** RECEIVE REQUEST FOR N/C PROGRAMME
- CAPE E120 → X10
- **X20** OBTAIN TOOL LIST(S) & LAYOUTS ← CAPE E330 (TOOL LIBRARY)
- **X30** OBTAIN FIXTURE LAYOUTS ← CAPE E490 (FIXTURE REGISTER)
- **X40** DETERMINE CONTENT OF EACH OPERATION
- **X50** DETERMINE SEQUENCE OF MACHINING
- **X60** DEFINE GEOMETRY OF RAW MATERIAL
- **X70** DEFINE ALL FINISH PART GEOMETRY
- **X80** DEFINE MACHINING & CLEARANCE PLANES
- **X90** DEFINE COLLISION ZONES
- **X100** DEFINE MACHINING CONDITIONS – SURFACE FINISH Etc. → **X110**

RELEASED DESIGN — CAD D320 D530 → X60, X70, X80

1. TOOL LIBRARY
2. FIXTURE REGISTER

CAPE E330 E490 → X90

CAPE E STD → X100

M/C PARAMETERS

X1.2   SUB SYSTEM FLOWCHART   Sheet 2 of 2

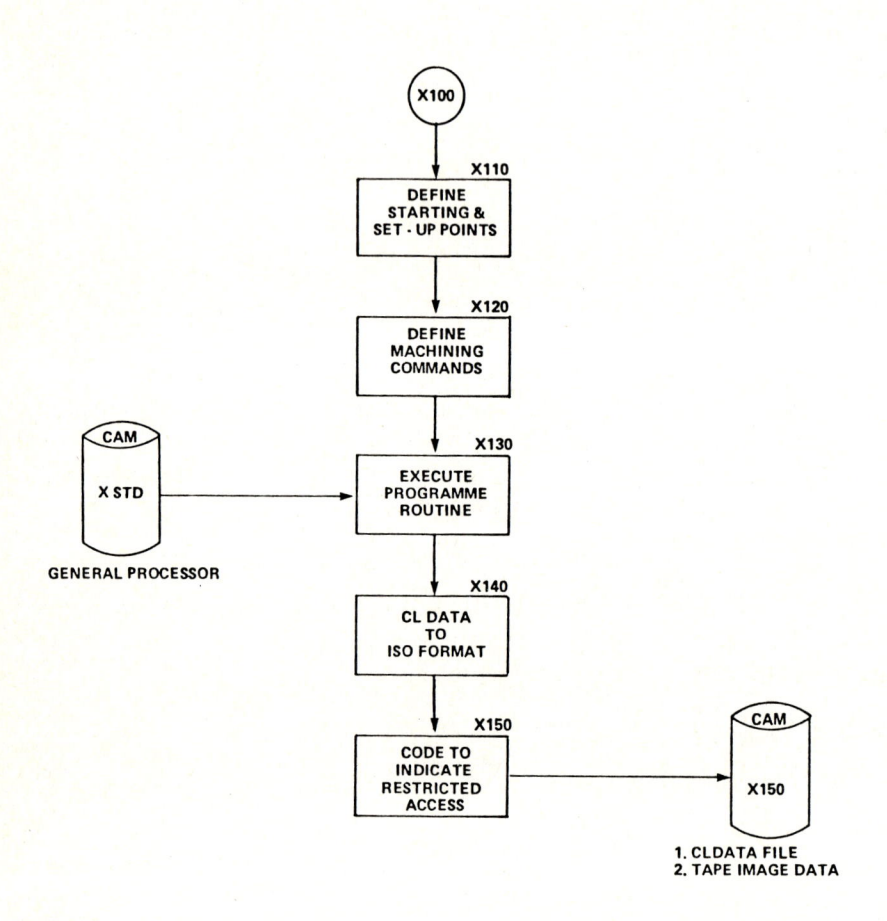

```
                          ( X100 )
                             │
                             ▼        X110
                        ┌─────────┐
                        │ DEFINE  │
                        │STARTING &│
                        │SET-UP POINTS│
                        └─────────┘
                             │
                             ▼        X120
                        ┌─────────┐
                        │ DEFINE  │
                        │MACHINING│
                        │COMMANDS │
                        └─────────┘
                             │
   ╔═══════╗                 ▼        X130
   ║ CAM   ║            ┌─────────┐
   ║ X STD ║──────────▶ │ EXECUTE │
   ╚═══════╝            │PROGRAMME│
                        │ ROUTINE │
 GENERAL PROCESSOR      └─────────┘
                             │
                             ▼        X140
                        ┌─────────┐
                        │ CL DATA │
                        │   TO    │
                        │ISO FORMAT│
                        └─────────┘
                             │
                             ▼        X150                    ╔═══════╗
                        ┌─────────┐                           ║ CAM   ║
                        │ CODE TO │                           ║       ║
                        │INDICATE │──────────────────────────▶║ X150  ║
                        │RESTRICTED│                          ╚═══════╝
                        │ ACCESS  │                       1. CLDATA FILE
                        └─────────┘                       2. TAPE IMAGE DATA
```

X1.3	**Design rules - N/C Part-programming**
X1.3.1	**Input data**
	The system must be capable of accepting the minimum data items listed below, for any detail:
X1.3.1.1	Data Relating to Detail:

- **Identity**

    The identity must be an unambiguous reference by which the detail can be selected and must include revision level.

- **Dimensioning**

    The dimensions, tolerances, geometric tolerances and datum points for both the finished machined detail and unmachined blank. This information must be quoted in suitable form to allow absolute determination of the precise shape, form, size and position of every feature.

- **Surface Finish**

    The surface finish or texture requirements of any element of the detail.

- **Material**

    The type of material from which the detail is to be made. Where appropriate this should include the grade or composition of the material, eg "high speed steel" rather than steel.

- **Material State**

    This refers to any material treatment, eg. heat treatment or surface hardening.

X1.3.1.2　The following data relating to the cutting tools to be used is required to be available:

- **Identity**

    A unique identity which distinguishes any tool from all others is required.

- **Tool Characteristics**

    Each characteristic of the tool must be individually identified such that its application may be verified by the N/C Part Programmer. Characteristics required include:

Tool Type	- eg. Twist Drill
Tool Material	- eg. High Speed Steel
Tool Length	- eg. 200 mm
Tool Free Length	- eg. 75 mm
Drill Flute Length	- eg. 60 mm
Tool Diameter	- eg. 20 mm
Tool Point	- eg. 118 degrees inc.

X1.3.1.3　The following data relating to the workholding fixture is required to be available:

- **Identity**

    An unambiguous reference to each fixture is required.

- **Location**

A finite description of the position of the location devices within the fixture is required together with their location relative to the machine location features.

- **Description of Clamping Features**

    A finite description of the location and size of clamping devices is required to enable a cutter location path to be constructed which avoids collisions between the fixture and the machine and/or its tools.

X1.3.1.4 **NC General Purpose Processor**

Facility is required for manipulation of the data produced within the Part-Programming sub-system in order to prepare the cutter path location data (CLDATA) for a detail without reference to the machine tool on which the machining will occur. This could be achieved by use of APT or its derivatives which could exist as a stand-alone facility or reside as a part of a CAD package.

X1.3.2 **Input re-formatting**

Data will be received in batch format   The sub-system must be capable of re-formatting, as required, input data into the form required by the sub-system. It is assumed that the input data will be structured in a disciplined and identified manner.

X1.3.3 **Output data**

Output data from this sub-system will be fully descriptive of the machining process and will be structured according to the accepted ISO standard format for NC machine CLDATA files.

The CLDATA file will also contain all those instruction sets which are necessary to enable the post processor to incorporate suitable commands within th e Tape Image version of the Programme, eg. coolant ON/OFF, etc.

N/C Part-programming management sub-system (X2)

X2.1 **Sub-system description**

The sub-system involves reviewing the production plan and checking the availability of those N/C Part-Programmes which are required to enable the plan to be satisfied. Liaison with the resource scheduler is required to enable the plan to be verified or changed. Conversion of the Cutter Location Data (CLDATA) to the format required by the selected machine tool is also a function of this sub-system. This involves the processing of the CLDATA within a computer process which is peculiar to that machine which is to be used.

Finally, when the N/C Part-Programme is available within the required form it is included within the file of allocated programmes which are required to satisfy the requirements of the Production Plan.

N/C Part programming management sub-system

X2.2 Flowchart

X2.2. SUB SYSTEM FLOWCHART

Sheet 1 of 1

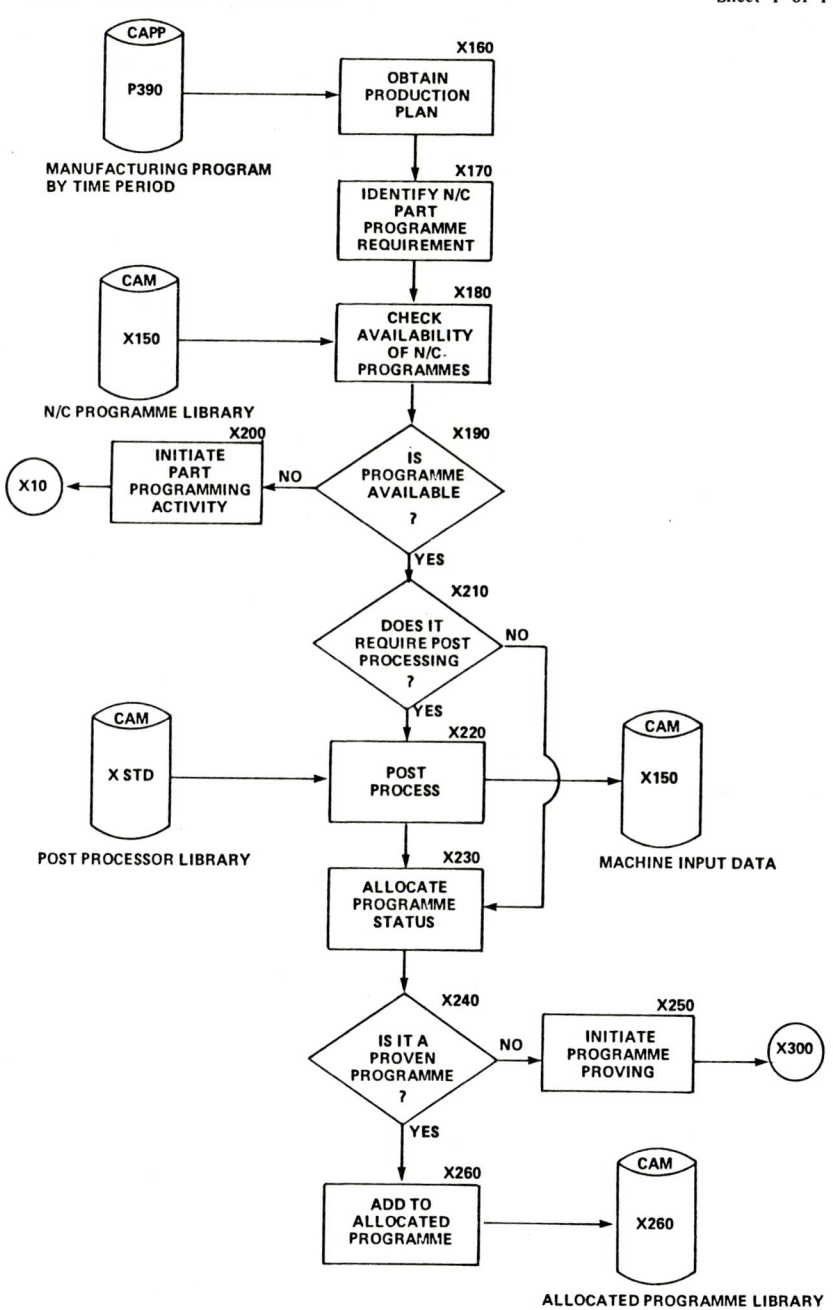

CAPP P390 — MANUFACTURING PROGRAM BY TIME PERIOD

- X160: OBTAIN PRODUCTION PLAN
- X170: IDENTIFY N/C PART PROGRAMME REQUIREMENT
- X180: CHECK AVAILABILITY OF N/C PROGRAMMES

CAM X150 — N/C PROGRAMME LIBRARY

- X190: IS PROGRAMME AVAILABLE?
  - NO → X200: INITIATE PART PROGRAMMING ACTIVITY → X10
  - YES ↓
- X210: DOES IT REQUIRE POST PROCESSING?
  - NO → (to X230)
  - YES ↓
- X220: POST PROCESS

CAM X STD — POST PROCESSOR LIBRARY

CAM X150 — MACHINE INPUT DATA

- X230: ALLOCATE PROGRAMME STATUS
- X240: IS IT A PROVEN PROGRAMME?
  - NO → X250: INITIATE PROGRAMME PROVING → X300
  - YES ↓
- X260: ADD TO ALLOCATED PROGRAMME

CAM X260 — ALLOCATED PROGRAMME LIBRARY

X2.3 Design rules - N/C programme management

X2.3.1 Input data

The sub-system should be capable of accepting the minimum data items listed below in respect of the prioritised production plan from P470.

- **Detail Identity**

    An unambiguous reference by which the detail can be identified. The reference must clearly identify the modification level or version.

- **Allocated Machine**

    The identity of the machine which has been allocated for the production of the detail.

- **Part-Programme Identity**

    An unambiguous reference to the N/C programme which will produce the part of the stated modification level on the allocated machine.

- **Part-Programme Status**

    A code which indicates whether this programme is the current version, an obsolete version or an improved version.

X2.3.2 Input re-formatting

Data will be received in batch format. The sub-system must be capable of re-formatting, as required, input data into the form required by the sub-system. It is assumed that the input data will be structured in a disciplined and identified manner.

X2.3.3 Output data

Output data relating to availability of part-programmes is required to be provided to production and resource schedulers.

The status of part-programmes is also an output which enables additional machine time to be allocated for activities such as part-programme prove-out.

X2.3.4 Post-processing

This sub-system must provide access to a post-processor library containing the post-processors for all the combinations of machine tool/NC systems that exist in the machine shops under consideration.

Post-processors are commonly written for implementation on a particular computer. If more than one computer is used for post-processing, then this sub-system must take into account that more than one version of a particular post-processor may exist.

N/C Part-programming prove-out sub-system (X3)

X3.1 Sub-system description

This sub-system is concerned with verifying the data which was generated within the part-programming sub section (X1). It is necessary to verify the geometric definitions together with the suitability of machining technology employed under dynamic conditions. It is also necessary to ensure that clearance zones have been observed prior to release of the N/C part-programme for production purposes.

Two stages are involved - "off" and "on" machine. "Off" machine checking can be accomplished by graphical techniques. However, "on" machine requires that the machine tool together with all other resources are available. A means of verifying the data with the minimum of machine time is essential particularly in companies which encounter frequent change of components. A number of management decisions are called for within this sub-section which are related to the necessity of maintaining edited data and also the need to maintain previous versions of an N/C part-programme.

N/C Part-programming prove-out sub-system

X3.2 Flowchart

X3.2. SUB SYSTEM FLOWCHART

Sheet 1 of 1

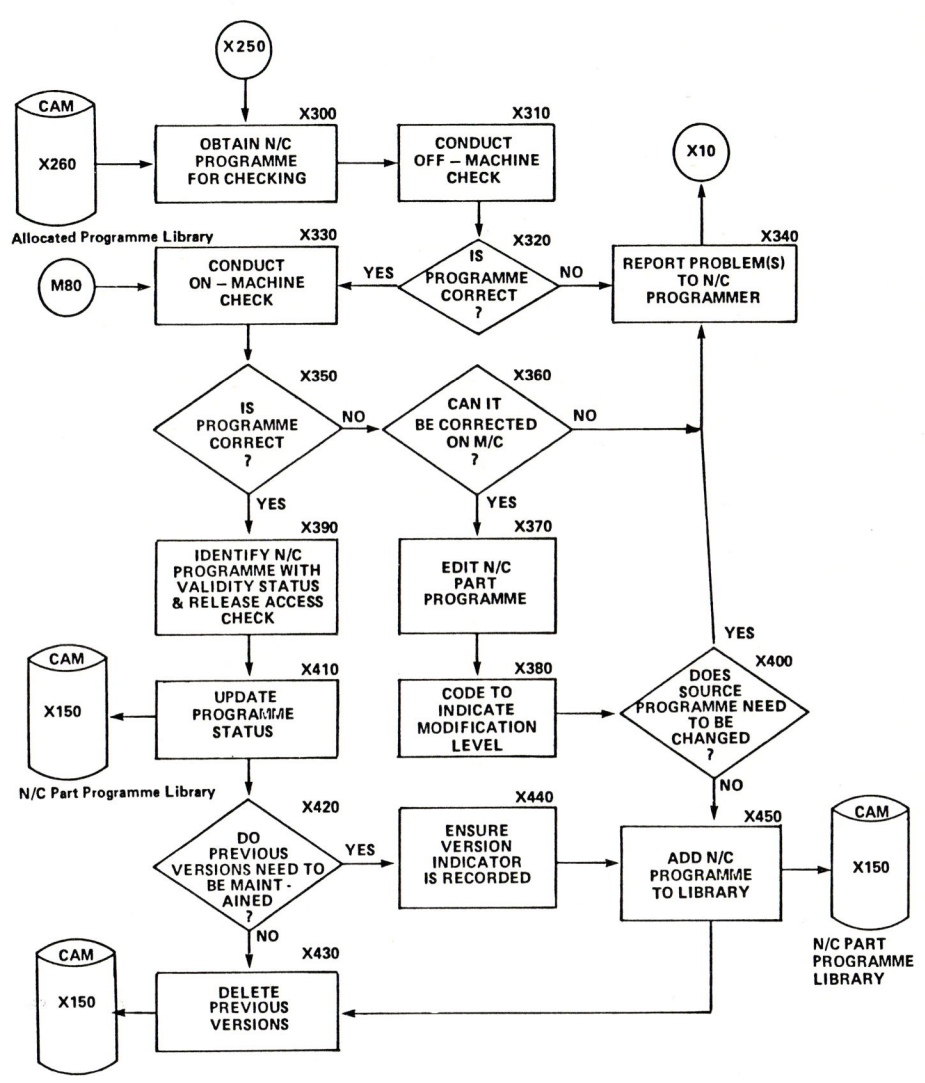

X3.3 Design rules - Part-programme prove-out

X3.3.1 Input data

The sub-system must be capable of accepting the minimum data items listed below:

Data is required relating to:

- The detail required to be proved.
- The data relating to the N/C programme from the library of allocated programmes.
- Instructions to proceed with part-programme prove-out from the resource scheduler (Node M80).

X3.3.2 Input re-formatting

Data will be received in a batch format. The sub-system must be capable of re-formatting, as required, input data into the form required by the sub-system. It is assumed that the input data will be structured in a disciplined and identified manner.

X3.3.3 Other design rules

X3.3.3.1 Identified Problems (X340) (X400)

Provision must be made within this sub-system to inform the Part Programmer of the problems encountered and preventing further use of the programme until corrections have been made.

X3.3.3.2 Previous Versions (X420)

When a programme has been proved provision must be made for maintaining or deleting previous versions if they exist.

X3.3.4 Output data

Provision must be made to permit the storage of completed part-programmes within (X150) and release of the Access Check.

The Part Programme identity must be able to incorporate the version indicator.

X3.3.5 Management decisions

Within this sub-system there emerges the need to provide a means of human intervention in the following activities:

- **Part-Programme Versions (X420)**

  Whenever a new part-programme is added to the library, the manager responsible must decide whether or not to retain any previous versions of that particular part-programme.

  A means must be provided, therefore, to allow deletion to be performed on instructions from the manager responsible.

  Safeguard must also be included to prevent unauthorised deletion or alteration to any part-programme residing in the library.

- **Source Programme Changes (X400)**

  When a part-programme is being finally proved on a machine tool and modifications are found to be necessary, the manager responsible must decide whether or not to modify the machine-independent source programme.

A means must be provided, therefore, to allow such modifications to be made on instructions from the manager responsible.

Safeguards must also be included to prevent unauthorised modifications being made to source programmes residing in the library.

- **Reporting Programme Changes (X340)**

Any changes in a part-programme which are found to be necessary during prove-out will need to be reported to the writer of the particular part-programme.

This sub-system must therefore provide an effective means of reporting such changes.

Production management

Introduction

Production Management is that part of CAM/CAST which is concerned with carrying out the processes and operations which are required to satisfy manufacturing orders once those orders are commenced.

Production Management is concerned only with current activities and planned activities within a very short time span, it does not include any form of forward planning of manufacturing orders.

The processes which are initiated, monitored and co-ordinated by Production Management sub-systems are:

1) The movement of allocated raw material and work-in-progress between stores, buffers and workcentres.
2) The delivery of tools and accessories to fixed facilities.
3) The delivery of process instructions to fixed facilities.
4) The mounting and fixturing or workpieces.
5) The machining of workpieces.
6) The measurement, testing and inspection of workpieces.
7) The packing of finished components

To manage production properly Production Management sub-systems must be constantly aware of the status of every facility, resource and workpiece within its scope. Production Management sub-systems therefore encompass the monitoring of all physical activity within the scope of a CIM system.

The generalised events which are monitored by the Production Management sub-systems are:

1) The acknowledgment or commencement of authorised tasks.
2) The successful completion of authorised tasks.
3) The failure to complete authorised tasks.
4) The delays to completion of authorised tasks.
5) The occurrence of unauthorised activities.

Computer Aided Manufacturing/Computer Aided Storage and Transportation

Production Management

Flowcharts

CAM/CAST PRODUCTION MANAGEMENT

CAPP P 470
1. PRIORITISED CURRENT PRODUCTION REQUIREMENTS

CAPE E120
1. OPERATION DETAILS

M 10 — RECEIVE PRIORITY-SEQUENCED PART ORDERS, MERGE WITH OPERATIONS LIST

M 20 — PROVIDE CAPP WITH LISTS OF ORDERS PRODUCED OR NOT ACTIONED REPORT ACHIEVEMENT

CAM M 20
1. COMPLETED PRODUCTION

M 30 — PROVIDE CAPP WITH RESOURCE AVAILABILITY CAPACITY & PERFORMANCE DATA

CAM M 30
1. AVAILABLE CAPACITIES

CAM M 10
1. ORDER STATUS
2. OPERATION STATUS

M 40 — MAINTAIN STATUS OF PART PROGRAMS FOR MACHINING OPERATIONS

CAM X260
1. ALLOCATED PROGRAM LIBRARY

M 50 — OPERATING CONTROLS:
– REPRIORITISE ORDERS
– REPRIORITISE MACHINES
– START/STOP USE OF M/Cs
– HOLD/INSPECT/REMOVE ORDER
– RELEASE MANUAL OPS.
– NOTIFY COMPLETION OF MANUAL OPERATIONS
– START/STOP FACILITY CONTROL

CAM M 50
1. FACILITY MAP
2. RESOURCE STATUS
3. STRATEGIES

M 60 — START INTERFACE TO DELIVERY MANAGEMENT

M 70 — START INTERFACE TO WORK CENTRE MANAGEMENT

(M80) (M540)

Sheet 1 of 11

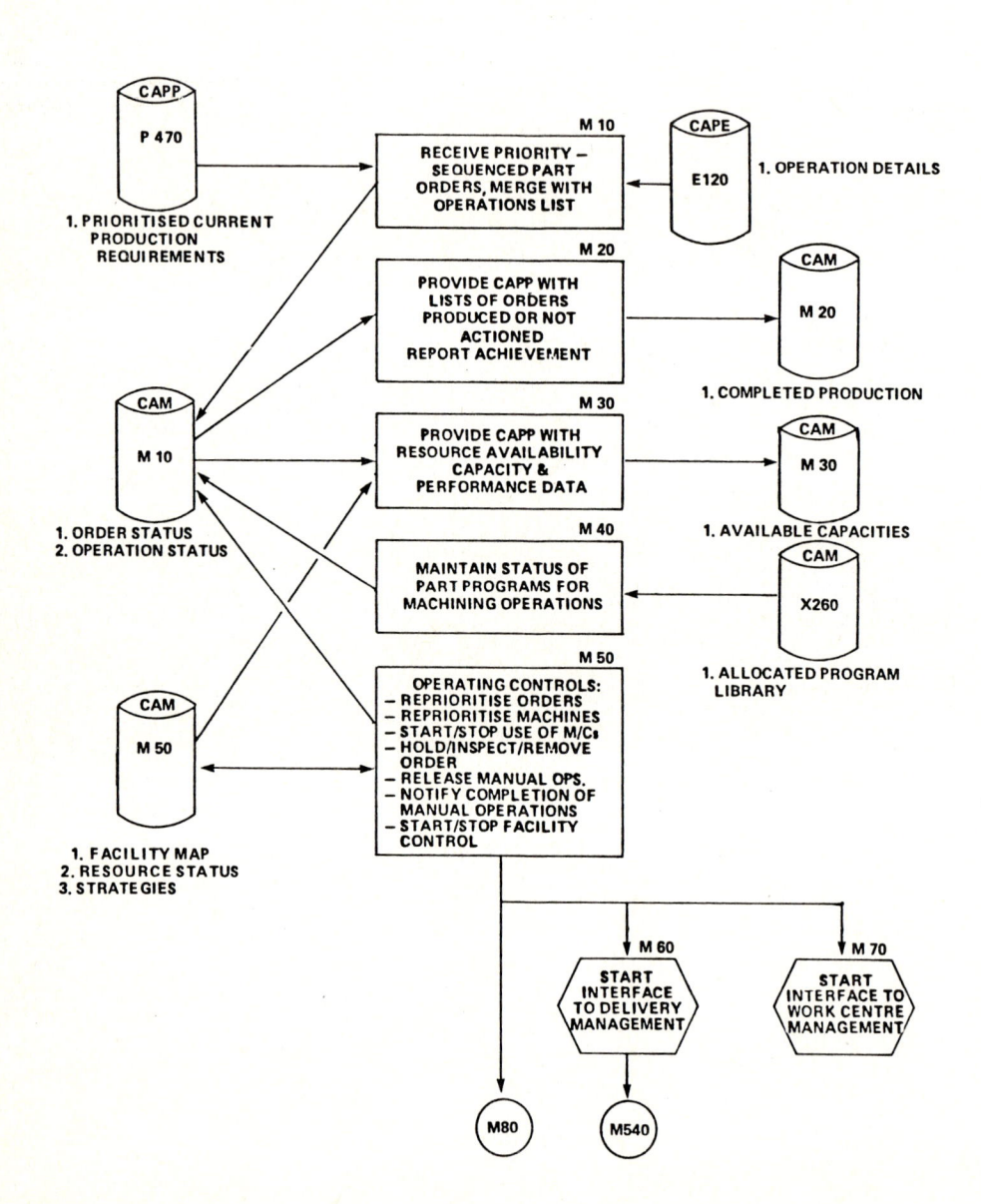

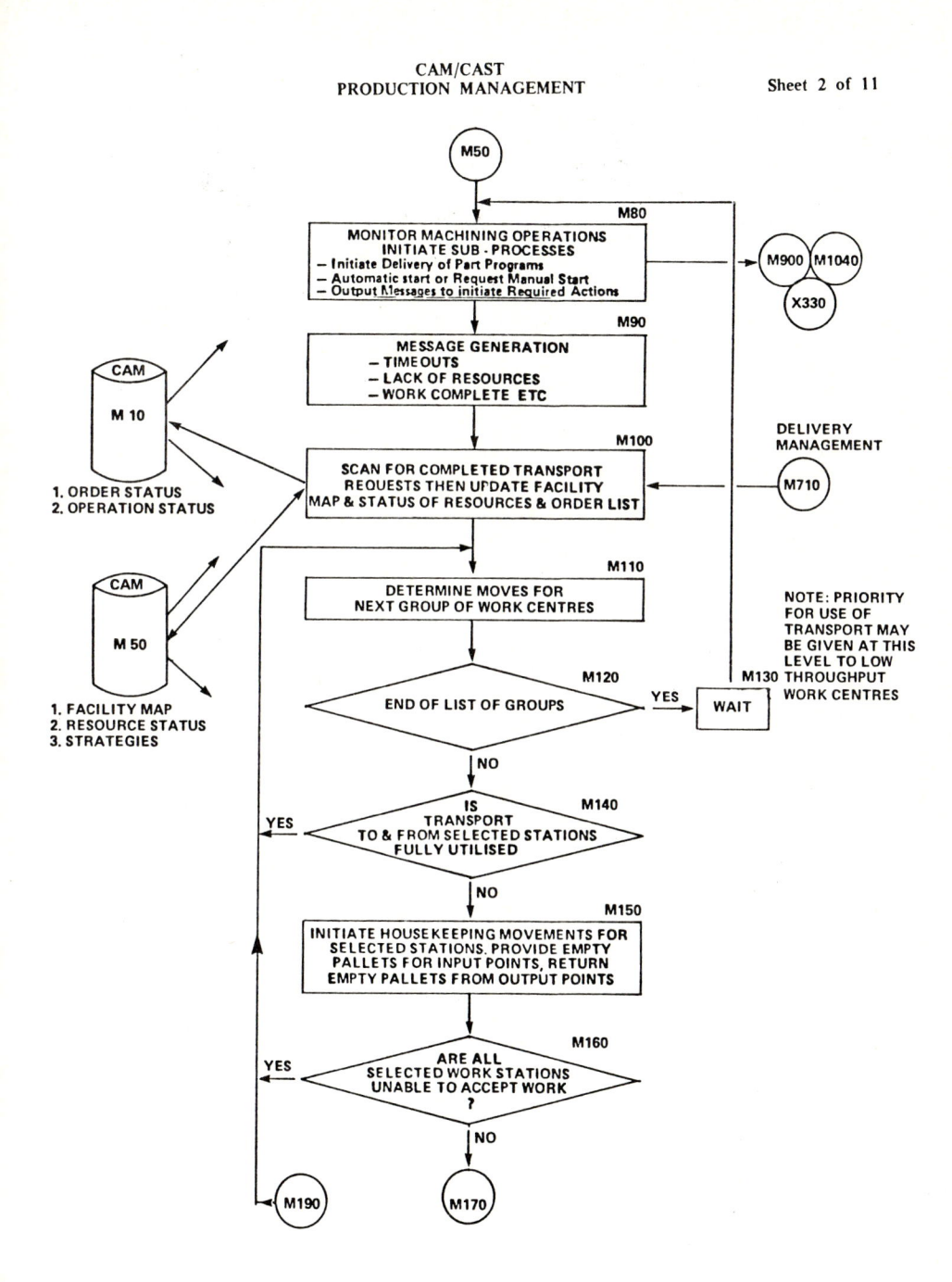

CAM/CAST PRODUCTION MANAGEMENT

Sheet 2 of 11

M50

M80 — MONITOR MACHINING OPERATIONS / INITIATE SUB-PROCESSES
- Initiate Delivery of Part Programs
- Automatic start or Request Manual Start
- Output Messages to initiate Required Actions

→ M900, M1040, X330

M90 — MESSAGE GENERATION
- TIMEOUTS
- LACK OF RESOURCES
- WORK COMPLETE ETC

M100 — SCAN FOR COMPLETED TRANSPORT REQUESTS THEN UPDATE FACILITY MAP & STATUS OF RESOURCES & ORDER LIST

DELIVERY MANAGEMENT ← M710

CAM M10
1. ORDER STATUS
2. OPERATION STATUS

CAM M50
1. FACILITY MAP
2. RESOURCE STATUS
3. STRATEGIES

M110 — DETERMINE MOVES FOR NEXT GROUP OF WORK CENTRES

M120 — END OF LIST OF GROUPS
- YES → **M130** WAIT
- NO ↓

NOTE: PRIORITY FOR USE OF TRANSPORT MAY BE GIVEN AT THIS LEVEL TO LOW THROUGHPUT WORK CENTRES

M140 — IS TRANSPORT TO & FROM SELECTED STATIONS FULLY UTILISED
- YES →
- NO ↓

M150 — INITIATE HOUSEKEEPING MOVEMENTS FOR SELECTED STATIONS. PROVIDE EMPTY PALLETS FOR INPUT POINTS, RETURN EMPTY PALLETS FROM OUTPUT POINTS

M160 — ARE ALL SELECTED WORK STATIONS UNABLE TO ACCEPT WORK?
- YES →
- NO ↓

M190   **M170**

238   CAM CAST sub-topics

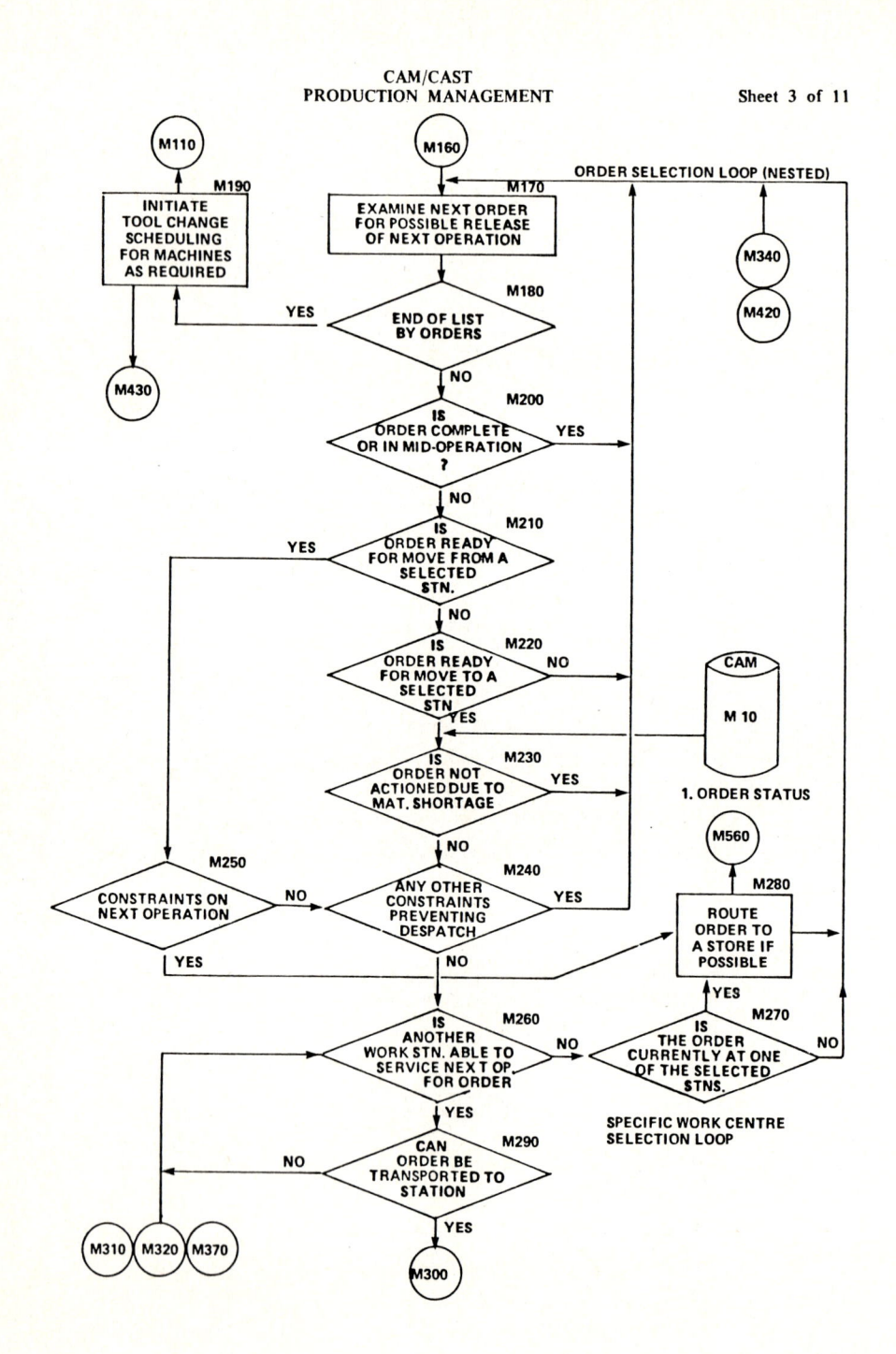

CAM/CAST PRODUCTION MANAGEMENT

Sheet 3 of 11

- **M110** → **M190**: INITIATE TOOL CHANGE SCHEDULING FOR MACHINES AS REQUIRED → **M430**
- **M160** → **M170**: EXAMINE NEXT ORDER FOR POSSIBLE RELEASE OF NEXT OPERATION
- **M180**: END OF LIST BY ORDERS — YES → M190; NO ↓
- **M200**: IS ORDER COMPLETE OR IN MID-OPERATION? — YES → (ORDER SELECTION LOOP, via M340/M420); NO ↓
- **M210**: IS ORDER READY FOR MOVE FROM A SELECTED STN. — YES → M250; NO ↓
- **M220**: IS ORDER READY FOR MOVE TO A SELECTED STN — NO → CAM M10; YES ↓
- **M230**: IS ORDER NOT ACTIONED DUE TO MAT. SHORTAGE — YES → 1. ORDER STATUS; NO ↓
- **M240**: ANY OTHER CONSTRAINTS PREVENTING DESPATCH — YES → M280; NO ↓
- **M250**: CONSTRAINTS ON NEXT OPERATION — NO → M240; YES → M260
- **M280**: ROUTE ORDER TO A STORE IF POSSIBLE → M560
- **M260**: IS ANOTHER WORK STN. ABLE TO SERVICE NEXT OP. FOR ORDER — NO → M270; YES ↓
- **M270**: IS THE ORDER CURRENTLY AT ONE OF THE SELECTED STNS. — YES → M280; NO → (ORDER SELECTION LOOP)
- **M290**: CAN ORDER BE TRANSPORTED TO STATION — NO → M310/M320/M370; YES → **M300**

CAM M10 — 1. ORDER STATUS

SPECIFIC WORK CENTRE SELECTION LOOP

ORDER SELECTION LOOP (NESTED) — M340, M420

CAM/CAST PRODUCTION MANAGEMENT

Sheet 4 of 11

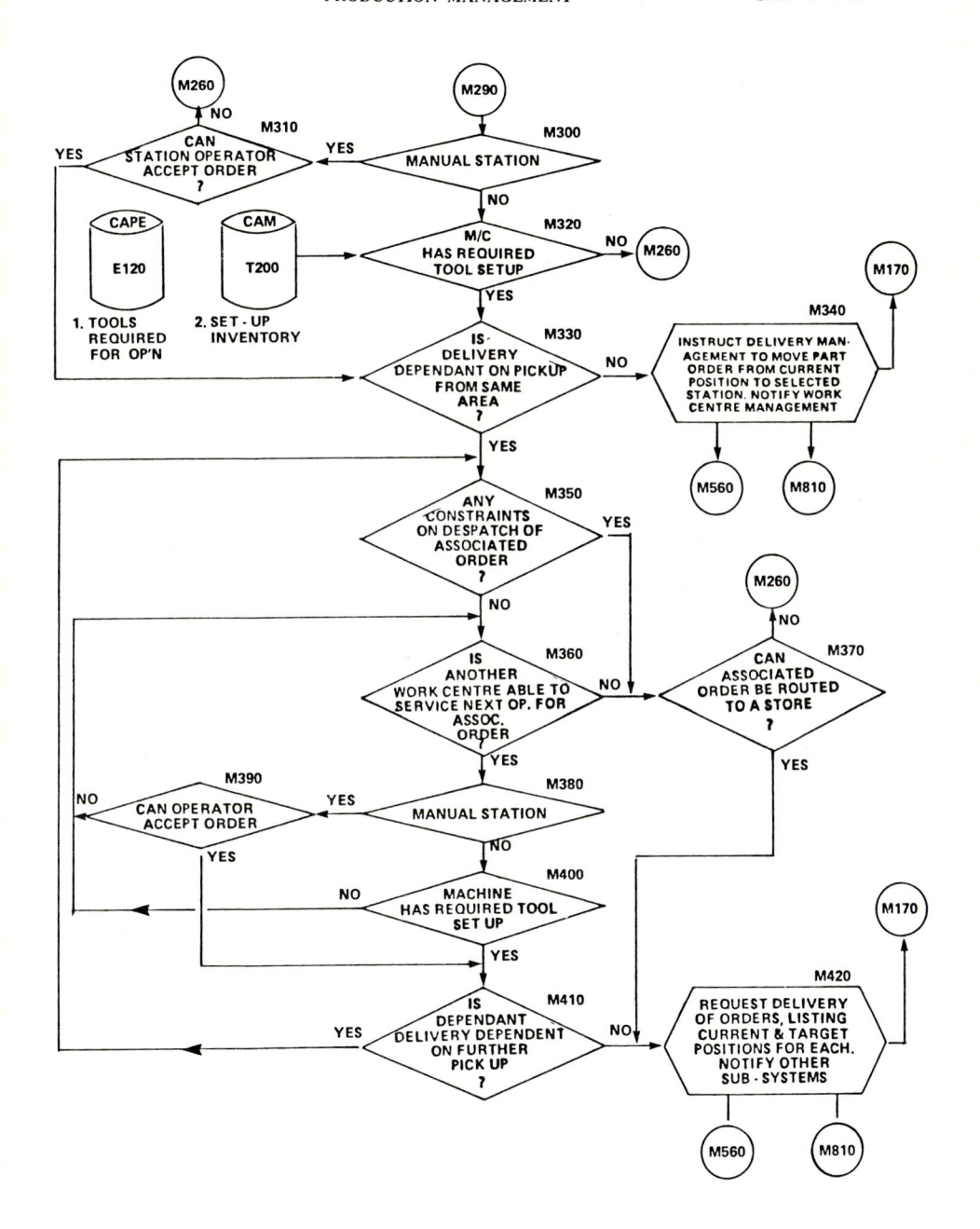

- **M290** — MANUAL STATION?
  - YES → **M310** CAN STATION OPERATOR ACCEPT ORDER?
    - NO → **M260**
    - YES → (to CAPE E120 / CAM T200 data: 1. TOOLS REQUIRED FOR OP'N ; 2. SET-UP INVENTORY)
  - NO → **M320** M/C HAS REQUIRED TOOL SETUP?
    - NO → **M260**
    - YES → **M330** IS DELIVERY DEPENDANT ON PICKUP FROM SAME AREA?
      - NO → **M340** INSTRUCT DELIVERY MANAGEMENT TO MOVE PART ORDER FROM CURRENT POSITION TO SELECTED STATION. NOTIFY WORK CENTRE MANAGEMENT → **M170**, **M560**, **M810**
      - YES → **M350** ANY CONSTRAINTS ON DESPATCH OF ASSOCIATED ORDER?
        - YES → **M360** IS ANOTHER WORK CENTRE ABLE TO SERVICE NEXT OP. FOR ASSOC. ORDER?
          - NO → **M370** CAN ASSOCIATED ORDER BE ROUTED TO A STORE?
            - NO → **M260**
            - YES → (down)
          - YES → **M380** MANUAL STATION?
            - YES → **M390** CAN OPERATOR ACCEPT ORDER?
              - NO → (loop back)
              - YES → (loop back)
            - NO → **M400** MACHINE HAS REQUIRED TOOL SET UP?
              - NO → (loop back)
              - YES → **M410** IS DEPENDANT DELIVERY DEPENDENT ON FURTHER PICK UP?
                - YES → (loop back)
                - NO → **M420** REQUEST DELIVERY OF ORDERS, LISTING CURRENT & TARGET POSITIONS FOR EACH. NOTIFY OTHER SUB-SYSTEMS → **M170**, **M560**, **M810**

CAM/CAST PRODUCTION MANAGEMENT

Sheet 5 of 11

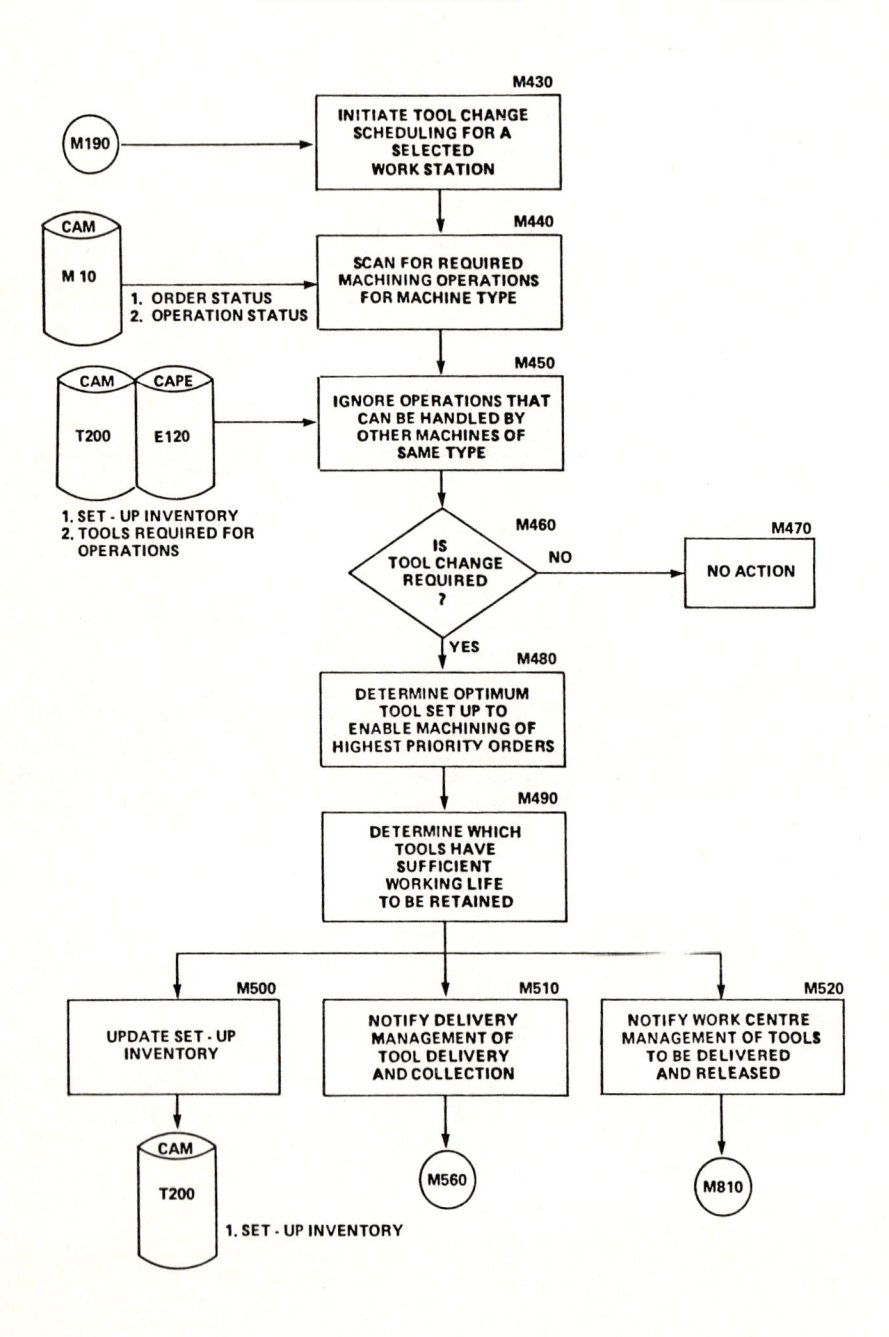

- **M190** → **M430**: INITIATE TOOL CHANGE SCHEDULING FOR A SELECTED WORK STATION
- **CAM M10** (1. ORDER STATUS, 2. OPERATION STATUS) → **M440**: SCAN FOR REQUIRED MACHINING OPERATIONS FOR MACHINE TYPE
- **CAM T200**, **CAPE E120** (1. SET-UP INVENTORY, 2. TOOLS REQUIRED FOR OPERATIONS) → **M450**: IGNORE OPERATIONS THAT CAN BE HANDLED BY OTHER MACHINES OF SAME TYPE
- **M460**: IS TOOL CHANGE REQUIRED?
  - NO → **M470**: NO ACTION
  - YES → **M480**: DETERMINE OPTIMUM TOOL SET UP TO ENABLE MACHINING OF HIGHEST PRIORITY ORDERS
- **M490**: DETERMINE WHICH TOOLS HAVE SUFFICIENT WORKING LIFE TO BE RETAINED
  - **M500**: UPDATE SET-UP INVENTORY → **CAM T200** (1. SET-UP INVENTORY)
  - **M510**: NOTIFY DELIVERY MANAGEMENT OF TOOL DELIVERY AND COLLECTION → **M560**
  - **M520**: NOTIFY WORK CENTRE MANAGEMENT OF TOOLS TO BE DELIVERED AND RELEASED → **M810**

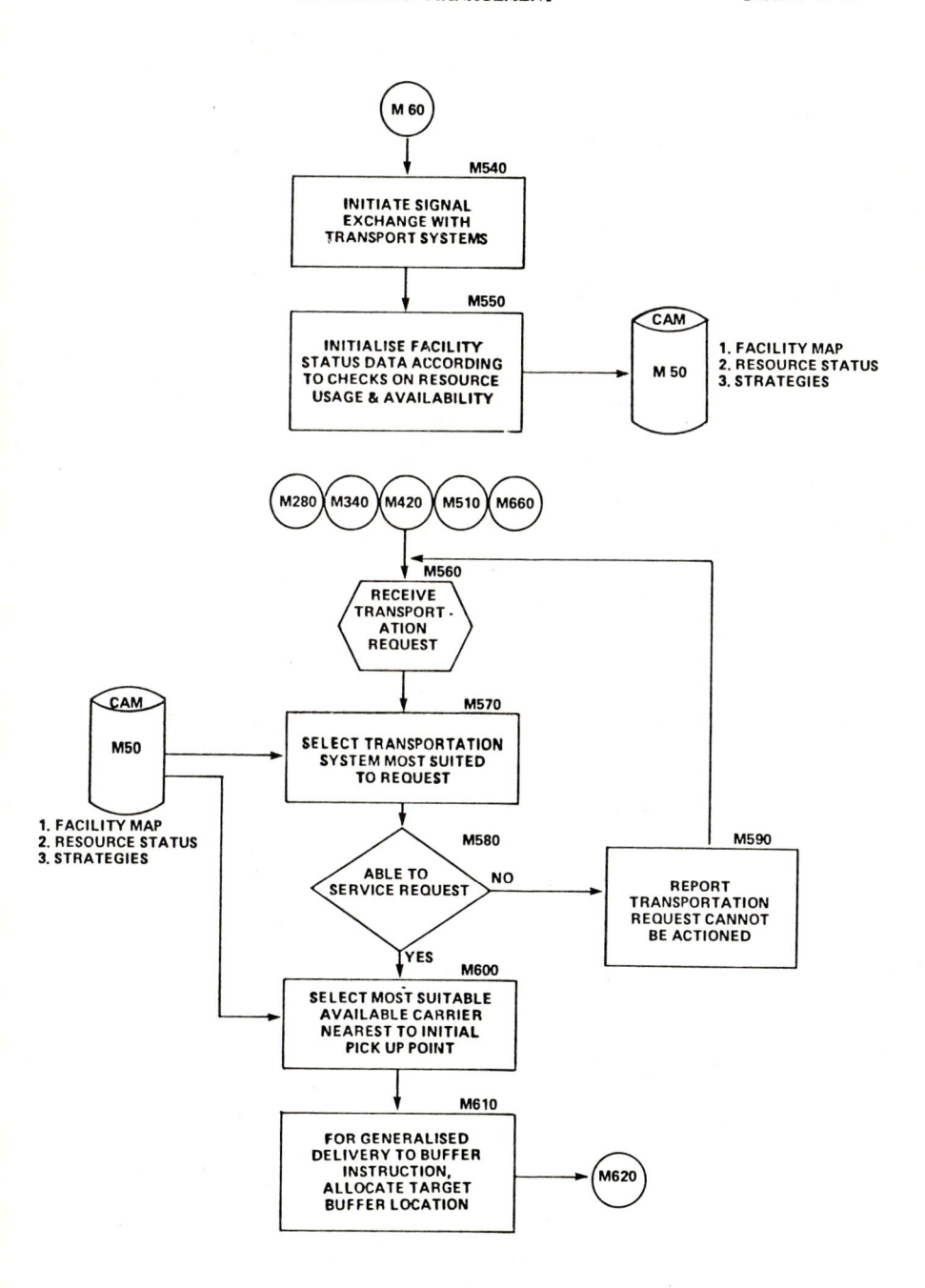

**CAM/CAST
PRODUCTION MANAGEMENT**

```
                    (M 60)
                       │
                       ▼
                     M540
         ┌──────────────────────────┐
         │   INITIATE SIGNAL        │
         │   EXCHANGE WITH          │
         │   TRANSPORT SYSTEMS      │
         └──────────────────────────┘
                       │
                       ▼
                     M550
         ┌──────────────────────────┐           ╔═══╗
         │  INITIALISE FACILITY     │           ║CAM║   1. FACILITY MAP
         │  STATUS DATA ACCORDING   │──────────▶║M50║   2. RESOURCE STATUS
         │  TO CHECKS ON RESOURCE   │           ╚═══╝   3. STRATEGIES
         │  USAGE & AVAILABILITY    │
         └──────────────────────────┘

      (M280)(M340)(M420)(M510)(M660)
                       │
                       ▼
                     M560
                  ╱────────╲
                 ╱ RECEIVE   ╲
                ╱ TRANSPORT-  ╲
                ╲    ATION    ╱
                 ╲  REQUEST  ╱
                  ╲─────────╱
                       │
        ╔═══╗          ▼           M570
        ║CAM║    ┌──────────────────────┐
        ║M50║───▶│  SELECT TRANSPORTATION│
        ╚═══╝    │  SYSTEM MOST SUITED   │
                 │  TO REQUEST           │
  1. FACILITY MAP└──────────────────────┘
  2. RESOURCE STATUS     │
  3. STRATEGIES          ▼       M580                      M590
                    ╱─────────╲                  ┌──────────────────┐
                   ╱  ABLE TO  ╲     NO          │    REPORT        │
                   ╲  SERVICE  ╱───────────────▶│  TRANSPORTATION   │
                    ╲ REQUEST ╱                  │  REQUEST CANNOT   │
                     ╲───────╱                   │  BE ACTIONED      │
                       │YES                      └──────────────────┘
                       ▼            M600
                 ┌──────────────────────┐
                 │ SELECT MOST SUITABLE │
                 │ AVAILABLE CARRIER    │
                 │ NEAREST TO INITIAL   │
                 │ PICK UP POINT        │
                 └──────────────────────┘
                       │            M610
                       ▼
                 ┌──────────────────────┐
                 │ FOR GENERALISED      │
                 │ DELIVERY TO BUFFER   │──────▶(M620)
                 │ INSTRUCTION,         │
                 │ ALLOCATE TARGET      │
                 │ BUFFER LOCATION      │
                 └──────────────────────┘
```

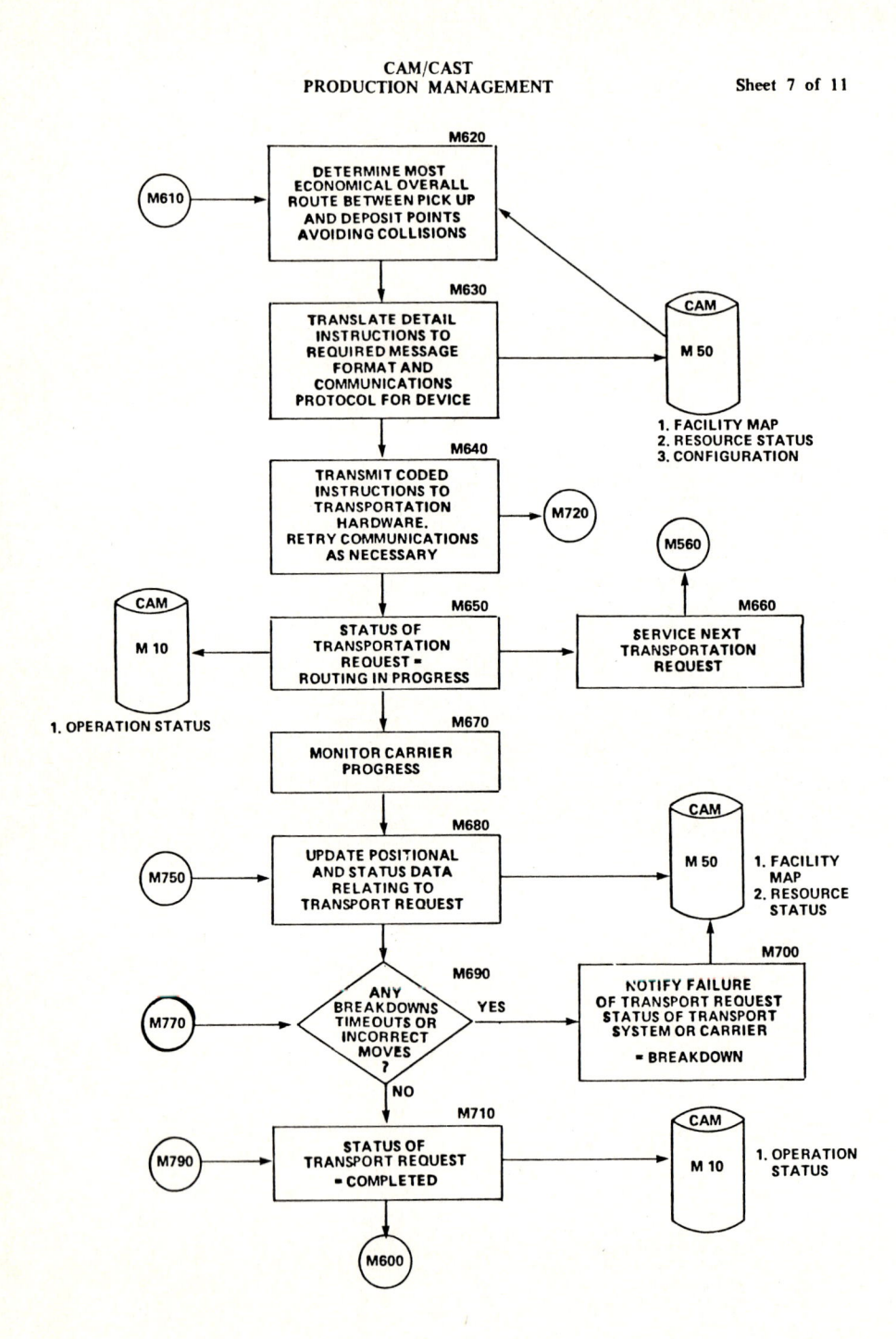

CAM/CAST
PRODUCTION MANAGEMENT

Sheet 7 of 11

```
                          M620
    ┌───────┐      ┌─────────────────────┐
    │ M610  │────▶│ DETERMINE MOST      │
    └───────┘      │ ECONOMICAL OVERALL  │
                   │ ROUTE BETWEEN PICK UP│
                   │ AND DEPOSIT POINTS  │
                   │ AVOIDING COLLISIONS │
                   └─────────────────────┘
                            │
                          M630
                   ┌─────────────────────┐         ┌──────┐
                   │ TRANSLATE DETAIL    │         │ CAM  │
                   │ INSTRUCTIONS TO     │────────▶│      │
                   │ REQUIRED MESSAGE    │         │ M 50 │
                   │ FORMAT AND          │         └──────┘
                   │ COMMUNICATIONS      │         1. FACILITY MAP
                   │ PROTOCOL FOR DEVICE │         2. RESOURCE STATUS
                   └─────────────────────┘         3. CONFIGURATION
                            │
                          M640
                   ┌─────────────────────┐
                   │ TRANSMIT CODED      │
                   │ INSTRUCTIONS TO     │       ┌───────┐
                   │ TRANSPORTATION      │──────▶│ M720  │
                   │ HARDWARE,           │       └───────┘
                   │ RETRY COMMUNICATIONS│        ┌───────┐
                   │ AS NECESSARY        │        │ M560  │
                   └─────────────────────┘        └───────┘
                            │                        ▲
   ┌──────┐               M650                     M660
   │ CAM  │      ┌─────────────────────┐   ┌─────────────────┐
   │ M 10 │◀─────│ STATUS OF           │──▶│ SERVICE NEXT    │
   └──────┘      │ TRANSPORTATION      │   │ TRANSPORTATION  │
                 │ REQUEST =           │   │ REQUEST         │
   1. OPERATION  │ ROUTING IN PROGRESS │   └─────────────────┘
      STATUS     └─────────────────────┘
                            │
                          M670
                   ┌─────────────────────┐
                   │ MONITOR CARRIER     │
                   │ PROGRESS            │
                   └─────────────────────┘
                            │
                          M680
   ┌───────┐      ┌─────────────────────┐         ┌──────┐
   │ M750  │─────▶│ UPDATE POSITIONAL   │────────▶│ CAM  │  1. FACILITY
   └───────┘      │ AND STATUS DATA     │         │ M 50 │     MAP
                  │ RELATING TO         │         └──────┘  2. RESOURCE
                  │ TRANSPORT REQUEST   │                      STATUS
                  └─────────────────────┘
                            │                            M700
                          M690                  ┌─────────────────────┐
                         ╱     ╲       YES      │ NOTIFY FAILURE      │
   ┌───────┐           ╱  ANY   ╲  ───────────▶ │ OF TRANSPORT REQUEST│
   │ M770  │─────────▶│BREAKDOWNS│               │ STATUS OF TRANSPORT │
   └───────┘           │TIMEOUTS OR│              │ SYSTEM OR CARRIER   │
                       │ INCORRECT │              │ = BREAKDOWN         │
                        ╲ MOVES ╱                └─────────────────────┘
                         ╲  ?  ╱
                          ╲___╱
                           │ NO
                          M710
   ┌───────┐      ┌─────────────────────┐         ┌──────┐
   │ M790  │─────▶│ STATUS OF           │────────▶│ CAM  │  1. OPERATION
   └───────┘      │ TRANSPORT REQUEST   │         │ M 10 │     STATUS
                  │ = COMPLETED         │         └──────┘
                  └─────────────────────┘
                            │
                        ┌───────┐
                        │ M600  │
                        └───────┘
```

CAM/CAST
PRODUCTION MANAGEMENT

Sheet 8 of 11

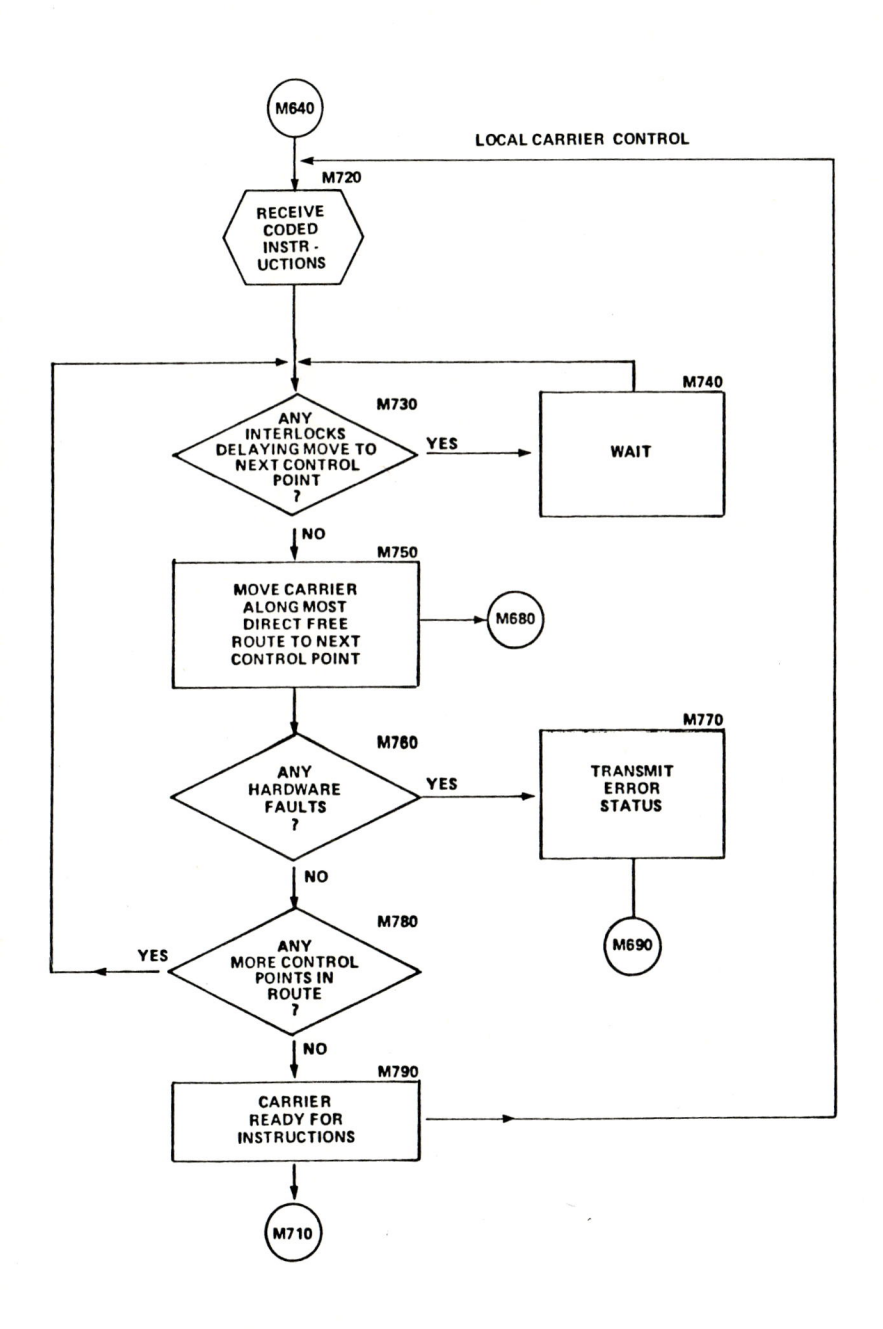

LOCAL CARRIER CONTROL

- M640
- M720 RECEIVE CODED INSTRUCTIONS
- M730 ANY INTERLOCKS DELAYING MOVE TO NEXT CONTROL POINT ?
  - YES → M740 WAIT
  - NO ↓
- M750 MOVE CARRIER ALONG MOST DIRECT FREE ROUTE TO NEXT CONTROL POINT → M680
- M760 ANY HARDWARE FAULTS ?
  - YES → M770 TRANSMIT ERROR STATUS → M690
  - NO ↓
- M780 ANY MORE CONTROL POINTS IN ROUTE ?
  - YES ↑ (loop back)
  - NO ↓
- M790 CARRIER READY FOR INSTRUCTIONS
- M710

**CAM/CAST
PRODUCTION MANAGEMENT**

Sheet 9 of 11

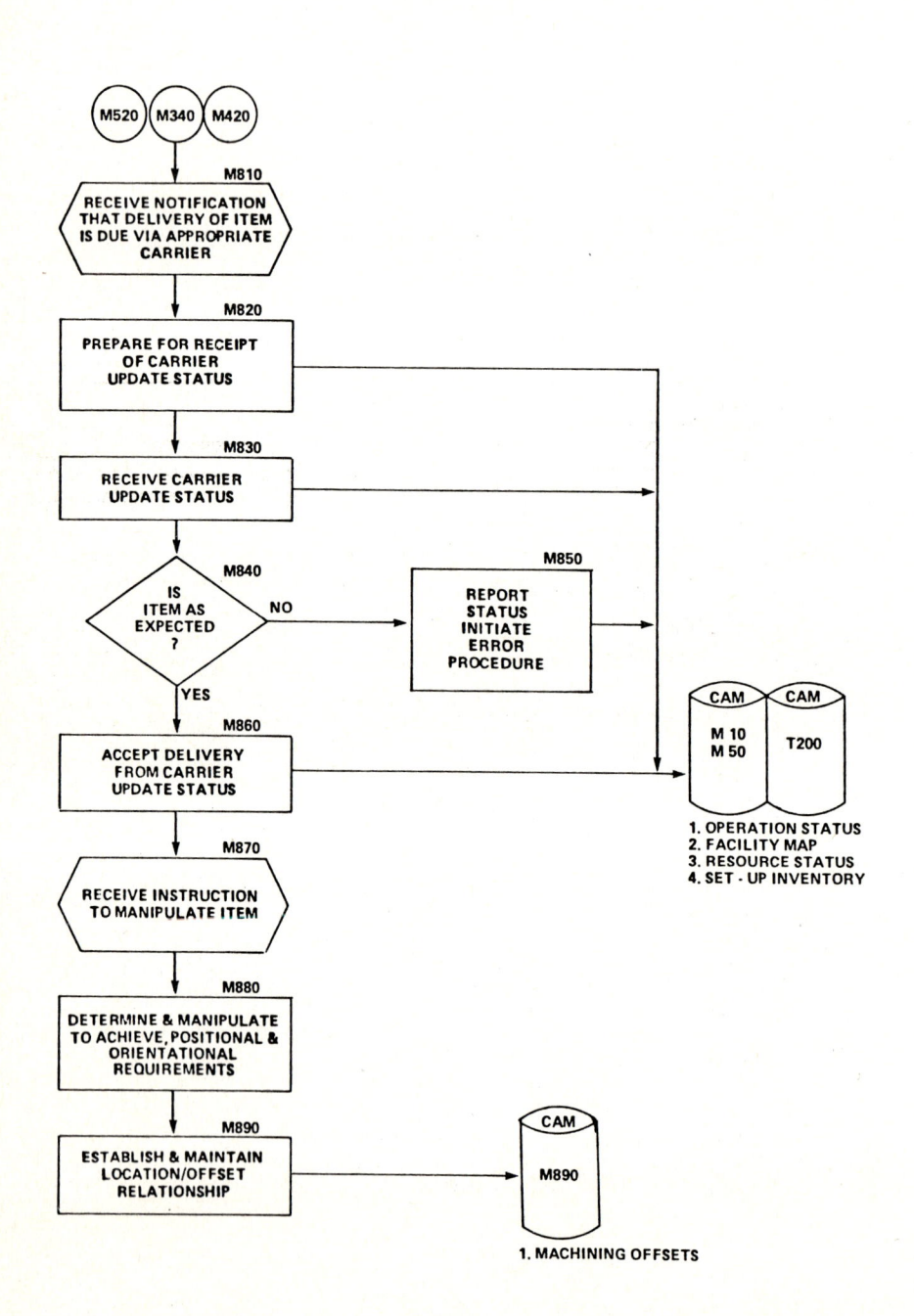

- M520, M340, M420

M810 RECEIVE NOTIFICATION THAT DELIVERY OF ITEM IS DUE VIA APPROPRIATE CARRIER

M820 PREPARE FOR RECEIPT OF CARRIER UPDATE STATUS

M830 RECEIVE CARRIER UPDATE STATUS

M840 IS ITEM AS EXPECTED ?
- NO → **M850** REPORT STATUS INITIATE ERROR PROCEDURE
- YES ↓

M860 ACCEPT DELIVERY FROM CARRIER UPDATE STATUS

CAM M10 M50 / CAM T200
1. OPERATION STATUS
2. FACILITY MAP
3. RESOURCE STATUS
4. SET-UP INVENTORY

M870 RECEIVE INSTRUCTION TO MANIPULATE ITEM

M880 DETERMINE & MANIPULATE TO ACHIEVE, POSITIONAL & ORIENTATIONAL REQUIREMENTS

M890 ESTABLISH & MAINTAIN LOCATION/OFFSET RELATIONSHIP

CAM M890
1. MACHINING OFFSETS

CAM/CAST
PRODUCTION MANAGEMENT

Sheet 10 of 11

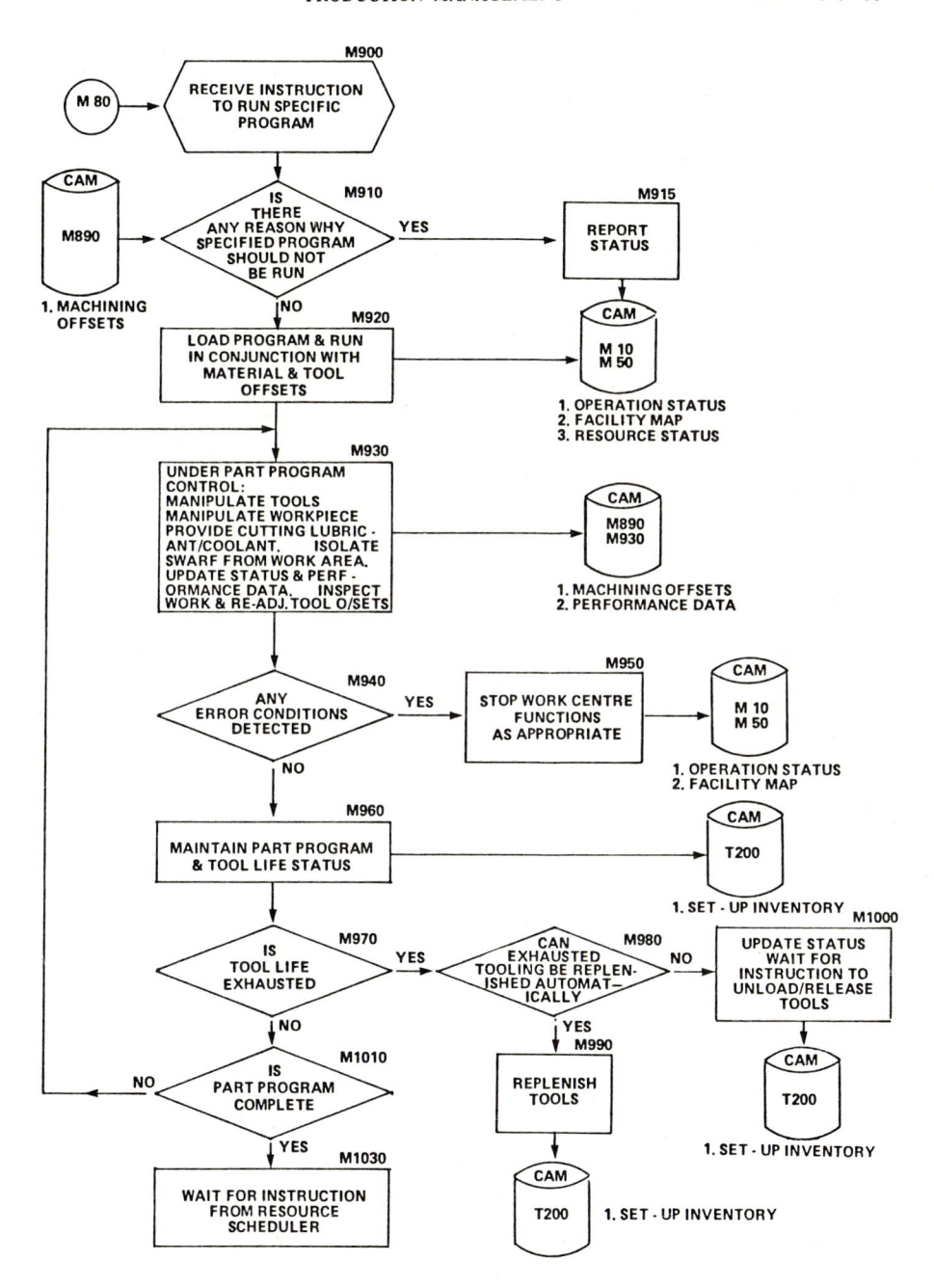

- M80
- M900 RECEIVE INSTRUCTION TO RUN SPECIFIC PROGRAM
- CAM M890 — 1. MACHINING OFFSETS
- M910 IS THERE ANY REASON WHY SPECIFIED PROGRAM SHOULD NOT BE RUN — YES → M915 REPORT STATUS
- CAM M10 M50 — 1. OPERATION STATUS 2. FACILITY MAP 3. RESOURCE STATUS
- NO ↓
- M920 LOAD PROGRAM & RUN IN CONJUNCTION WITH MATERIAL & TOOL OFFSETS
- M930 UNDER PART PROGRAM CONTROL: MANIPULATE TOOLS MANIPULATE WORKPIECE PROVIDE CUTTING LUBRICANT/COOLANT. ISOLATE SWARF FROM WORK AREA. UPDATE STATUS & PERFORMANCE DATA. INSPECT WORK & RE-ADJ. TOOL O/SETS
- CAM M890 M930 — 1. MACHINING OFFSETS 2. PERFORMANCE DATA
- M940 ANY ERROR CONDITIONS DETECTED — YES → M950 STOP WORK CENTRE FUNCTIONS AS APPROPRIATE → CAM M10 M50 — 1. OPERATION STATUS 2. FACILITY MAP
- NO ↓
- M960 MAINTAIN PART PROGRAM & TOOL LIFE STATUS → CAM T200 — 1. SET-UP INVENTORY
- M970 IS TOOL LIFE EXHAUSTED — YES → M980 CAN EXHAUSTED TOOLING BE REPLENISHED AUTOMATICALLY — NO → M1000 UPDATE STATUS WAIT FOR INSTRUCTION TO UNLOAD/RELEASE TOOLS → CAM T200 — 1. SET-UP INVENTORY
- YES ↓ M990 REPLENISH TOOLS → CAM T200 — 1. SET-UP INVENTORY
- NO ↓ M1010 IS PART PROGRAM COMPLETE — NO (loop back)
- YES ↓ M1030 WAIT FOR INSTRUCTION FROM RESOURCE SCHEDULER

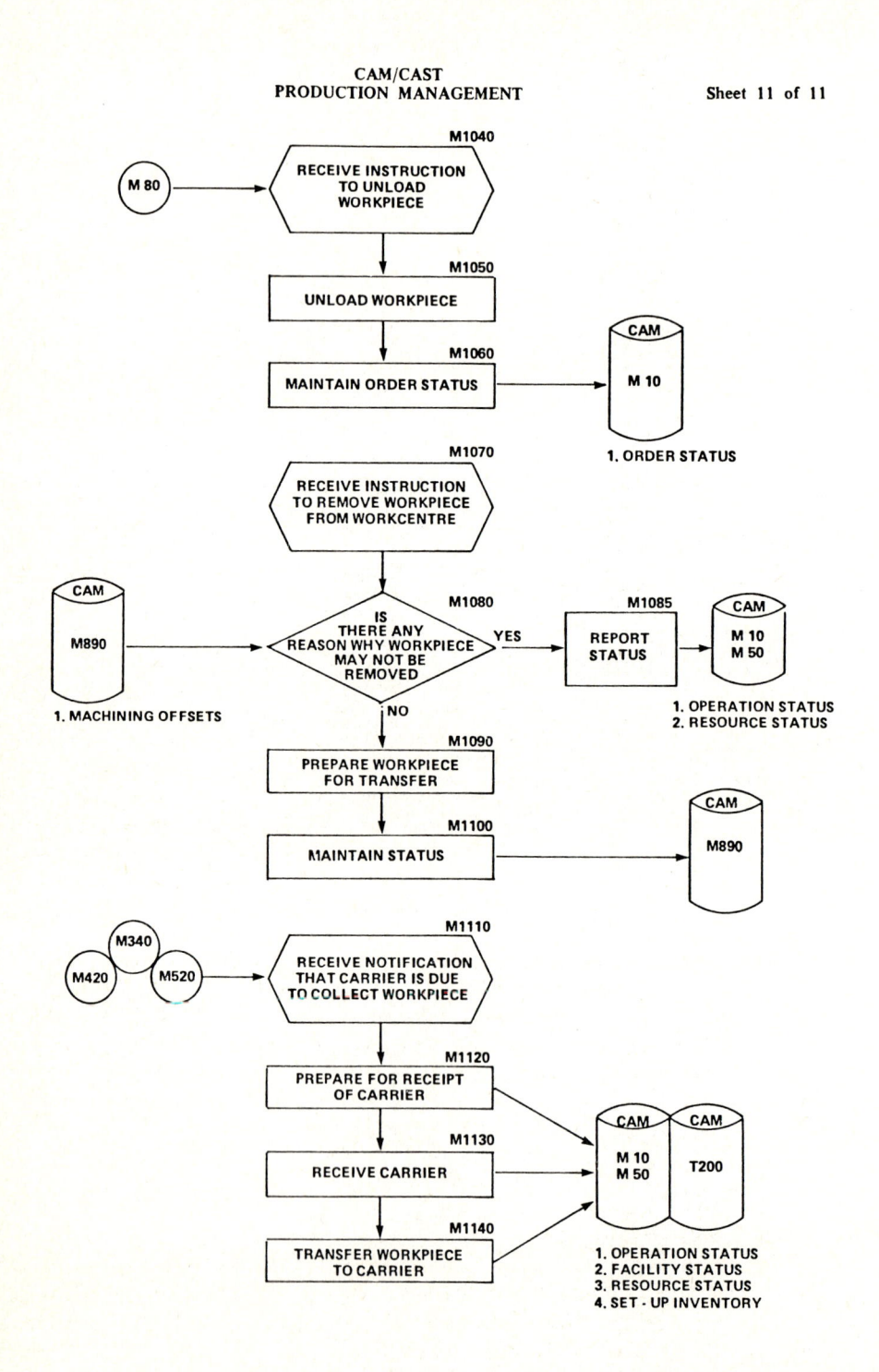

**CAM/CAST
PRODUCTION MANAGEMENT**

Sheet 11 of 11

- M 80 → M1040 RECEIVE INSTRUCTION TO UNLOAD WORKPIECE
- M1050 UNLOAD WORKPIECE
- M1060 MAINTAIN ORDER STATUS → CAM M 10
  1. ORDER STATUS
- M1070 RECEIVE INSTRUCTION TO REMOVE WORKPIECE FROM WORKCENTRE
- CAM M890 (1. MACHINING OFFSETS) → M1080 IS THERE ANY REASON WHY WORKPIECE MAY NOT BE REMOVED
  - YES → M1085 REPORT STATUS → CAM M 10 M 50
    1. OPERATION STATUS
    2. RESOURCE STATUS
  - NO → M1090 PREPARE WORKPIECE FOR TRANSFER
- M1100 MAINTAIN STATUS → CAM M890
- M420, M340, M520 → M1110 RECEIVE NOTIFICATION THAT CARRIER IS DUE TO COLLECT WORKPIECE
- M1120 PREPARE FOR RECEIPT OF CARRIER
- M1130 RECEIVE CARRIER → CAM M 10 M 50 | CAM T200
- M1140 TRANSFER WORKPIECE TO CARRIER
  1. OPERATION STATUS
  2. FACILITY STATUS
  3. RESOURCE STATUS
  4. SET - UP INVENTORY

Narrative description of Production Management

M10 The CAPP system will produce a priority-sequenced list of part orders to be manufactured within a time period such as a shift. This data will be merged with information from the CAPE system, describing the list of operation required to produce each part. The resource scheduler will then control the manufacturing facility to produce the parts in the order list, updating the order list after each completed operation.

M20 The CAM system will have the capability of reporting achievement against the order list. Whenever required it will provide the CAPP system with details of orders produced or not produced, as an input to future scheduling.

M30 Whenever required, the manufacturing system will provide the production planning system with information about resource availability, manufacturing capacity, recent and other performance data, so that realistic future schedules may be generated.

M40 Whenever part-programmes become available for machining operations the part-programme management system will request a modification to the status of planned machining operations on the orders file.

M50 The CAM system will incorporate a comprehensive range of operating controls to enable the people responsible for the manufacturing facility to have full control over the system.

Some applications may require facilities to reprioritise orders or reprioritise work flow to specific machine. It will be necessary to include a facility to stop the scheduling of selected machines to allow for tool changes, maintenance, or night shift operation.

Overrides on orders may be needed to stop further work on selected parts or to route them for inspection or removal from the facility. Facilities may be required to control the flow of orders to manual stations and to inform the resource scheduler of completed manual operations. An interface to CAD is possible here, as instructions for manual operations may be presented in graphical form.

It will also be necessary to provide a method of starting and stopping automatic control of the facility.

M60 The resource scheduler will require the services of a delivery management module to interface to transport and storage facilities. This module will ensure that routing instructions are carried out and that completion of routing instruction is notified to the resource scheduler. The module must be active and ready for instructions at all times, when the resource scheduler has overall control.

M70 The resource scheduler will require the services of work centre management to interface to CNC machines. The interface module will interchange signals relating to deliveries and pickups, the loading of part-programmes, the use of tools and automatic start of scheduled operations. Work centre management must be active and ready to exchange data, whenever the resource scheduler has overall control.

M80 The Resource Scheduler is a loop structured continuous process. At regular intervals, for example every two seconds, the scheduler will examine the status of all resources and the order list, and decide what to do next.

The short time delay between scans can be replaced by a wait until any event that may enable a movement or operation to be started.

The decision process takes in the entire manufacturing facility, even if initiated in response to an event. This is because further progress directly relating to event may not be possible due to lack of resources but progress will become possible after some future event.

The first thing the resource scheduler will do every scan is to examine the current status of every work centre currently in mid-operation.

If the operation is just beginning it may be necessary to request the down-line load or delivery of a part-programme or the transmission of control signals to initiate events such as programme start or transfer from input buffer.

Where manual action is required the scheduler will initiate message output. This could be anything from lighting a bulb to interfacing to a CAD data base to generate detailed instructions in graphical form. Sometimes CNC centres will require a manual start if the part-programme does not have a proven status.

The scheduler will have access to information describing the detail of all operations to be carried out on parts at all work centres and is responsible for co-ordination of sub-processes affecting parts currently at work centres.

M90   The resource scheduler should initiate the output of warning messages when appropriate. Examples are timeouts, lack of work to maintain full utilisation and lack of resources needed to meet production requirements. Alarm messages and prompts for action must be generated when any problem affects the overall functioning of the facility.

M100   Any notification from delivery management of a completed movement by a transport facility should be recognised, After completed movements the system will update the facility map to reflect where all parts are situated, and the availability status of all resources. The parts order list is updated to reflect completed operations.

M110   When the data base is fully updated the final requirement for the main processing loop of the resource scheduler is to initiate new operations to be started for ordered parts, by routing the parts to the next station.

The logic flow allows for any application dependent prioritisation of groups of machines, for example to resolve contention for the allocation of transport resources.

M120   The decision processing loops through each group of work centres scanning for any possible movement of parts to or from any work centres in the group. Priority may be given to low throughput work centres through this mechanism.

M130   At the end of all routing decisions the resource scheduler will suspend activity until the next scan, as described in M80.

M140   The selected work centres will be serviced by transport systems for delivery and pick up of parts. If all these transport systems are fully loaded then there is no need to select parts for routing to or from the work centres at this priority level.

M150   For some of the selected work centres, housekeeping movements may be necessary to keep those work centres operational, for example the provision of empty pallets for input points or the return of empty pallets from output points. These movements are released by the scheduler prior to scanning the part order list, as they are movements that may not relate to a specific part order.

M160   If all the selected work centres are either in mid-operation, withdrawn from use or broken then they do not need to be supplied with work. The configuration of the facility will have an influence on whether or not it is possible to route a part to or from a machine. If a work station has an input buffer or pallet changer it may be possible to route a part to the machine even if the machine is currently busy with

CAM CAST sub-topics

another part. For other work centres it may be required to delay delivery of a part until a completed part can be collected as a paired movement.

M170   In order to select orders for routing to or from work centres in the group, it is necessary to scan the priority sequenced part order list. Each order is considered for possible selection for routing (M170 - M340).

M180   When all orders have been considered, the next group of work centres can be processed. Note that there is no need to scan to the end of the list of orders if resources such as transport for the work centre group are exhausted.

M190   If no further work can be allocated for any machines within a group of work centres because of tooling problems then a mechanism can be included to automatically start up a tool change scheduling decision process (M430 - M520) for each machine without work.

M200   Within the part order scan, each part is considered for possible routing to or from any work centre in the group. If a part is completed or in mid-operation it need not be considered further.

M210   If the part order is present at any selected work station, and it has completed an operation and is ready for direct collection by a transport system, then it can be considered for collection (see M250). If the work centre has an output buffer or pallet changer then the part must be in a position where it can be collected.

Note that in prioritising work flow to work centres of selected types it is just as important to clear finished parts to make way for new parts, as it is to deliver those new parts.

M220   If the next required operation for the part order is for a work centre of the selected type it can be considered in detail. Otherwise the part is not selected, though it may be picked up in a later scan for a lower priority work centre. Part overrides may determine where the part must go next e.g. for inspection or removal.

M230   The part cannot be selected to start its first operation if data from the material management system indicates that necessary material (including fixturing material) is not available.

M240   Other constraints may prevent selection of the part for routing such as unavailability of a part-programme for the next machining operation, or an operator request to stop operations for the part.

Other application dependent constraints may be applied according to local operating strategies.

M250   Similar constraints may be applied to parts which are ready for routing from work centres of the selected type. If these parts cannot be routed to start the next operation they should be routed to a storage or buffer area if possible (See M280).

M260   Whether the selected part is ready for routing from or to any work centre of the selected type, it is now necessary to select a specific work centre for the next operation for the part. Note that overrides such as 'inspect' and 'remove' may determine where the part must go next.

It is the next operation for the part that determines the work centres scanned in this inner loop and these are considered one by one. When routing from a work centre of the selected type (determined at M110) this level of scan may consider work centres of another type.

Work centres are eliminated from consideration if they are in mid-operation, withdrawn from use or broken. Sometimes input buffering arrangements will permit delivery of the part even if the work centre is busy.

M270   If after detailed consideration, no work centre is available for the next operation for the part then if the part is ready for collection from a work centre of the selected type (determined at M110) then the part should be cleared to a storage or buffer area if possible.

Priority parts are stored only if they cannot be routed directly to the next operation.

This is why the storage decision is invoked at the end of a scan for target work centres for a part, within a scan of the part order list.

M280   A part may be cleared to a storage or buffer facility if it cannot be directly routed to the next operation. This is possible when a transport facility is available and it connects to a storage area with free capacity from which the part may later be routed to its next operation. The scan continues with the next part order (M170).

M290   As part of the inner loop for selecting a work station for the next operation on a specific part order it is necessary to consider transport availability i.e. a specific transport system must be ready to carry the part from where it is now to the chosen work centre.

M300   The final checks on work station suitability will vary according to the type of work station, for example manual stations may be scheduled differently.

M310   The operating controls (See M50) may include mechanisms for selecting which parts may be routed to manual stations therefore it may only be acceptable to route a part to a manual stations if it is not blocked by operator request.

The next check (M320) may apply to manual stations where tools are in use.

M320   The final check is to determine if the work centre has the required tool setup for the next operation for the part.

Previous checks will ensure that any complex checks on tool availability are only carried out when all other criteria indicate a movement decision is probable.

The check may include configuration and tool life criteria and matching all tools required for the operation against tools currently held by the machine.

If the machine does not have the required tool setup then the next suitable machine is considered.

M330   The resource scheduler has now determined that a movement of a part to a delivery point may now take place. However some applications may include operating strategies that dictate that deliveries to certain areas should be paired with a pick up by the same carrier, of a part from the same area.

M340   For single movement requests the delivery management system is instructed to transport the part from its current position to the selected station. Appropriate updates to the status of the transport system and the allocated work station will affect later decision scans. The work centre management system may be instructed to accept the delivery.

Processing continues at M170 so that other parts may be routed to or from other stations of the selected type if further transport resources are available.

M350 - M420
       When the delivery is dependent on an associated pick up, the resource scheduler will check for any despatch constraints on the part awaiting pickup. The scheduler will scan for a work centre ready to accept the associated part, or if none are available it will try to route the part to a store. It is possible that the dependent delivery is itself dependent on a further delivery.

CAM CAST sub-topics

	A chain of movements to be serviced by a single transport resource is passed as a request to delivery management. For each movement a pick up and delivery point is specified.
M430	Tool change scheduling for a machine may be initiated by manual request or within a time-based scan or by the resource scheduler as appropriate.
M440	The part order list is scanned to identify parts that will need to be routed to machines of the type being considered for planned machining operations.
M450	Some manufacturing facilities may include several similar machines. When scheduling tool changes for a specific machine, the planned tooling need not meet the requirements of machining operations that are within the capacity of the current tool setup on other machines.
M460	The scan will determine if the current tool setup on a specific machine can be retained for further planned operations, or if a tool change is required.
M470	If no tool change is required, no action is taken.
M480	The optimum tool set up is determined to enable the machine to accommodate the highest priority orders in the parts order list.
M490	Some tools already held by the machine may be retained if they have sufficient remaining working life.
M500	The tool management system is notified which tools are to be used and returned.
M510	The delivery management system is instructed to deliver the new tools and collect the tools to be removed.
M520	The work centre management system is provided with details of tools to be delivered and released.
M540	Delivery Management is a service to the resource scheduling sub-system and to any other source of transportation requests. It will generally be started at the same time as the resource scheduler (M60).
	Delivery Management is responsible for the transmission of routing instructions to transportation hardware, therefore it will initiate an exchange of signals with the transport and storage systems out in the manufacturing facility.
M550	The exchange of signals may include information received describing the current status of transportation devices or indication of which facilities, including buffer storage stations, are occupied or in use. These incoming signals are cross checked against data shared with the resource scheduler i.e. the facility map and resource status. This information must be updated as required.
M560	The delivery management sub-system is now ready to process transportation requests. These requests can come from any source and can be generated either manually or automatically for example by resource scheduler software. Requests for transport may be processed within a time-based scan or as a callable or interrupt-driven routine.
M570	Some applications may require the delivery management sub-system to select a transport system to service a request, from alternatives e.g. AGV or conveyor or rail car. Other applications may request movements by specific transport systems.
M580	Incoming requests must be validated against the limitations of available transport systems.
M590	An error reporting procedure will be called for transport requests which can not be actioned.

M600	A transport system may include several carriers e.g. AGV's perhaps half of which are currently in use. In some systems specific types of carrier may be requested, perhaps relating to the size or weight of the load. In factory-wide systems the distance of the carrier to the initial pick up point may influence the choice of carrier.
M610	In some systems it may be necessary to allocate a specific location within a storage buffer area as part of the processing of a generalised 'deliver to buffer' instruction.
M620	In some systems there may be alternative routes between pick up and deposit points. Unnecessary movement should be avoided although longer routes may be selected to avoid collisions with other carriers moving in other directions. A detailed route is planned, including the final return movements of the carrier to await further instructions, and a possible alternative route to be followed in the event of an error such as the rejection of a delivery by a work station.
M630	The detailed route must be translated into a message format and communications protocol that the selected transportation hardware or manual operator can understand.
M640	Coded instructions are transmitted to the transportation hardware. Alternatively for manually controlled systems detailed instructions may be displayed. Communication errors can be handled using transmission retry and breakdown reporting procedures.
M650	Facility status data and the status of the transport request is updated to indicate that routing is in progress.
M660	In systems with many carriers, other transport requests can be serviced to fully utilise all resources.
M670	All movements in progress must be monitored.
M680	Incoming signals from the transport facilities may be used to note the progress of the carrier through each part of the route, in servicing the transport request.
M690	Any breakdowns, timeouts or incorrect moves are recognised.
M700	An error reporting procedure will be called for any transport request that does not complete successfully. Where necessary the transport system or carrier is assigned a breakdown status so that the system will not try to use it until a repair is notified.
M710	On successful completion of a movement a notification that the transport request is completed should be passed to the resource scheduler or source of the request.
M720	Local control of carriers may be considered separately as this processing may be manually controlled or resident in an intelligent facility perhaps incorporating a programmable controller or microprocessor. The local control facility must be capable of receiving coded instructions.
M730	In some systems the local control logic must be capable of handling interlocks to prevent collisions between carriers.
M740	The carrier may need to wait before entering sections of the route.
M750	When the route section is free of carriers moving in other directions then the carrier may move to the next decision point or destination. In some systems local intelligence may have the capability of following an alternative route such as a rail siding. Pick up and deposit sequences should be handled automatically at the lowest level of control logic.
M760 to M790	Hardware faults are detected and breakdown procedures in the monitoring module are activated. After each successfully completed movement the next movement is started until all coded instructions are carried out. The carrier is then ready for

CAM CAST sub-topics

further instructions.

M810 — The workcentre will be advised that delivery of an item is imminent - the item may be material, tool, fixture, or NC Programme; for housekeeping purposes it may be a container or receptacle for any of these items. The notification should include the item key so that the workcentre may, upon receipt of the carrier, validate that the correct item has been delivered. In the case of tools, the notification should also include the tool life. The intended point of delivery at the workcentre boundary should also be identified unless it is safe to adopt a default value in line with the item being delivered.

M820 — The point of delivery should be checked to determine whether the carrier can be accommodated, and the area into which the item will be loaded should be checked for vacancy. Any local preparation necessary to accommodate the carrier in the correct position relative to the workcentre should be carried out. The workcentre status should then be updated so that the resource scheduler will know that the workcentre awaits the carrier.

M830 — The carrier may now interface with the workcentre, any positional accuracy requirements should be addressed by the delivery system and checked by the workcentre before proceeding. Once the positional/stability requirements have been satisfied, the workcentre status should be updated so that the resource scheduler is aware that the carrier is correctly positioned for any subsequent transfer.

M840 — The workcentre should now validate that the carrier contents are the same as notified in M810, validation should include confirmation of item identity, and if possible, physical characteristics such as size, orientation, surface condition, etc. If the carrier contents are as expected, the workcentre will take action as in M860, if not as in M850. Via the established interface, the workcentre rejects the item, and the carrier is released. The workcentre status is updated and the error reported.

M860 — The carrier contents should be accepted into the workcentre. Upon acceptance of the item the workcentre status should be update to reflect the transfer. Via the established interface the carrier is notified of a successful transfer, the carrier can then be released.

M870 — In workcentres where a choice of items are to be manipulated, the workcentre will be unaware of any priority between the items. The resource scheduler will therefore instruct the workcentre to manipulate items in accordance with its requirements.

M880 — Material will be secured in jigs and fixtures, tools will be mated to holders and/or loaded to carousels, pallets will be located as required to accurately receive material after an operation has been concluded, and NC programmes may have to be post-processed to adapt them for use on a particular workcentre.

M890 — Having secured material, tools, pallets or NC programmes in relation to datum-points, it is now necessary to establish the finite deviations from these data; these offsets will be required for up to three planes and will be dependent upon the item. Workcentre management should maintain these offsets for every item along with its location within the workcentre.

M900 — In workcentres where a choice of part-programmes exists, the workcentre will be instructed to run a specific part-programme by the resource scheduler.

M910 and M915
    The workcentre then applies a series of tests to determine if there is any reason why the specified programme should not be run. The tests will include a check that the specified programme, and associated tools and materials, are located within the

workcentre and are ready for work; a check that it is safe to load the programme to the run area so as not to interfere with any other currently running programme, and a check whether the workcentre is in manual or automatic mode. The workcentre should progress the instruction to run a programme as far as it can, and in the event that it is unable to run, a concise status report should be generated.

M920 The programme will call for material and tools by their key, the workcentre will perform a reconciliation exercise to match material and tool offsets to the programme requirements. Material and tools will be manipulated in the workstation to the compensated dimensions. The workcentre status should be updated to show programme running.

M930 The programme now has effective local control of the workstation and will make demands of the workcentre and its services. The workstation will be required to manipulate both tools and materials, again in accordance with any prevailing offset conditions, any re-orientation of either tool or material will be reflected in the relevant status. Under programme control the workstation will be provided with cutting lubricant and blasts of air to maintain a swarf-free working area.

Throughout the duration of the programme it should update status and performance data. Some data is required by the resource scheduler and some will be used to verify or establish synthetics values for use in CAPE activities.

Increasingly, in-progress inspection and adaptive control, will be a feature of both workcentres and programmes. The process may be contact or non-contact. Whichever is selected only the programme will be aware of the parameter to be measured, its dimensions and tolerances. The programme will therefore direct the workcentre to establish the actual workpiece dimensions, the programme will then compare them with the desired dimensions and re-adjust tool offsets as required.

M940 Throughout the period that the workcentre is active it will be checking for error symptoms. These must be sufficient to full describe any condition which may degrade workcentre performance. The conditions include tool failure, programme failure, inadequate air supply, inadequate cutting fluid supply, inadequate servo response, inadequate material location, power overload, et al.

M950 Monitored activities that deviate from specified standards should be reported. The workcentre should distinguish between functions that are in use and those that are unutilised; failed elements that are unutilised should be inhibited from use, but should not cause the workcentre to stop unless it has need of the failed resource. Naturally, functions that are in use, and fail, must result in the cessation of work in that area.

M960 The workcentre must maintain part-programme and tool status so that it can decide its next action.

M970 Tooling with an exhausted tool life expectancy should not be used to commence a cutting operation. The test is made during the operation so that the tool may be withdrawn from service as soon as is desirable. Sometime it will be permissible to allow the tool to finish the operation, but occasionally the cutting cycle time will be too great, with a consequent risk of producing below specification work; in this case a tool change must precede further cutting action, though this will usually be under programme control.

M980 The feasibility of exhausted tool replenishment, without bringing the workcentre to a standstill, will be dependent upon the tooling in use, the workcentre configuration, and the availability of suitable replacements. In some instances it will be quite straightforward to index a tool tip and continue as before, in other cases sister tooling will be required. Some configurations will demand that the workcentre must be brought to a complete halt whilst whole magazines or carousels are changed, certain other configurations support a 'pick andplace' fully automatic tool exchange facility to the workstation. In the 'pick and place' environment the workcentre may hold its own stock to replenish those exhausted, on the other hand it may have access, via the resource scheduler and delivery system, to a shared tool store from which tools may be drawn on a 'needs' basis.

M990   The workcentre should initiate and perform the tool replacement procedure as applicable; the workcentre status should reflect the new tool situation with respect to tool key, location, offset and life expectancy.

M1000   If tool replenishment requires the workcentre to stop, the tool status should be updated to reflect its exhausted condition and the workcentre should continue. The tool should not be used again in this state. The workcentre will receive a separate command (incorporated in M1070) to change the tools.

M1010 to M1030

Upon programme completion the workcentre can start to progress the material out of the workstation and towards the workcentre boundary if machining is complete. If further machining is to take place within the workcentre, then further instruction will be required from the resource scheduler to indicate new positional and orientational requirements.

M1040   In workcentres where there is a choice of material to be prepared for transfer, the resource scheduler will instruct the workcentre to commence preparation.

M1050   The material will have to be unloaded from the machine tool, de-fixtured and loaded to a pallet for subsequent transfer to the delivery system. The nature of the preparation activity will depend upon the workcentre configuration, the nature and quantity of the material, and the requirements of any subsequent processes.

M1060   On each occasion that the material changes location, orientation or state within the workcentre, its status should reflect the changed condition, this will include completion of preparation for transfer.

M1070   The workcentre will receive an instruction from the resource scheduler to remove a specific item from its boundaries. This may occur to satisfy many conditions and may be applied to any item. For example, material may be found to be unfit for use after it has been accepted, or unfit for further work because of a tool breakage; tool removal from the workcentre boundary will always be via this instruction; programmes subjected to proving runs and perhaps altered at the workcentre will require to be fed back, though in normal use most programmes are likely to be overwritten or erased and the memory re-initialised prior to the next programme load; pallets or other housekeeping items will also be subject to this instruction.

M1080   The workcentre must determine if there is any reason why the item will be inhibited from reaching the location from which it will be unloaded. This will usually be a problem of conflict with other similar items that are in the path between the item to be removed and the location from which it will be unloaded. Further examples that might lead to the workcentre being unable to service the instruction would be the breakdown of a service necessary to progress the item, or the non-availability of a pallet, a tool holder or clamps.

If the workcentre cannot comply with the instruction to remove, the workcentre should fully report its status so that the resource scheduler can issue alternative instructions as required.

M1090   The item will have to undergo a change of state prior to subsequent transfer. In all cases it will have to be removed from the workstation, and any positional media, and manipulated into a different media to be transferred to the carrier. The nature of the preparation activity will depend upon the workcentre configuration, the nature and quantity of the item and the requirements of any subsequent processes.

M1100   On each occasion that the item changes location, orientation or state within the workcentre, its status will reflect the changed condition, this will include completion of preparation for transfer.

M1110   The workcentre will be advised that the arrival of a carrier is imminent to collect an item, this item may be material, tool, programme or a housekeeping associated item such as a pallet. This notification may arise in one of two circumstances; the normal sequence of workcentre production will be that the workcentre will update its status to show that material is ready for transfer, the resource scheduler will then instruct the delivery system to send a suitable carrier. In this case the carrier should be due to arrive at at an unload point associated with the final location in M1060.

Alternatively, the workcentre will receive instructions to remove items from its boundary as depicted in node M1070. This will usually result in an item being made available for transfer from a specific location shown in M1100. The reasons for the alternative initiation of a removal operation have been discussed in M1070.

The intended point of collection should form part of the collection notification unless a default position applies.

M1120   The point of collection should be checked to determine whether the carrier can be accommodated; any local preparation necessary to accommodate the carrier in the correct position relative to the workcentre should be carried out. An example of this is a machine tool being served by AGV's of varying dimensions, the machine tool might be required to guide the AGV into different physical locations in light of this. The workcentre status should be updated so that the resource scheduler will know that the workcentre awaits the carrier.

M1130   The carrier may now interface with the workcentre, any positional accuracy requirements should be addressed by the delivery system and checked by the workcentre before proceeding. Once the positional/stability requirements have been satisfied the workcentre status should be updated.

M1140   The transfer of the item from workcentre boundary to the carrier may now proceed. Upon completion, the workcentre status should be updated to reflect the transfer. Via the established interface the carrier is notified of a successful transfer, the carrier can then be released.

**Computer Aided Manufacturing/
Computer Aided Storage and Transportation**

Production Management sub-systems

The Flowchart for the Production Management activities in the ESPRIT Pilot Study has been broken down into the following sub-systems:

1   Resource scheduling
        (Nodes M10 to M520)

2   Delivery management
        (Nodes M540 to M790)

3   Workcentre management
        (Nodes M810 to M1140)

Resource scheduling sub-system (M1)

M1.1 Sub-system description

This sub-system is responsible for controlling the manufacturing facility to produce parts to meet priority-sequenced orders from the Computer Aided Production Planning System. Achievement reporting back to CAPP will influence future scheduling.

The resource scheduler will co-ordinate the activities of all work stations and transport facilities through interfaces to the work centre management and delivery management sub-systems. The system will be capable of fully automatic control in scheduling resources, taking into account operator controls and local operating strategies.

As operations at work centres are completed, the status of orders is maintained, along with the status of all scheduled facilities.

The sub-system requires information from other CAM sub-systems as an input to scheduling decisions.

The resource scheduler will initiate operations at work centres, instruct delivery management to transport parts and initiate the down-line loading or delivery of part-programmes to CNC machines.

It will also produce messages as required to warn staff responsible for the facility, of any critical situations.

The flowchart illustrates the general requirement to scan real time data, react quickly to changing situations and initiate the highest priority required actions as soon as the necessary resources are available. In practice the structure of a resource scheduling module must be tuned to optimise the flow of part orders in a particular manufacturing facility and will to some extent be determined by the physical layout of that facility.

Applications will also need a facility for scheduling tool changes. This is considered as a separate module as it may or may not be automated and it may require a limited degree of forward planning.

Production Management

Resource scheduling - sub-system

M1.2 Flowcharts

M1.2 SUB SYSTEM FLOWCHART

Sheet 1 of 5

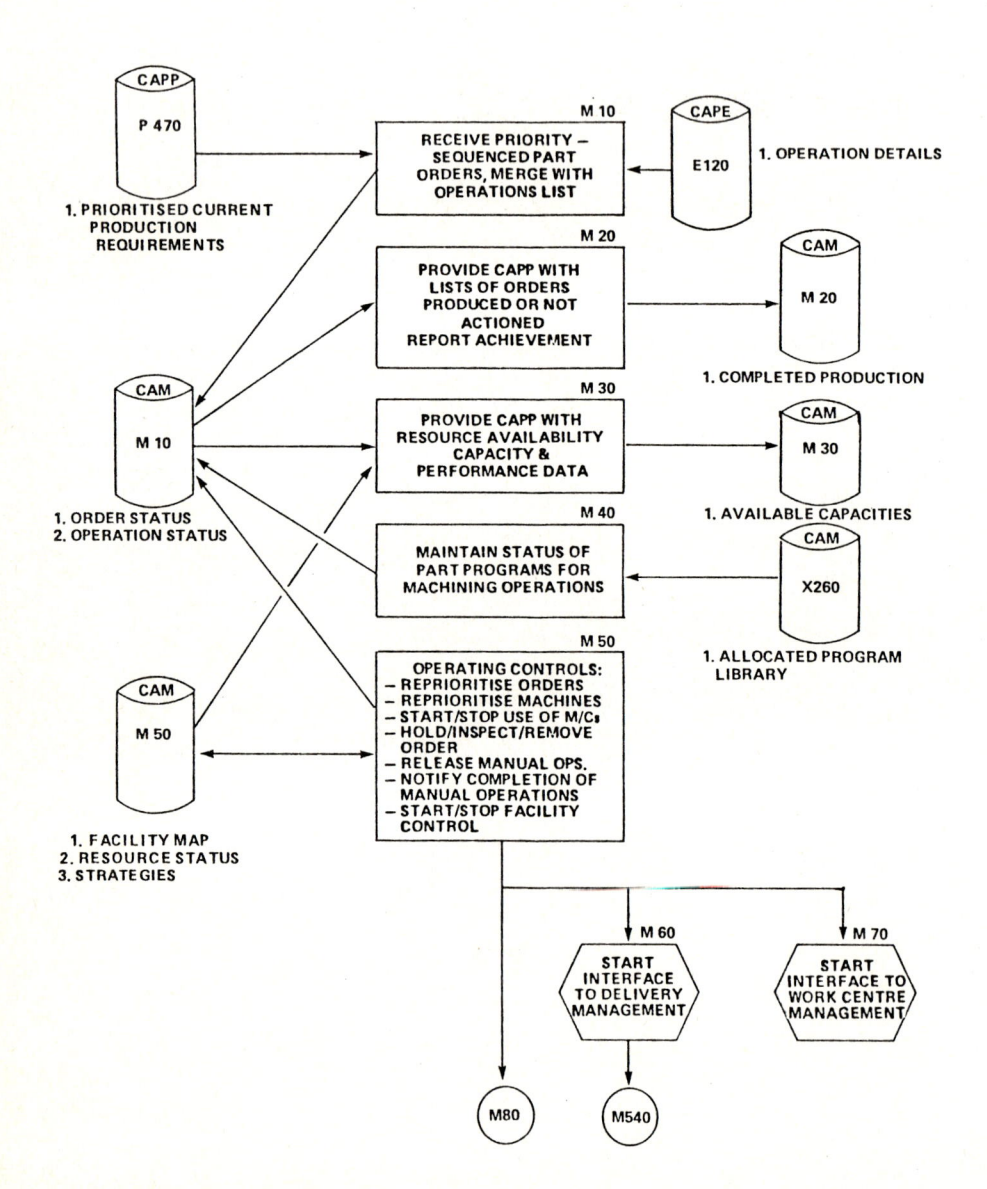

CAPP / P 470
1. PRIORITISED CURRENT PRODUCTION REQUIREMENTS

CAPE / E120
1. OPERATION DETAILS

CAM / M 10
1. ORDER STATUS
2. OPERATION STATUS

CAM / M 20
1. COMPLETED PRODUCTION

CAM / M 30
1. AVAILABLE CAPACITIES

CAM / X260
1. ALLOCATED PROGRAM LIBRARY

CAM / M 50
1. FACILITY MAP
2. RESOURCE STATUS
3. STRATEGIES

M 10 RECEIVE PRIORITY – SEQUENCED PART ORDERS, MERGE WITH OPERATIONS LIST

M 20 PROVIDE CAPP WITH LISTS OF ORDERS PRODUCED OR NOT ACTIONED REPORT ACHIEVEMENT

M 30 PROVIDE CAPP WITH RESOURCE AVAILABILITY CAPACITY & PERFORMANCE DATA

M 40 MAINTAIN STATUS OF PART PROGRAMS FOR MACHINING OPERATIONS

M 50 OPERATING CONTROLS:
– REPRIORITISE ORDERS
– REPRIORITISE MACHINES
– START/STOP USE OF M/Cs
– HOLD/INSPECT/REMOVE ORDER
– RELEASE MANUAL OPS.
– NOTIFY COMPLETION OF MANUAL OPERATIONS
– START/STOP FACILITY CONTROL

M 60 START INTERFACE TO DELIVERY MANAGEMENT

M 70 START INTERFACE TO WORK CENTRE MANAGEMENT

(M80)   (M540)

M1.2. SUB SYSTEM FLOWCHART

Sheet 2 of 5

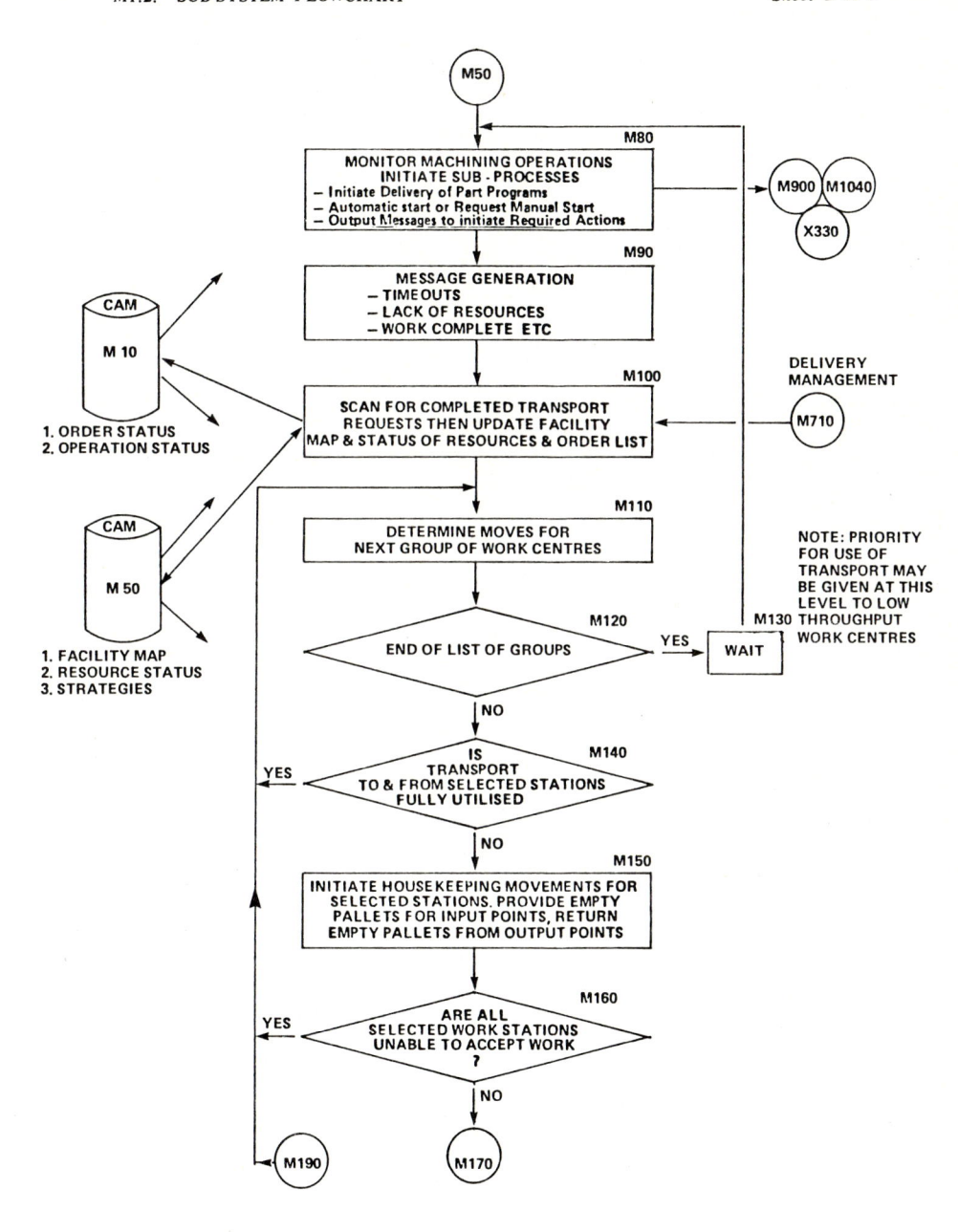

M50

M80 MONITOR MACHINING OPERATIONS INITIATE SUB - PROCESSES
- Initiate Delivery of Part Programs
- Automatic start or Request Manual Start
- Output Messages to initiate Required Actions

→ **M900**, **M1040**, **X330**

M90 MESSAGE GENERATION
- TIMEOUTS
- LACK OF RESOURCES
- WORK COMPLETE ETC

CAM M10
1. ORDER STATUS
2. OPERATION STATUS

M100 SCAN FOR COMPLETED TRANSPORT REQUESTS THEN UPDATE FACILITY MAP & STATUS OF RESOURCES & ORDER LIST

DELIVERY MANAGEMENT ← **M710**

CAM M50
1. FACILITY MAP
2. RESOURCE STATUS
3. STRATEGIES

M110 DETERMINE MOVES FOR NEXT GROUP OF WORK CENTRES

NOTE: PRIORITY FOR USE OF TRANSPORT MAY BE GIVEN AT THIS LEVEL TO LOW THROUGHPUT WORK CENTRES

M120 END OF LIST OF GROUPS — YES → **M130** WAIT

NO ↓

M140 IS TRANSPORT TO & FROM SELECTED STATIONS FULLY UTILISED — YES →

NO ↓

M150 INITIATE HOUSEKEEPING MOVEMENTS FOR SELECTED STATIONS. PROVIDE EMPTY PALLETS FOR INPUT POINTS, RETURN EMPTY PALLETS FROM OUTPUT POINTS

M160 ARE ALL SELECTED WORK STATIONS UNABLE TO ACCEPT WORK ? — YES →

NO ↓

M190   **M170**

M1.2. SUB SYSTEM FLOWCHART

Sheet 3 of 5

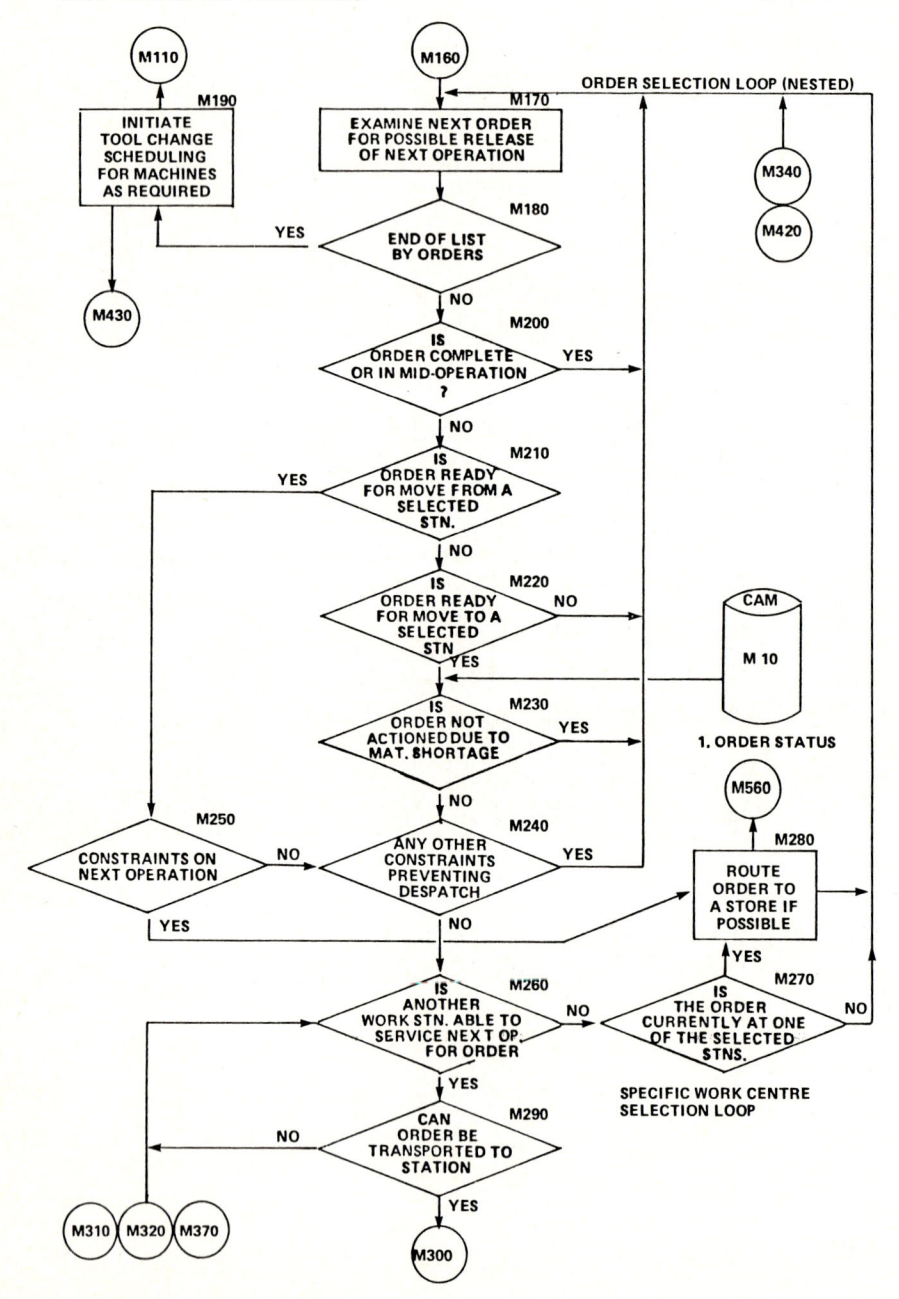

- **M110**
- **M190** — INITIATE TOOL CHANGE SCHEDULING FOR MACHINES AS REQUIRED
- **M430**
- **M160**
- **M170** — EXAMINE NEXT ORDER FOR POSSIBLE RELEASE OF NEXT OPERATION
- **M180** — END OF LIST BY ORDERS? (YES → M190; NO ↓)
- **M200** — IS ORDER COMPLETE OR IN MID-OPERATION? (YES →; NO ↓)
- **M210** — IS ORDER READY FOR MOVE FROM A SELECTED STN. (YES →; NO ↓)
- **M220** — IS ORDER READY FOR MOVE TO A SELECTED STN (NO →; YES ↓)
- **M230** — IS ORDER NOT ACTIONED DUE TO MAT. SHORTAGE (YES →; NO ↓)
- **M250** — CONSTRAINTS ON NEXT OPERATION (NO →; YES ↓)
- **M240** — ANY OTHER CONSTRAINTS PREVENTING DESPATCH (YES →; NO ↓)
- **M280** — ROUTE ORDER TO A STORE IF POSSIBLE
- **M560**
- **M340**
- **M420**
- CAM M10 — 1. ORDER STATUS
- **M260** — IS ANOTHER WORK STN. ABLE TO SERVICE NEXT OP. FOR ORDER (NO →; YES ↓)
- **M270** — IS THE ORDER CURRENTLY AT ONE OF THE SELECTED STNS. (YES ↑; NO →)
- **M290** — CAN ORDER BE TRANSPORTED TO STATION (NO →; YES ↓)
- **M300**
- **M310**, **M320**, **M370**

ORDER SELECTION LOOP (NESTED)

SPECIFIC WORK CENTRE SELECTION LOOP

CAM CAST sub-topics

M1.2. SUB SYSTEM FLOWCHART

Sheet 4 of 5

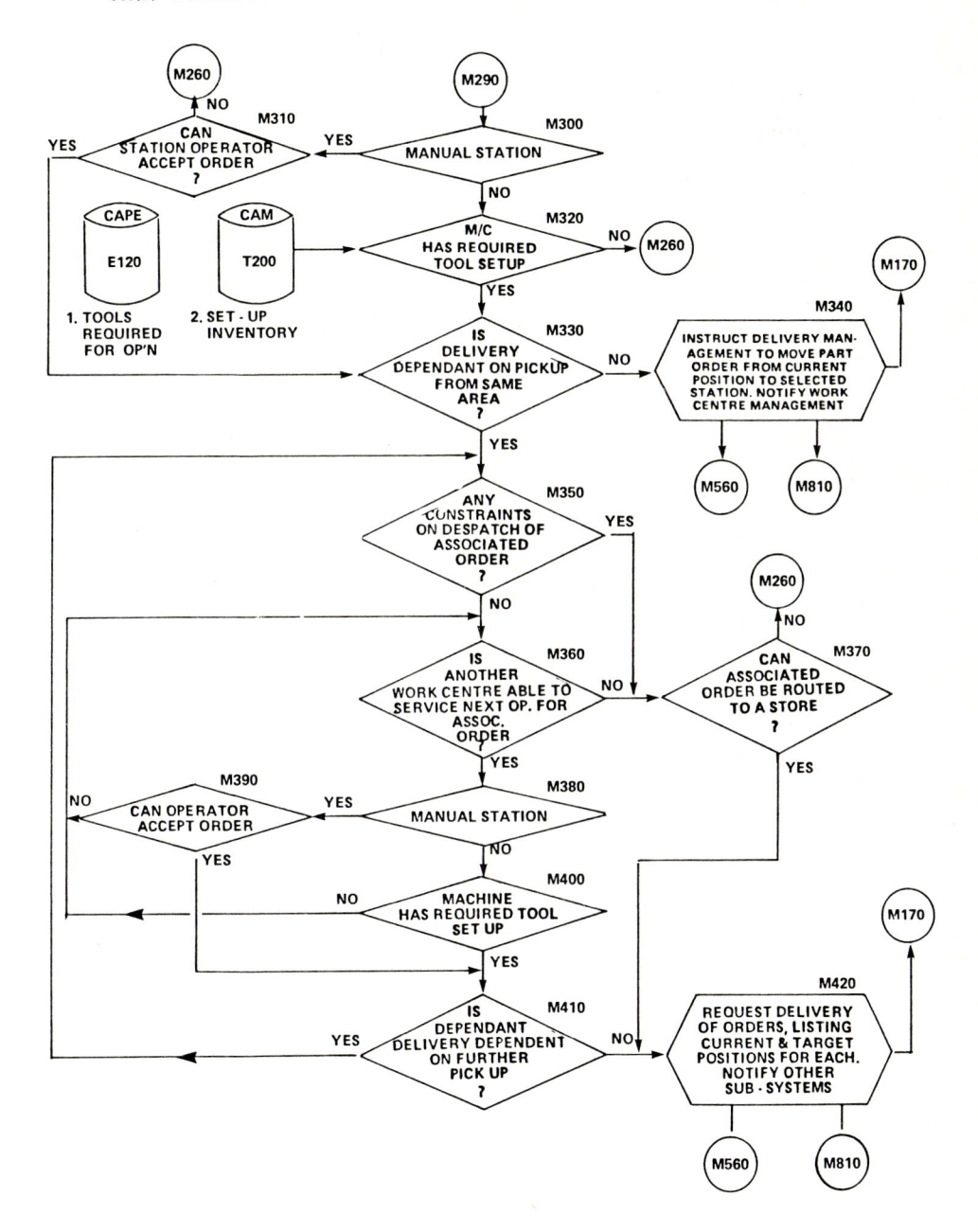

M1.2. SUB SYSTEM FLOWCHART   Sheet 5 of 5

M190 → **M430** INITIATE TOOL CHANGE SCHEDULING FOR A SELECTED WORK STATION

CAM M10
1. ORDER STATUS
2. OPERATION STATUS

→ **M440** SCAN FOR REQUIRED MACHINING OPERATIONS FOR MACHINE TYPE

CAM T200 / **CAPE E120**
1. SET - UP INVENTORY
2. TOOLS REQUIRED FOR OPERATIONS

→ **M450** IGNORE OPERATIONS THAT CAN BE HANDLED BY OTHER MACHINES OF SAME TYPE

M460 IS TOOL CHANGE REQUIRED ? — NO → **M470** NO ACTION

YES ↓

M480 DETERMINE OPTIMUM TOOL SET UP TO ENABLE MACHINING OF HIGHEST PRIORITY ORDERS

M490 DETERMINE WHICH TOOLS HAVE SUFFICIENT WORKING LIFE TO BE RETAINED

M500 UPDATE SET - UP INVENTORY

M510 NOTIFY DELIVERY MANAGEMENT OF TOOL DELIVERY AND COLLECTION

M520 NOTIFY WORK CENTRE MANAGEMENT OF TOOLS TO BE DELIVERED AND RELEASED

CAM T200
1. SET - UP INVENTORY

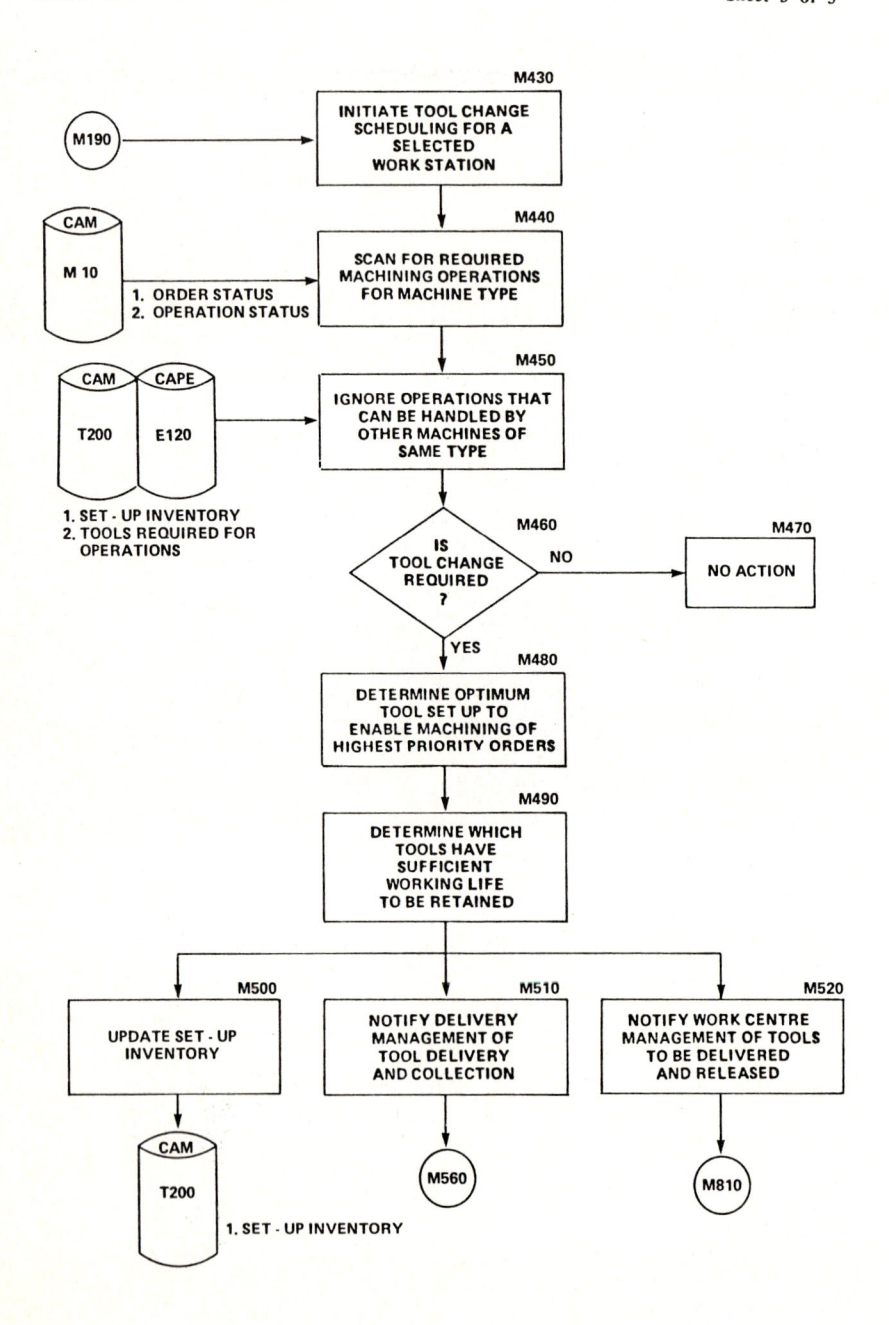

(M560)   (M810)

M1.3 Design rules - resource scheduling

M1.3.1 Input data

The sub-system must be capable of accepting the following minimum input data, although not every realisation of a resource scheduling sub-system will use all the data elements.

Input data is required in respect of:
- Manufacturing Orders
- Operations and Operational Dependencies

M1.3.1.1 Manufacturing order data

- **Manufacturing Order Reference**

  A unique and unambiguous reference by which the order is identified and reported.

- **Type of Order**

  A code which indicates whether an order is for production, development, programme proving or any other user-defined order category.

- **Rough Component Identity**

  The part no. of the rough component (or blank).

- **Raw Material Specification**

  The dimensions and type of material of the rough component.

  NB: Normally only one of the previous two data will be available.

- **Finished Component Identity**

  The part no. of the component(s) for which the order was raised (including level or version no. if appropriate).

- **Order Quantity**

  The quantity of the finished component that is required.

- **Due Date**

  The time by which the order is required to be complete, expressed in appropriate units (hour, shift, day, week, month, quarter, etc.).

  NB: This is not necessarily the due date of sales order (or orders) which gave rise to the manufacturing order.

- **Priority**

  The criterion for resolving contention for an available facility between operations with similar "Latest Release Times".

- **Total Process Time Outstanding**

  The sum of the time allowances for all outstanding operations required to complete the order (including transportation, fixturing/de-fixturing, machining, inspection/testing and packing operations).

  NB: When the order data is first received this datum represents the total "make time" of the order.

- **Order Status**

  One of a user-defined set of codes which indicate:
  - an operation is in process

- an operation is waiting to be released
- all operations are completed
- an order is temporarily suspended during an operation
- an order is temporarily suspended between operations

- **Operation Number**

  Depending on the Order Status this is the current operation, or the next operation to be released.

M1.3.1.2 **Operation and dependency data**

- **Operation Number**

  A reference which is unique and unambiguous within a finished component part no., and identifies a particular process, or set sequence of processes, which occur at a particular facility, or at one of a particular user-defined group of facilities.

  NB: An Op. No. can refer to an automated process such as NC machining, or to a manual process such as manual inspection.

- **Operation Type**

  A code which identifies to which of a user-defined range of classifications an operation belongs.

  eg. Codes may be defined to distinguish between transportation, fixturing/de-fixturing, machining, inspection/testing and packing operations.

- **Operation Quantity**

  The number of workpieces normally involved in a single execution of this operation. This datum may take different values for different operations within a routing.

  These data within one routing are required to calculate outstanding processing time on an order, and to determine how an operation may be despatched after the previous operation in the routing has been executed, or to calculate how many times the previous operation must be executed before a single instances of an operation can be despatched.

- **Operational Dependency**

  The Op. No. (or Op. No's) whose completion is necessary before this operation can be released.

  eg. If a component is made in four operations, of which Op.1 and Op.4 must always be first and last respectively, but Op's 2 and 3 may be performed in either order; the dependencies would be as follows:

Op. No.	Dependencies
1	0 - implies no dependency
2	1) either may directly
3	1) follow Op.1
4	2 & 3 - no need to specify Op.1

- **Time Allowance**

  A synthetic value for normal Op. processing time.

- **Workcentre or Group of Workcentres**

  Either an unambiguous reference to a particular machine, robot or fixturing station, or a code which represents a user-defined group of machines, robots or fixturing stations.

- **Programme Reference**

  An unambiguous reference which identifies the programme (or procedure) to be used on this operation. The programme could be a part-programme for an N/C machine tool, a programme for a robot, or hard copy operating instructions for a manual procedure.

- **Programme Status**

  A code which represents the user-defined status of the programme.

  eg. It may need to be obtained from a central source, then post-processed, then down-lined to a workcentre.

  eg. It may be permanently held in usable form at the workcentre where the programme is required.

  In the first example three different status apply, in the second example only one.

- **Tool List Reference**

  An unambiguous reference to the list of tool types and their associated usage times required for this operation.

  Either the complete tool requirement should appear in the list, on only those which are not standard items at the machine(s) identified above.

- **Tool Kit Status**

  A code which represents the user-defined availability of the tool kit for this operation. The status indicated by the code could include the following:

  - tools available in machine tool magazine
  - tools available within workcentre
  - tools available within setting area
  - tools not available

- **Mounting and Fixturing Reference**

  An unambiguous reference to the list of jigs, fixtures, pallets, etc., required to mount this component for this operation.

  Either the complete mounting/fixturing requirement should appear in the list, or only those which are not standard items at the fixturing station(s) identified above.

- **Carrier Type**

  A code which identifies which type (or types) of transportation system may be used for this operation.

  NB: Some of the data are only applicable to particular operations types as defined above, and the interpretation of certain data depends on the "operation type".

M1.3.2   **Input re-formatting**

The sub-system must be capable of validating and re-formatting input data and creating cross linkages between data to meet its own access requirements. It must be possible to describe the format of the input data to the sub-system (it is assumed that the input data will be structured in a consistent manner). The sub-system must be capable of receiving new order and operation order in batch mode, ideally it will be capable of receiving new order and operation data in interactive mode. The sub-system must be able to receive facility status amendments interactively.

M1.3.3 **Order amendment**

The sub-system must be capable of accepting amendments to priorities and due-dates at any stage of an order, ideally the sub-system should accept a deletion or an amendment to order quantity for an order which has not been commenced.

M1.3.4 **Facility status amendments**

The sub-system must be capable of accepting manually entered amendments to facility status in interactive mode (normally this ability will be restricted to re-establishing the availability of previously unavailable facilities - other status amendment input will normally be generated automatically).

M1.3.5 **Output data**

The sub-system must be capable of reporting achievements and current order and facility status. The sub-system must allow the system-user(s) to specify the format and frequency of the outputs.

M1.3.6 **Despatching tasks**

The sub-system must be capable of despatching tasks to delivery management, workcentre management and lower level resource scheduling sub-systems. The sub-system must be capable of despatching any type of task (see "Operation Type" in M1.3.1.2)

The sub-system must be capable of despatching tasks to human operators, robot operators, N/C machinery or to computer or PLC controlled sub-systems.

M1.3.7 **Transfer of process instructions**

The sub-system must be able to initiate the transfer of programmes to robots and machine controllers. The sub-system must be able to initiate the transfer of textual or graphical instructions to human operators.

M1.3.8 **Facility monitoring**

The sub-system must be aware of the current status of every facility within its scope. The sub-system must be able to automatically update the status of facilities in respect of newly despatched tasks.

The sub-system must be able to update the status of facilities in respect of events reported to it by delivery management, workcentre management and lower-level resource scheduling sub-systems.

Awareness of facility status must include knowledge of:
- operational status of every facility
- capacity and current load of every facility
- availability of manpower (at manual stations)
- availability of pallets, fixtures, jigs (at fixturing stations)
- availability of uncommitted tool life (at work centres)
- existence of spent and totally committed tools (at workcentres)

M1.3.9 **Order and operation status**

The sub-system must be aware of the current status of every order and operation within its scope, and the location of every workpiece (or batch of workpieces) in any condition for the orders within its scope. The sub-system must be able to establish which operations may be released to a workcentre that is ready to accept a new workpiece (or batch of workpieces).

Awareness of order and operation status must include knowledge of:
- location and condition of every workpiece
- all operations currently in process
- all operations ready for release
- requirements for tool types and associated usage times
- availability of uncommitted tool life in tool magazines
- availability of prepared tools outside tool magazines
- availability of part-programmes

M1.3.10 **Task selection and initiation**

The sub-system must be able to despatch tasks whenever a favourable combination of facility status and order/operation status exists.

The sub-system must be capable of recognising favourable conditions by applying user-defined scheduling algorithms and decision rules to its knowledge of the various status.

Ideally the sub-system will be capable of applying different user-defined rules and algorithms according to its perception of the current situation.

eg. The sub-system may despatch tasks in such a manner that reduced manning is possible for night-shift operations.

eg. The sub-system may bias its selection in favour of tasks with long machining times if it recognises that the transportation system(s) is(are) operating at less than normal capacity.

M1.3.11 **Facility selection**

The sub-system must be able to incorporate user-defined bias in favour of designated facilities when contention for transportation arises, or when alternative routings for an order exist.

Facility bias logic should be incorporated in the task selection rules.

Ideally the sub-system would be capable of dynamically adjusting bias in accordance with its perception of the current situation.

M1.3.12 **Operational control**

The sub-system must be capable of accepting changes to task selection rules and algorithms. Ideally the sub-system would accept changes interactively without suspending its supervisory role.

The sub-system must be able to accept manually generated overrides and exceptions. These either force a deviation from the choice of task that would otherwise derive naturally from the task selection rules, or delay (or bring forward) the despatching of a selected task.

M1.3.13 **Task validation**

The sub-system must not despatch a transportation task to move a workpiece (or batch of work) unless it knows from its status data that the intended destination can physically accommodate the workpiece and has the necessary resources to perform the subsequent operation. Depending on the operation the resources may include man-power, jigs, fixtures, pallets, tools, programmes or machining capability. Task validation should be incorporated in the task selection rules.

M1.3.14 **Housekeeping**

The sub-system must be able to despatch any tasks necessary to maintain the continuous operation of the total system, which will not be explicitly included in the operations data inputs.

- eg. Empty pallets may need to be transported from a de-fixturing station to a fixturing station.
- eg. Spent tools may need to be replaced in tool magazines.

M1.3.15 **Forward planning**

The sub-system should not attempt to forward plan at one time all the operations implied by the uncompleted orders within its scope.

The sub-system should attempt to despatch tasks in accordance with an optimum sequence determined by the production planning system, but should override this sequence when it conflicts with the result of using the task selection rules. Ideally the task selection rules can incorporate the position of an order/operation within the optimum sequence as a factor affecting the selection of tasks.

M1.3.16 **Tool replenishment**

The sub-system must be able to despatch instructions both to replace tools and machine tool magazines, and to deliver new tool kits to workcentres.

The sub-system must be able to predict future events to determine a need for tool renewal, this rule is therefore an exception to M1.3.15.

The sub-system must be able to access tool identities and their remaining committed and uncommitted lives in respect of tools at workcentres, both for tools already loaded to machine tool magazines and for those not yet loaded.

M1.3.17 **Error detection and reporting**

The sub-system must be able to detect error conditions that need to be reported to a higher authority, such as instances of facility breakdowns, and orders whose due dates become unachievable.

Delivery management sub-system (M2)

M2.1 Sub-system description

This sub-system is responsible for controlling all transport systems in the manufacturing system. This could include AGV systems, conveyor systems, rail cars, monorails, overhead slings, robot pick and place, roller beds and any future material handling system.

This sub-system may be implemented as a single monolithic entity, a hierarchy of processes, or as independent sub-systems managing different physical transport systems.

Also included are manual transfers and fork lift trucks where the operators respond to generated instructions. Hardware for automatic transportation could be obtained from a variety of suppliers, and the interface to each type of hardware is at present subject to a wide degree of variation.

More than one module may be needed to exchange signals with different transport systems because of the variations in message formats and communications protocol.

The system will generate and transmit detailed coded instructions to transport systems. For manually operated systems this could take the form of displayed instructions.

The system must also monitor movements in progress and inform the source of the transport request of completed movements.

Other interfaces may be necessary e.g. to work centre management to exchange signals relating to deliveries and collections. Also store control modules may be needed for high volume storage.

The sub-system should include facilities as required for selecting carriers and planning a detailed route to avoid collision with other carriers. The lowest level of traffic control will hold or divert carriers at control points to avoid collisions.

The sub-system will frequently be implemented with distributed intelligence; typically a central co-ordinating intelligence, intelligence on-board discrete mobile units, and local intelligence controlling the use of junctions and other shared facilities.

Production Management

Delivery management sub-system

M2.2 Flowcharts

CAM CAST sub-topics

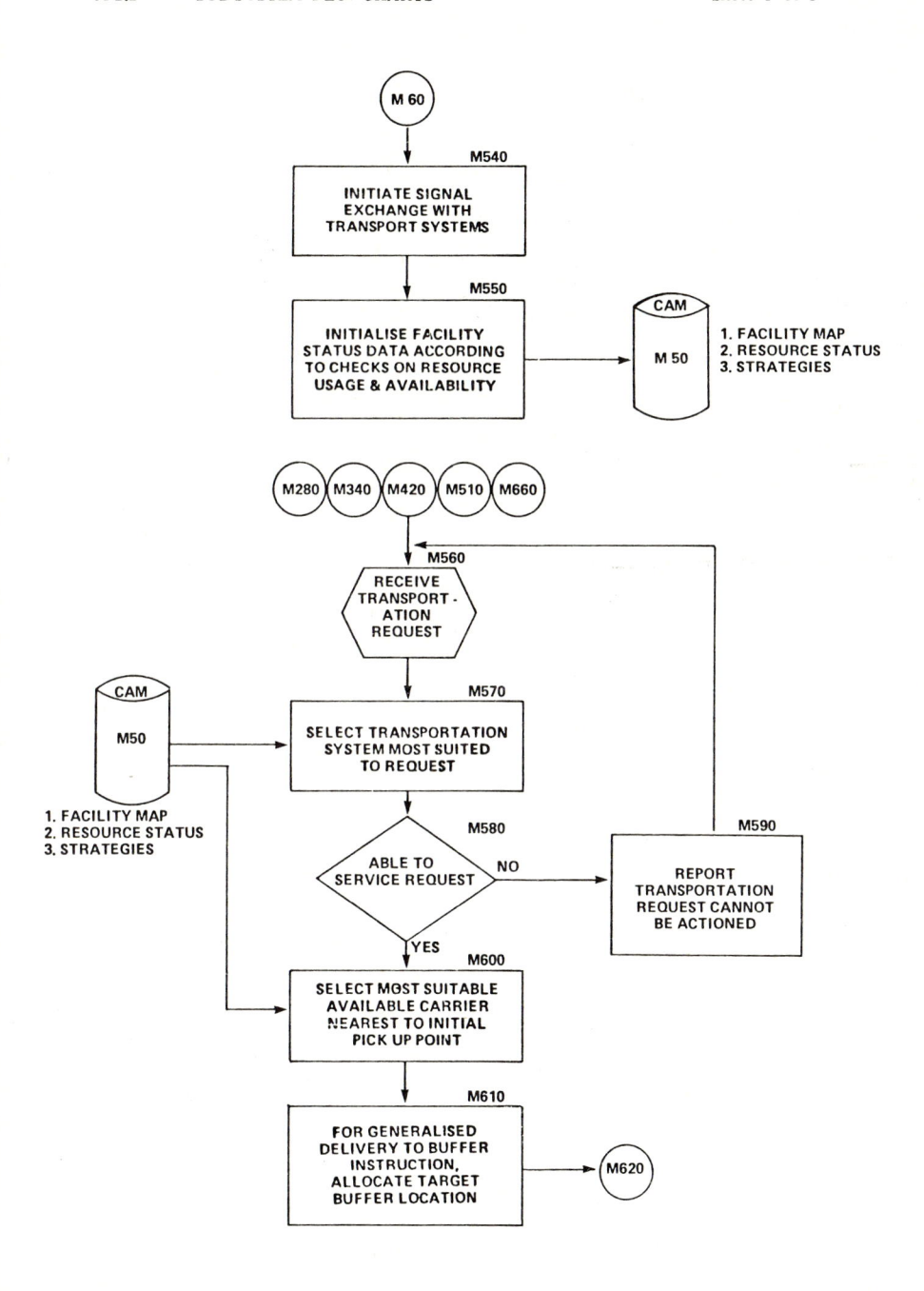

M 2.2   SUB SYSTEM FLOWCHARTS   Sheet 1 of 3

```
                          ( M 60 )
                             │
                             ▼
                    ┌──────────────────┐ M540
                    │ INITIATE SIGNAL  │
                    │ EXCHANGE WITH    │
                    │ TRANSPORT SYSTEMS│
                    └──────────────────┘
                             │
                             ▼
                    ┌──────────────────┐ M550        ╔════════╗
                    │ INITIALISE       │             ║  CAM   ║   1. FACILITY MAP
                    │ FACILITY STATUS  │────────────▶║  M 50  ║   2. RESOURCE STATUS
                    │ DATA ACCORDING   │             ║        ║   3. STRATEGIES
                    │ TO CHECKS ON     │             ╚════════╝
                    │ RESOURCE USAGE   │
                    │ & AVAILABILITY   │
                    └──────────────────┘

         (M280)(M340)(M420)(M510)(M660)
                       │
                       ▼
                    ┌─────────┐  M560
                    │ RECEIVE │
                    │TRANSPORT│◀──────────────────────┐
                    │  ATION  │                       │
                    │ REQUEST │                       │
                    └─────────┘                       │
  ╔════════╗              │                           │
  ║  CAM   ║              ▼                           │
  ║  M50   ║      ┌──────────────────┐ M570           │
  ║        ║─────▶│ SELECT TRANS-    │                │
  ╚════════╝      │ PORTATION SYSTEM │                │
                  │ MOST SUITED TO   │                │
 1. FACILITY MAP  │ REQUEST          │                │
 2. RESOURCE      └──────────────────┘                │
    STATUS                │                           │
 3. STRATEGIES            ▼           M580            │  M590
                       ◇ ABLE TO ◇ ─── NO ───▶ ┌────────────┐
                       ◇ SERVICE ◇             │ REPORT     │
                       ◇ REQUEST ◇             │TRANSPORTATION
                           │                   │ REQUEST    │
                          YES                  │ CANNOT BE  │
                           │                   │ ACTIONED   │
                           ▼          M600     └────────────┘
                  ┌──────────────────┐
                  │ SELECT MOST      │
                  │ SUITABLE AVAIL-  │
                  │ ABLE CARRIER     │
                  │ NEAREST TO       │
                  │ INITIAL PICK     │
                  │ UP POINT         │
                  └──────────────────┘
                           │
                           ▼          M610
                  ┌──────────────────┐
                  │ FOR GENERALISED  │
                  │ DELIVERY TO      │──▶ (M620)
                  │ BUFFER INSTRUC-  │
                  │ TION, ALLOCATE   │
                  │ TARGET BUFFER    │
                  │ LOCATION         │
                  └──────────────────┘
```

M 2.2 SUB SYSTEM FLOWCHARTS

M610 → **M620** DETERMINE MOST ECONOMICAL OVERALL ROUTE BETWEEN PICK UP AND DEPOSIT POINTS AVOIDING COLLISIONS

↓

M630 TRANSLATE DETAIL INSTRUCTIONS TO REQUIRED MESSAGE FORMAT AND COMMUNICATIONS PROTOCOL FOR DEVICE → **CAM M 50**
1. FACILITY MAP
2. RESOURCE STATUS
3. CONFIGURATION

↓

M640 TRANSMIT CODED INSTRUCTIONS TO TRANSPORTATION HARDWARE. RETRY COMMUNICATIONS AS NECESSARY → **M720**

↓

CAM M 10 ← **M650** STATUS OF TRANSPORTATION REQUEST = ROUTING IN PROGRESS → **M660** SERVICE NEXT TRANSPORTATION REQUEST → **M560**

1. OPERATION STATUS

↓

M670 MONITOR CARRIER PROGRESS

↓

M750 → **M680** UPDATE POSITIONAL AND STATUS DATA RELATING TO TRANSPORT REQUEST → **CAM M 50**
1. FACILITY MAP
2. RESOURCE STATUS

↓

M770 → **M690** ANY BREAKDOWNS TIMEOUTS OR INCORRECT MOVES ? —YES→ **M700** NOTIFY FAILURE OF TRANSPORT REQUEST STATUS OF TRANSPORT SYSTEM OR CARRIER = BREAKDOWN

↓ NO

M790 → **M710** STATUS OF TRANSPORT REQUEST = COMPLETED → **CAM M 10**
1. OPERATION STATUS

↓

M600

Sheet 2 of 3

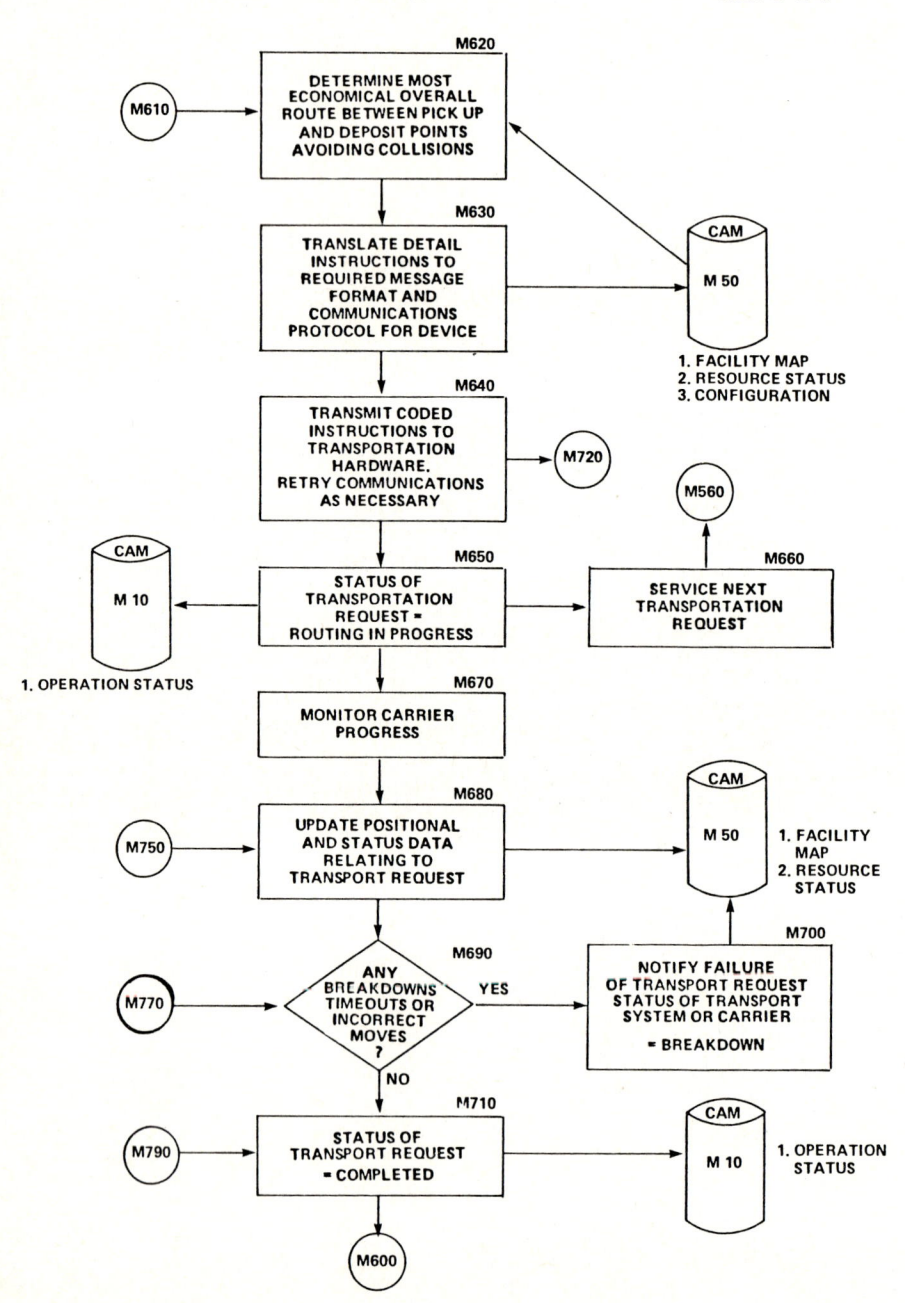

CAM CAST sub-topics 275

M 2.2   SUB SYSTEM FLOWCHARTS   Sheet 3 of 3

LOCAL CARRIER CONTROL

- M640
- M720 RECEIVE CODED INSTRUCTIONS
- M730 ANY INTERLOCKS DELAYING MOVE TO NEXT CONTROL POINT ?
  - YES → M740 WAIT
  - NO ↓
- M750 MOVE CARRIER ALONG MOST DIRECT FREE ROUTE TO NEXT CONTROL POINT → M680
- M760 ANY HARDWARE FAULTS ?
  - YES → M770 TRANSMIT ERROR STATUS → M690
  - NO ↓
- M780 ANY MORE CONTROL POINTS IN ROUTE ?
  - YES → (loop back to M730)
  - NO ↓
- M790 CARRIER READY FOR INSTRUCTIONS
- M710

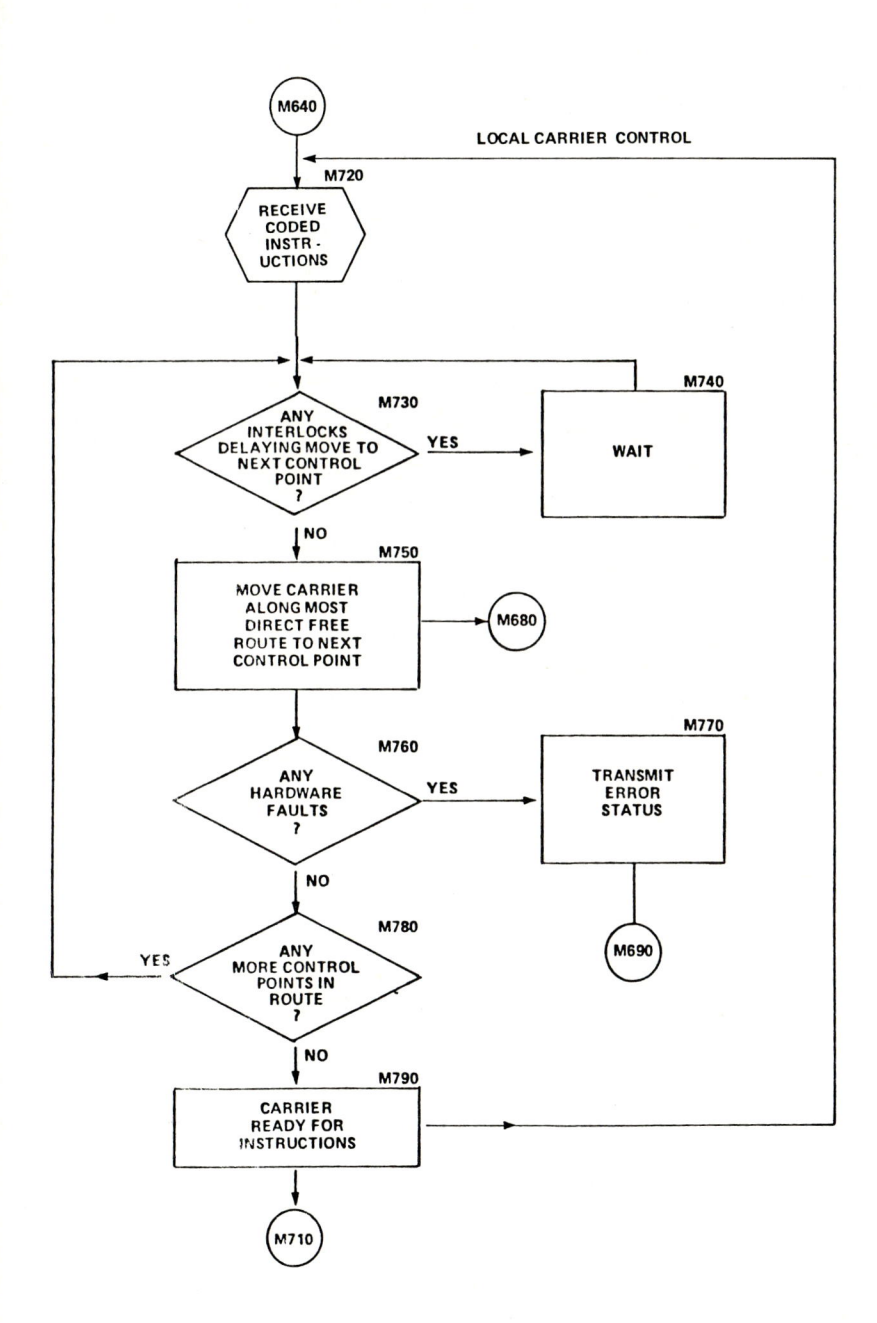

M2.3	**Design rules - delivery management**
M2.3.1	**Input data**

The sub-system must be capable of accepting movement instructions interactively.

The sub-system must be capable of accepting the following minimum input data:

- **Message Identifier**

    A code generated within the sub-system which is the source of message. The identifier must be an unambiguous reference to the message and its source.

- **Load Categories**

    Either a user defined code which specifies the range of physical characteristics of the load or actual physical characteristics of the load - size, weight, shape, etc. These data may be used to select between alternative docking and transfer routines, and to select a specific carrier if the physical system employs discrete mobile units with differing capacities, and to select a particular physical system (eg. monorail or AGV) as necessary.

- **Movement Type**

    A user-defined code that identifies what operation is to occur at the specified address - movement types will include the repositioning of carriers, the collection of loads, and the delivery of loads.

- **Address**

    Either a reference to a specific address or to a general area containing several specific addresses. Each address must be paired with a movement type datum.

- **Error Address**

    A reference to a specific point served by the transport system to which the carrier or load is sent if the physical system is prevented from completing the required movement.

M2.3.2	**Input re-formatting**

The sub-system must be able to reformat input data to meet its own requirements for data access.

The sub-system must be able to accept specification of re-formatting procedures.

Ideally the system will accept different re-formatting specifications which are peculiar to individual sources of input messages (it is assumed that all messages from a particular source will have consistent formats).

M2.3.3	**Multiple movements**

The sub-system must be able to receive input messages incorporating paired movements (eg. deliver a rough component to a machine and collect a finished component from that machine).

Ideally the sub-system will be able to receive input messages incorporating more than two individual movements.

Ideally the sub-system will accept input messages which include a datum specifying the number of individual movements incorporated in the message.

Ideally the sub-system will allow the input message to include either a single error address, or a specific error address for each individual movement.

M2.3.4 **Validation**

The sub-system must be able to validate input messages and to interactively issue acknowledgments and rejections to the sources of the input messages.

The sub-system must be able to include a user-defined diagnostic in rejection messages which identifies the reason for rejection.

M2.3.5 **System selection**

Ideally the sub-system will be able to manage several different physical transport systems. In this case the sub-system must be able to select a physical system in respect of an input message depending on the specified addresses and/or the specified load categories, (and if more than one physical system can meet all the requirements, the delivery management sub-system must be able to base a selection on the current status of the contending physical systems).

M2.3.6 **Carrier selection**

The sub-system must be able to select a mobile unit to execute a single or multiple movement.

Ideally the sub-system will be able to manage carriers within mixed characteristics in one physical transport system. The sub-system must then be able to select a carrier with the characteristics appropriate to the specified load category (or categories), and to the specified addresses.

M2.3.7 **Docking and load transfer selection**

The sub-system must be capable of initiating different docking and transfer procedures depending on the choice of transport system, the choice of carrier, the docking address, and the load category.

M2.3.8 **Final destination**

If the physical transport system includes discrete mobile units with on-board intelligence which are not in permanent communication the central co-ordinating intelligence, the delivery management sub-system must be able to append a final address to a movement instruction so that the mobile unit involved is able to communicate at the end of a movement sequence. (If the last delivery point specified in an input message in a communication point, no other address needs to be appended).

M2.3.9 **Error destination**

The sub-system must be able to generate an error address if one is not specified in an input message (trivially, the error address may be the same as the specified delivery address).

M2.3.10 **Route planning**

The system must be able to generate a complete route for a carrier or unmounted load from its current (resting) position, through the specified addresses in the correct sequence to the final destination.

Ideally the system will be aware of obstructions and congestion and will generate the optimum route in respect of the current situation.

M2.3.11 **Tracking**

The sub-system must be aware of what movements are in progress, and where in the physical system any idle carriers are waiting.

Ideally the sub-system will track carriers (or unmounted loads) past user-defined control points and thus be aware of the progress of current movements.

M2.3.12 **Navigation**

Mobile units with on-board intelligence must be able to navigate in accordance with the planned routes.

Ideally the on-board intelligence will be able to recognise temporary obstructions and make deviations from the planned route without altering its arrival sequence at the specified addresses.

M2.3.13 **Generalised destinations**

Ideally the delivery management sub-system will be able to accept input messages and plan routes to a general delivery address which encompasses several specific addresses. In this case the specific address will be determined when the carrier (or unmounted load) reaches the general area, normally by local and/or on-board intelligence.

M2.3.14 **Collision avoidance and contention for resources**

The sub-system must be able to resolve contention for use of resources during simultaneous movements, such as may occur at track junctions.

The sub-system must be able to prevent head-on or side-on collisions of carriers (or unmounted loads), normally this will be achieved by local and on-board intelligence controlling traffic at sidings and junctions.

The sub-system must be able to prevent or allow head to tail impacts of carriers (or unmounted loads) at the discretion of the system user.

M2.3.15 **Docking**

The sub-system must be able to initiate docking routines when a mobile unit (or unmounted load) reaches a specified address.

The sub-system must be able to confirm that docking has been accurately performed.

Ideally the sub-system can initiate and confirm alternative docking routines depending on the physical system, particular carrier, and load category.

M2.3.16 **Load transfers**

The sub-system must be able to determine whether or not it is safe to initiate a load transfer.

The sub-system must be able to interface with workcentre management sub-systems to initiate and monitor transfers.

The sub-system must be able to determine when it is safe to release a mobile unit after a transfer has occurred (or has been attempted but failed to occur).

Ideally the sub-system can initiate and monitor alternative transfer routines depending on the physical system, particular carrier, and load category.

M2.3.17 **Error detection**

The sub-system must be able to detect when a movement may not be completed without the risk of damage or injury.

The sub-system must be able to determine whether the error prevents the use of the error address.

The sub-system must be able to alert a higher authority whenever an error occurs.

The sub-system must allow the system user to specify what conditions constitute an error, and what conditions constitute an acceptable delay to the completion of a movement.

M2.3.18 **Output messages**

The sub-system must be able to send messages interactively to the source of an input message when the specified movement(s) is complete, or when the sub-system recognises that an error has prevented the completion of the movement within a user-defined time span.

The sub-system must be able to positively identify the original input message, the result of the movement (success or failure), and in the event of a failure it must also identify the individual movement and the cause or symptom of the failure.

280   CAM CAST sub-topics

Workcentre management sub-system (M3)

M3.1 Sub-system description

In order that we might examine the nature of the Workcentre Management function we must first define a workcentre. The working definition we have evolved is contained in the following three paragraphs.

A single machine routing obtains for every workpiece processed within a workcentre. For example, if we consider two similar machines being loaded by a robot, and the robot is placing parts at both machines as they become available, and they are performing identical operations, then they are considered to be two workcentres. If the robot is serving two dissimilar machines and placing parts at both machines, they are again considered to be two workcentres.

A workcentre includes those machines whose use, in a specific fixed sequence, is always implied by the selection of the first machine in the sequence, to perform an initial operation on a workpiece (eg. a transfer line). Using the simple robot and two machines model above, we can say that if there is no operational relationship between the two machines, they are two workcentres. If, however, the operation demands that the material pass through one machine, then the other, in a fixed sequence involving no other machine, and with no possibility of an alternative route between them, then they must be considered as a single workcentre.

A workcentre encompasses all necessary functions to bring about a required change of state. Thus, where fixturing/de-fixturing or other processes specific to a particular workcentre/item combination are necessary, these activities are considered to be a workcentre responsibility. If the process leaves the item in a state such that it may be generally applied then it need not be a general workcentre responsibility, though it may be a workcentre activity in its own right. For example, a machine might have its own fixturing requirements for any, or a specific, workpiece, then fixturing should be a responsibility of the workcentre to retain full flexibility. If, however, fixturing of the workpiece leaves it available to more than one machine, then it could form a workcentre activity in its own right, though it may be part of any workcentre in which it may be subsequently machined.

Over and above the definition, we may use the following characteristics to further identify the role of the workcentre in CAM.

The workcentre boundary commences at the end of the Delivery Management boundary, the interface between the two results in an exchange of information and, subject to item validation, the transfer of an item to the workcentre.

Workcentre management and Delivery management interface only upon instruction, this instruction will be generated by the resource scheduler, usually a computer, though it may be manual.

The workcentre contains no scheduling/next item selection intelligence.

A workcentre requires a combination of material(s), tool(s) and instructions, each capable of being uniquely recognised, to reside within its boundaries before it may do useful work. Within the context of this document, instructions must be deemed to be as comprehensive as NC programmes as a minimum requirement.

Upon command, the workcentre will progress the item into a workable condition and maintain location/offset from standard information and life expectancy as necessary.

Upon command from an external authority, the workcentre provides the motivation to cause the tool(s) to do work on the material(s) in accordance with the instruction set.

Upon completion of the instruction set, any activities necessary to enable transfer of material out of the workcentre boundary will be carried out.

Tools and instruction sets each merit separate consideration from material. Tools will wear during use and in time become exhausted, at this point they serve no useful purpose to the workcentre and should, application permitting, be readied for transfer from the workcentre.

Upon request the workcentre should ready any item for transfer to the delivery system, circumstances permitting. This will accommodate tool transfers, programme proving runs, defective materials and materials that contain broken tools.

The workcentre will notify the resource scheduler that an item is ready for collection before the resource scheduler will notify the workcentre that a carrier is due to collect the item.

Dependent upon application complexity the workcentre may have a single load/unload point, it may have separate load and unload stations, it may have buffer stations either side of the workstation or a single buffer either before or after the workstation. The buffer stations may serve as preparations areas for manipulation to workstation orientation/location requirements in the case of incoming items, or for removing the orientation/location media and preparing for the delivery system in the case of outgoing items. These load/unload point and buffer station permutations may apply irrespective of the item being progressed through the workcentre. The above is only intended to indicate the nature of a workcentre, it is not intended to be exclusive.

The workcentre must maintain various status. These will relate to material, tools and programmes within the workcentre boundary, and the workcentre and its services. These status will be application dependent, but should reflect for each workpiece, tool and programme, the location and relevant offsets; tool data should include the tool life. The workcentre should maintain an awareness of item residence, unused space and relative locations. It should further include all status to indicate its extent of serviceability, such as air suppliers, cutting fluid suppliers, power overloads, et al.

The following narrative and the associated network describes the logic to progress a single item through a workcentre. In practice there may be a multiplicity of events occurring simultaneously within and at the workcentre boundaries. It is conceivable that material, tool and programme exchanges might all be taking place concurrently, but independently, via their respective carriers, whilst the workstation is running a programme, cutting material with a tool; and whilst the workcentre is simultaneously preparing the next piece of material for the workstation, preparing another workpiece for despatch from the workcentre and also treating tools, singularly or in batches, in a similar manner. No two of these items need be similar. In practice, the desirability of any permutation of concurrent events will be determined by the projected profitability of the workcentre when satisfying its forecast workload. Thus, this compromise for the business should be seen as a Production Planning activity.

Production Management

Workcentre management sub-system

M3.2 Flowcharts

M3.2. SUB SYSTEM FLOWCHART

Sheet 1 of 3

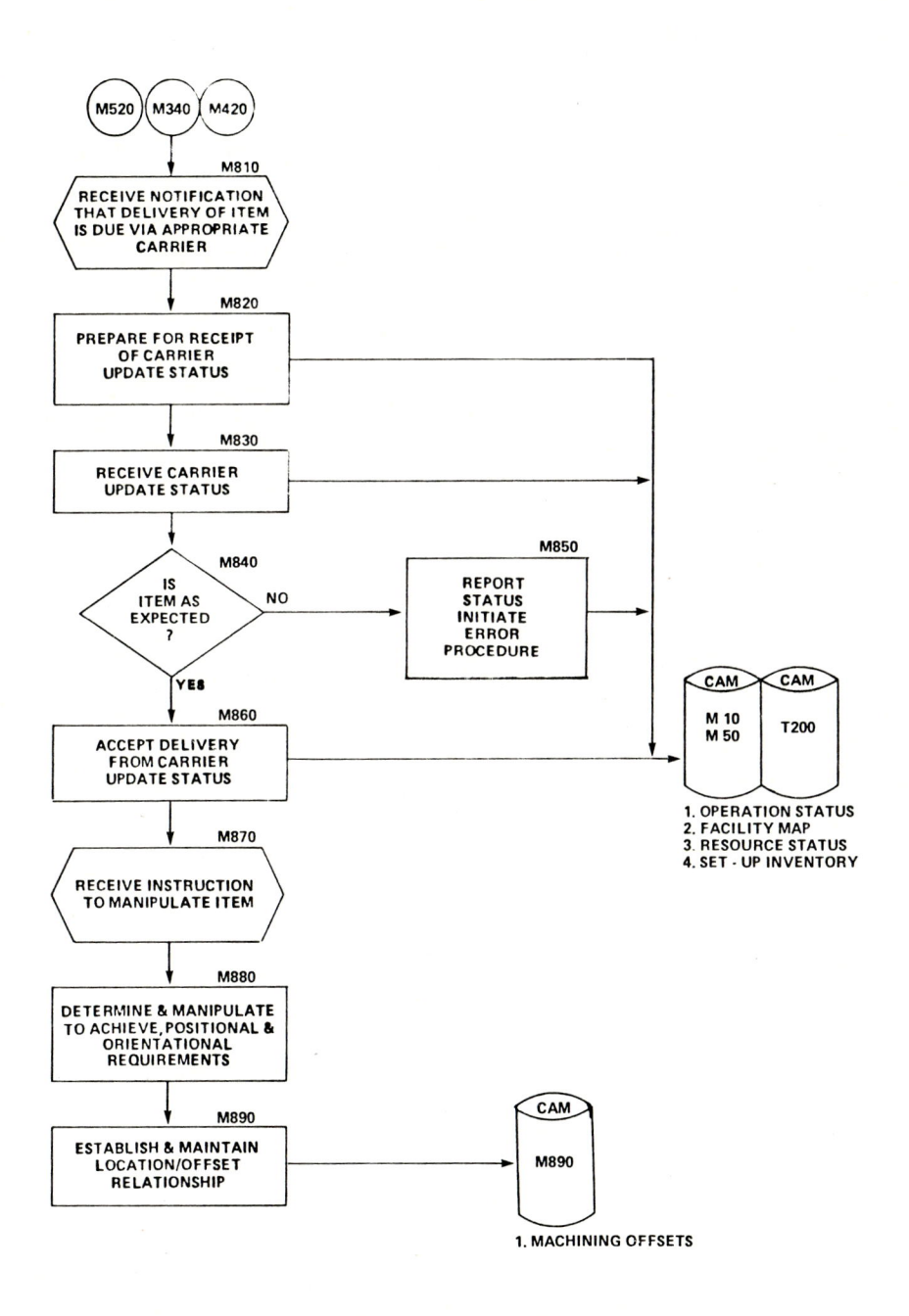

M520, M340, M420

M810 RECEIVE NOTIFICATION THAT DELIVERY OF ITEM IS DUE VIA APPROPRIATE CARRIER

M820 PREPARE FOR RECEIPT OF CARRIER UPDATE STATUS

M830 RECEIVE CARRIER UPDATE STATUS

M840 IS ITEM AS EXPECTED ?
- NO → **M850** REPORT STATUS INITIATE ERROR PROCEDURE
- YES ↓

M860 ACCEPT DELIVERY FROM CARRIER UPDATE STATUS

M870 RECEIVE INSTRUCTION TO MANIPULATE ITEM

M880 DETERMINE & MANIPULATE TO ACHIEVE, POSITIONAL & ORIENTATIONAL REQUIREMENTS

M890 ESTABLISH & MAINTAIN LOCATION/OFFSET RELATIONSHIP

CAM M10 M50, CAM T200
1. OPERATION STATUS
2. FACILITY MAP
3. RESOURCE STATUS
4. SET - UP INVENTORY

CAM M890
1. MACHINING OFFSETS

M3.2. SUB SYSTEM FLOWCHART

Sheet 2 of 3

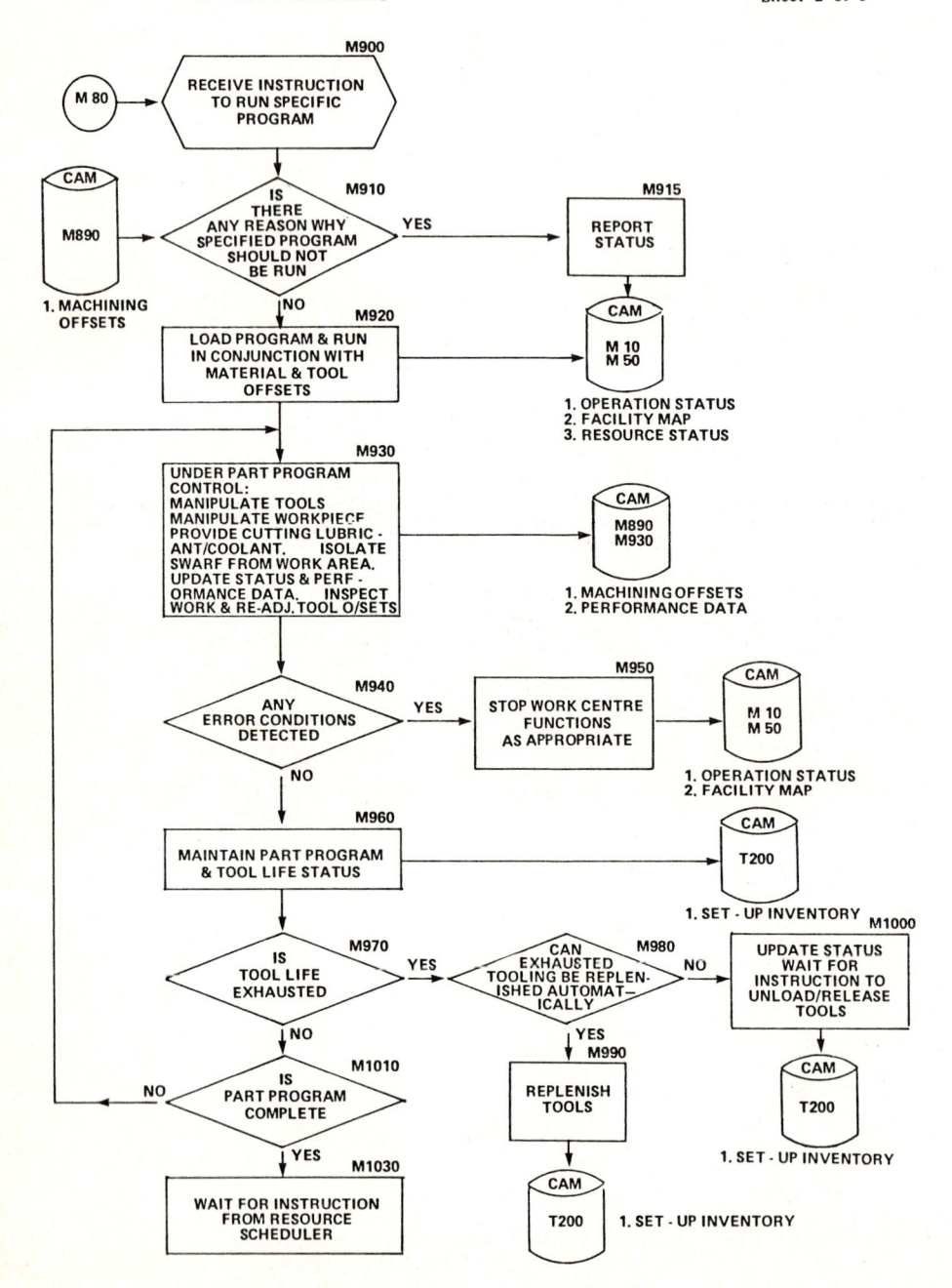

M80 → **M900** RECEIVE INSTRUCTION TO RUN SPECIFIC PROGRAM

CAM M890
1. MACHINING OFFSETS

M910 IS THERE ANY REASON WHY SPECIFIED PROGRAM SHOULD NOT BE RUN — YES → **M915** REPORT STATUS → **CAM M10 M50**
1. OPERATION STATUS
2. FACILITY MAP
3. RESOURCE STATUS

NO ↓

M920 LOAD PROGRAM & RUN IN CONJUNCTION WITH MATERIAL & TOOL OFFSETS

M930 UNDER PART PROGRAM CONTROL: MANIPULATE TOOLS, MANIPULATE WORKPIECE, PROVIDE CUTTING LUBRICANT/COOLANT, ISOLATE SWARF FROM WORK AREA, UPDATE STATUS & PERFORMANCE DATA, INSPECT WORK & RE-ADJ. TOOL O/SETS

CAM M890 M930
1. MACHINING OFFSETS
2. PERFORMANCE DATA

M940 ANY ERROR CONDITIONS DETECTED — YES → **M950** STOP WORK CENTRE FUNCTIONS AS APPROPRIATE → **CAM M10 M50**
1. OPERATION STATUS
2. FACILITY MAP

NO ↓

M960 MAINTAIN PART PROGRAM & TOOL LIFE STATUS → **CAM T200**
1. SET-UP INVENTORY

M970 IS TOOL LIFE EXHAUSTED — YES → **M980** CAN EXHAUSTED TOOLING BE REPLENISHED AUTOMATICALLY — NO → **M1000** UPDATE STATUS WAIT FOR INSTRUCTION TO UNLOAD/RELEASE TOOLS → **CAM T200**
1. SET-UP INVENTORY

YES ↓ **M990** REPLENISH TOOLS → **CAM T200**
1. SET-UP INVENTORY

NO ↓ **M1010** IS PART PROGRAM COMPLETE — NO → (loop back)

YES ↓ **M1030** WAIT FOR INSTRUCTION FROM RESOURCE SCHEDULER

M3.2. SUB SYSTEM FLOWCHART

Sheet 3 of 3

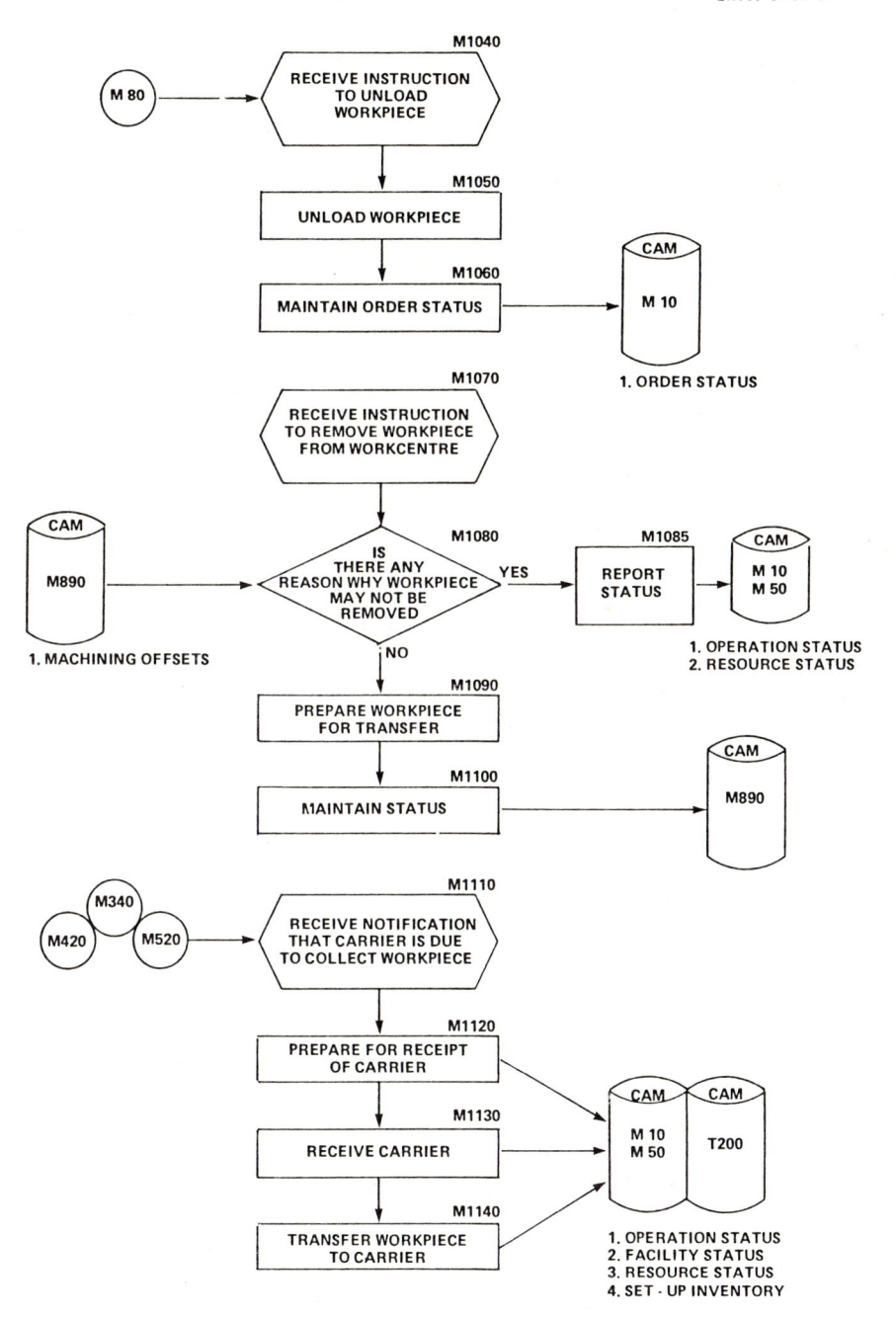

M1040 RECEIVE INSTRUCTION TO UNLOAD WORKPIECE ← M 80

M1050 UNLOAD WORKPIECE

M1060 MAINTAIN ORDER STATUS → CAM M 10
1. ORDER STATUS

M1070 RECEIVE INSTRUCTION TO REMOVE WORKPIECE FROM WORKCENTRE

CAM M890
1. MACHINING OFFSETS
→ **M1080** IS THERE ANY REASON WHY WORKPIECE MAY NOT BE REMOVED

YES → **M1085** REPORT STATUS → CAM M 10 M 50
1. OPERATION STATUS
2. RESOURCE STATUS

NO ↓

M1090 PREPARE WORKPIECE FOR TRANSFER

M1100 MAINTAIN STATUS → CAM M890

M340, M420, M520 → **M1110** RECEIVE NOTIFICATION THAT CARRIER IS DUE TO COLLECT WORKPIECE

M1120 PREPARE FOR RECEIPT OF CARRIER

M1130 RECEIVE CARRIER → CAM M 10 M 50 | CAM T200

M1140 TRANSFER WORKPIECE TO CARRIER →
1. OPERATION STATUS
2. FACILITY STATUS
3. RESOURCE STATUS
4. SET - UP INVENTORY

M3.3 **Design rules - workcentre management**

M3.3.1 **Input data**

The sub-system must be able to accept input messages interactively.

The sub-system must be able to accept the following minimum input data:

- **Message Identity**

    An unambiguous reference to the message which is generated by the source of the message.

- **Task Type**

    A user-defined code which identifies the action required of the sub-system, and defines the interpretation of the remaining data. The types of task which may be despatched to a workcentre management sub-system include:

    - Prepare for delivery of a workpiece (or batch).
    - Prepare for collection of a workpiece (or batch).
    - Fixture component(s).
    - De-fixture component(s).
    - Load component(s) to machine.
    - Unload component(s) from machine.
    - Prepare for delivery of part-programme.
    - Send amended part-programme.
    - Prepare to accept tool offset data.
    - Prepare for delivery of new tools or fixture.
    - Despatch worn/unwanted tools or fixtures.
    - Change content of tool magazine.
    - Begin machining.
    - Change chuck jaws.
    - Inspect component(s).

- **Task-Specific Data**

    This will vary according to the type of task. The following data will each be required for some of the above tasks:

    - **Load Identity**

        An unambiguous reference to a specific workpiece or a specific batch of work.

    - **Load Category**

        Either a user-defined code which specifies the range of physical characteristics of the load, or the actual physical characteristics independently specified (size, weight, shape, etc.).

    - **Address**

        An unambiguous (within the workcentre) reference to a specific location to which a load will be moved, or from which a load will be removed, or to a group of specific locations to one of which a load will be moved.

- **Process Identity**

  An unambiguous reference to manual or mechanical procedure for which a detailed specification exists.

- **Process Instruction Type**

  A user-defined code which identifies the type of specification and medium on which it is recorded. This datum may either be stated explicitly or the sub-system must be able to infer it from the task type.

- **Process Instructions**

  A file on computer storage media, or hard copy which may be in the form of a computer programme, test instructions or diagrams.

- **Resource Set Identity**

  An unambiguous reference to a specific combination of tools or fixtures.

- **Resource Identity**

  An unambiguous reference to a specific tool, fixture, pallet, etc.

- **Resource Category**

  Either a user-defined code that specifies a combination of characteristics, or the actual physical characteristics independently specified (eg. length, diameter, taper, hardness, etc.).

- **Offset**

  A value which specifies the variation between an actual dimension and a synthetic value in respect of tool setting or component fixturing.

- **Tool Life Expectancy**

  A value which expresses the remaining use that may be made of a tool before it requires re-sharpening.

- **Tool Life Commitment**

  A value that expresses the planned use of a particular tool on a particular job. This datum must be expressed in the same units as the previous datum (contact time, no. of surface breaks, volume of material removed, etc.).

- **Facility Identity**

  An unambiguous (within the workcentre) reference to the static facility on which or by which particular process us to be performed (eg. fixturing station, machine tool, robot, etc.).

M3.3.2   **Input re-formatting**

The sub-system must be able to re-format input data to meet its own requirements for data access.

The sub-system must be able to accept specification of re-formatting procedures.

Ideally the sub-system will accept alternative re-formatting specifications which are peculiar to particular input message sources (it is assumed that all messages of the same type from a particular source will have consistent formats).

M3.3.3   **Validation**

The sub-system must be able to validate input messages, and to interactively issue acknowledgments and rejections to the source of the input messages.

The sub-system must be able to include a user-defined diagnostic in a rejection message which identifies the reason for rejection.

M3.3.4   **Achievement/failure reporting**

The sub-system must be able to send a message interactively to the source of an instruction to confirm the successful execution of that instruction, or to report any failure to complete the task.

The sub-system must be able to include positive identifications of the input message and (if relevant) the cause or symptom of any failure.

M3.3.5   **Deliveries and collections**

The sub-system must be able to recognise when a transport system is ready to make a delivery to or a collection from a particular location within the workcentre and must be able to permit or prevent a load transfer according to whether or not the operation may be safely performed.

Ideally a location may be served by more than one physical transport system, and more than one type of carrier, in this case the workcentre management sub-system must be able to recognise the system and carrier utilised for each movement, and must be able to initiate the appropriate load transfer procedure.

M3.3.6   **Load verification**

The sub-system must be able to accept confirmation of load identity before initiating a load transfer to within the sub-system boundary.

Ideally the sub-system will be able to accept confirmation of load characteristics (batch size, orientation, condition of workpiece(s), etc.).

The sub-system must be able to permit or prevent the load transfer depending on the results of the load verification.

M3.3.7   **Load transfers**

The sub-system must be able to recognise when a transfer has been successfully completed and to authorise the release of the carrier.

The sub-system must be able to recognise when a transfer has not completed successfully and must be able to report the failure interactively to a user-specified sub-system.

M3.3.8   **Delivery notification**

The sub-system must be able to accept notification of a planned delivery or collection.

The sub-system must be able to report unnotified arrivals interactively to a user-specified sub-system.

M3.3.9   **Component fixturing**

The sub-system must be able to accept instructions to fixture or de-fixture components.

The sub-system must be able to accept a fixturing specification reference, ideally the sub-system will accept full fixturing specifications including offsets, and fixture requirements.

The sub-system must be able to report interactively to the source of the fixturing instruction message that a specific fixturing operation is complete.

The sub-system must be able to report to the source of the fixturing instruction message that a specific fixturing operation cannot be completed, and must be able to quote a user-defined diagnostic that identifies the reason for the failure.

M3.3.10 **Part-programmes**

The sub-system must be able to accept post-processed part-programmes.

The sub-system must be able to accept an instruction to load a specific part-programme to a specific machine controller.

Ideally the sub-system will be able to post-process a part-programme to a form suitable for the appropriate machine controller within the workcentre.

M3.3.11 **Programme proving**

Ideally the sub-system must be able to accept a user-defined code which indicates that a particular execution of a part-programme is a proving run which should be stepped through under manual control.

Ideally the sub-system will accept manually entered amendments to a part-programme and will produce a correspondingly modified version of the programme.

Ideally the sub-system will be able to transmit a modified programme to a user-determined sub-system.

M3.3.12 **Loading machines with work**

The sub-system must be able to accept instructions to load specified work-pieces to specified machines.

The sub-system must be able to accept instructions to unload specified work-pieces from specified machines.

The sub-system must be able to initiate the loading and unloading of work-pieces to and from machines within the workcentre.

The sub-system must be able to recognise when a load or unload operation has been successfully completed.

The sub-system must recognise when a load or unload command cannot be completed.

The sub-system must be able to report interactively to the source of the instruction the successful completion of a load or unload operation.

The sub-system must be able to report interactively to the source of the instruction the failure to complete a load or unload operation, and must be able to quote a user-specified diagnostic which identifies the cause or symptom of the failure.

M3.3.13 **Loading machines with tools**

The sub-system must be able to accept instructions to load a specifically identified tool into a specific location in a machine tool magazine.

The sub-system must be able to acknowledge tool load instructions if the specified tools are available, and to reject tool load instructions if the specified tools are not within the workcentre boundary.

The sub-system must be able to accept instructions to remove a specified tool from a specified location in a machine tool magazine.

The sub-system must be able to acknowledge tool unload instructions which can be executed, the sub-system must be able to reject tool unload instructions when the specified tool is not in the specified magazine location, or when the tool magazine cannot be accessed.

The sub-system must be able to send a message interactively to the source of the tool load/unload instruction to report the success or failure of the task.

M3.3.14 **Programme initiation**

The sub-system must be able to accept an instruction to commence a particular programme on a particular facility.

The sub-system must be able to acknowledge programme commencement instructions if all the required resources and services are available.

The sub-system must be able to reject a programme commencement instruction if any of the required services or the specified programmes are not available to the specified facility.

Ideally the sub-system will reject the instruction if the appropriate workpiece (properly fixtured) and the required tools (with sufficient life expectancy) are not resident on the specified facility.

M3.3.15 **Resource identification**

Ideally the sub-system will be able to recognise individual tools, fixture, pallets, etc.

Ideally the sub-system will be able to associate temporary attributes (such as tool offsets) with individual resources.

M3.3.16 **Tool life monitoring**

The sub-system must be able to prevent the selection of an exhausted tool for a new operation within a machining programme.

Ideally the sub-system will prevent the selection for a new operation of a tool whose remaining life expectancy is less than a synthetic tool life commitment value associated with the new operation.

M3.3.17 **Error reporting**

The sub-system must be able to detect any unauthorised activity which could interfere with the workcentres ability to execute authorised tasks.

The sub-system must be able to send a message interactively to a user-specified sub-system to report an unauthorised activity, the sub-system must be able to include the location and the nature of the unauthorised activity within the message contents.

NB: Unauthorised activities include failures of the control/management sub-systems such as an unnotified arrival of a carrier with a workpiece at an address, and mechanical failures such as the loss of power or coolant at a machine tool.

Chapter 8

General Interface Rules

8.1. Background

For Computer Integrated Manufacturing to be a reality, the sub-systems identified in the preceding sections must be integrated as shown in the overall Flowcharts. Therefore, each sub-system has effectively an interface to another sub-system. This interface performs the function of transferring information from one sub-system to another, or of initiating subsequent sub-systems. It effectively is the link between sub-systems that enables them to act in a coherent way and allows the flow of processing in the manufacturing environment to continue.

Rather than specifically describing the interface between each set of interacting sub-systems, it is possible to identify general interface principles that hold true for all sub-system to sub-system interfaces. Specific areas that apply to interfacing between individual sub-systems are dealt with in the description of the relevant sub-systems themselves. Therefore, if sub-system A passes control to sub-system B then the output from sub-system A will be identified in terms of the functional contents and the frequency of issuance within sub-system A design rules, and likewise the input requirement of sub-system B will be identified within that sub-system's design rules definitions. Therefore, by a combination of general interface principles and specific sub-system input or output requirements, it is possible to describe the interface for any sub-system to any other sub-system.

8.2. General principles

In the majority of cases normal process flow will be in a single threading operation where sub-system A leads to sub-system B, and then to sub-system C, etc. Occasionally however we will have a situation where sub-system A, B and C may be considered as parallel processes leading to sub-system D.

Irrespective of the flow of process we believe that there are three general principles of interface between any two sub-systems. It is true that at any one time there is only one transmitter and one receiver in any sub-system to sub-system pair and that, regardless of whether sub-system A, B and C interface to sub-system D, at any one point in time, only one of the sub-systems will interface to another sub-system.

Principle 1 is where the information from the first sub-system is transmitted to the second sub-system in 'batch mode'. This is where all the information is collected, gathered up and stored in the first sub-system and then transmitted 'en block' to the second sub-system.

The second principle is where the information from the first sub-system is again transmitted to the second sub-system in batch mode but, over a lengthy period of time. This we have termed 'trickle feed' principle. The trickle feed principle is where the data is fed bit by bit from the first sub-system to the second sub-system. The only response from the second sub-system probably will be an acknowledgment that it has received the information or an acknowledgment when error conditions occur, ie. when the information is corrupted.

The third principle of interfacing is that of interactive interfacing. This is effectively conversation mode where information is passed between the two sub-systems in true conversation format where sub-system A will pass information to sub-system B; sub-system B will respond, and vice-versa. This is very typical of interactive computer systems where enquiries on file information are being undertaken.

Having identified the three general principles in which information can be exchanged between sub-systems, it is useful to see how this can be interpreted. When two sub-systems, sub-system A and B need to interface we have already stated that sub-system A will define its output that it is going to create in terms of the contents of that output, and to some extent the frequency of that output, and the actions that will be taken on issuance of that output. Sub-system B will identify its input requirements, (ie. effectively that output from sub-system A) and will identify again the format, the frequency and the actions that are to be taken upon receiving that information. By associating a general principle of interfacing (this may be the batch, trickle or interactive principle) we can identify the way in which the data is to be passed between the two sub-systems, and hence the full interface rules for that sub-system to sub-system interaction.

However, if sub-system A is designed and supplied by one computer supplier, and sub-system B is created by another computer supplier, there could well be compliance with the design rules in terms of format frequency and general interface principle. However, the actual format of the data itself may be entirely different, as could the physical media used to transmit that information. It is the responsibility of the overall communications strategy and indeed the communications strategy and infrastructure of the company in question as to how the information is transmitted in physical terms. We believe that in order to gain some conformity of interfacing it is necessary to transmit the information from one sub-system to another sub-system in a standard format and under some standard protocol.

We therefore believe that, irrespective of how the data is created within a sub-system, or how it is used after being accepted by another sub-system, it is important that it be transmitted from sub-system to sub-system in ASCII format and under the HDLC protocol. It is therefore the responsibility of the issuing sub-system to create the data in this format and protocol before transmission and not the responsibility of the receiving sub-system to do anything with the format. The receiving sub-system will always receive information in the standard format and therefore is at liberty to change that format if desired. We have arbitrarily chosen this format and protocol because we believe it to be the most widely used and also the easiest for the majority of manufacturers to comply with.

Chapter 9

Development of Strategies

Computer Integrated Manufacturing (CIM), as the name implies, is the art and science of automation by the integration of Information Technology (IT) into the manufacturing process. Its beginnings can be traced to the demonstration of the first NC-machine in 1952 at MIT, developed under a USAF contract. Soon thereafter USAF launched the ICAM (Integrated Computer Aided Manufacturing) program to rationalize the burgeoning problem of parts supply and production, using the rudimentary computing facilities then available. In due course as IT developed, CIM emerged as an important activity in its own right. Transfer of IT to CIM has, however, occurred largely in an ad hoc manner without established standards or even the accepted norms of good practice. The absence of such rules or guidelines has seriously impeded CIM. This has allowed IT vendors to produce a bewildering array of products (modules) which suffer from two serious drawbacks. In the first place, IT modules from different vendors are difficult, if not impossible, to integrate with each other. Secondly, IT modules are seldom optimized for CIM needs and often suffer from under- or over-capacity.

This is further compounded by the fact that there does not exist a standard modularization of CIM in the context of IT goals. To address these problems and thereby facilitate IT integration into CIM, two concurrent approaches.
have been adopted in this study.

In the first approach, CIM is divided into modules and sub-modules. Each module (or sub-module) can be identified as a discrete CIM-activity of a generic nature, for example machine selection is a sub-module of a larger module CAM. The functional scope of these modules is analysed in terms of its input and output. For example, a machine selection module needs such input as size, shape and mass of the material to be machined, as also the machining characteristics such as grooves, holes, finish etc. The output of the machine selection module would then be the identity of the machine appropriate for such a job. The analysis of the functional scope of each module is carried out in a sufficiently general manner to ensure that the I/O specifications so elicited are of universal validity and not typical for a particular CIM task. A given CIM module is thus defined by a set of design rules, which in turn, determine the I/O specifications of the modules. It is hoped that the design rules so derived will be applicable to all the CIM modules that purport to fulfil a certain generic task, for example, machine selection modules for turning and milling share the same design rules. A modularization of CIM and the development of corresponding design rules has already been carried out in the preceding chapters.

The second approach concentrates on three IT activities of the most general nature, namely data, processing and communication. The role of data in CIM refers to such factors as data integrity, validity, authorship etc.. For processing, one would like to know what sort of processors one needs and how to distribute processing among them, for example should there be a single mainframe which processes the various needs in a time-sharing mode, or should there be a number of small dedicated processors for local processing? Similarly for communication one would like to know how to organize communication among various CIM-modules in an optimal manner, for example should there be a hierarchical star configuration, a bus oriented network etc.?

An attempt to answer such questions has been made in the following chapters. The overall role of data-activity in CIM can be stated in terms of the following query:

To automate a given manufacturing system, what sort of data management, processing units and communication organization are required and how best to implement them?

A complete answer to this, for a specific manufacturing task, would require a highly 'mission-oriented' detailed analysis on a case by case basis. Since such an approach lacks general validity, it is of little value as a methodology and is thus not attempted here. A more desirable alternative would be to develop a 'strategy' of wider validity. Analogous strategies can be found in games where a set of 'rules of success'; as distinct from 'rules of game'; collectively form a strategy and guide one in choosing the next move.

As an illustration, in a game like chess we first have a set of rules to define allowed moves. However, just adherence to these rules does not guarantee a victory, which depends on the ability of the players to apply them correctly in a given situation. To win a game one needs to develop a strategy. This strategy, in turn, may itself comprise of general guidelines, rules of good housekeeping, precepts of defensive game, precepts of offensive game etc. Formulation of such game strategies requires a deep understanding of chess, honed with long experience.

In a restricted sense, such strategies can also be likened to an expert system. Though the development of such a strategy or an expert system for CIM would be highly desirable, this task, however, still requires a lot of fundamental research and is beyond the scope of this study.

As a first step to reduce the complexity of CIM, it might be desirable to seek the general guidelines or rules of good housekeeping for data-activity in CIM. It may be noted here that data-activity permeats the entire CIM and therefore the guidelines sought above are not for a particular CIM module. On the contrary, these guidelines should address the role of data-activity in CIM as a whole. In keeping with this spirit of module-independent data-activity, we shall develop guidelines or rules of good housekeeping of general validity which, if followed, would greatly aid in optimizing data-activity for various manufacturing tasks.

Rules for data-activity will be developed in the form of maxims that are applicable to data-activity in all of CIM. A maxim specifies, in general, the optimal or desired form of an data-activity in CIM. There will be data maxims, processing maxims and communication maxims, which will indicate, in a general way, the direction one should follow to arrive at an optimal design of these activities in CIM. The set of maxims would be collectively called a strategy.

The design of sub-systems as obtained by applying general maxims may be further optimized by applying more specific maxims that have a restricted validity. The justification of first using the general maxims is that adhering to a general strategy results in a reduction of the size of the solution space. This size can be further reduced by applying the specific maxims, valid in that space. Trying to find the solution in one step by applying both general maxims and specific maxims may effectively reduce the size of the solution space to such an amount, that it becomes very difficult (if not completely impossible) to find a solution. Furthermore, it will be almost impossible to develop specific maxims if the solution space is not reduced before, by applying general maxims. In the following chapters, we have restricted ourselves to general maxims, because they are considered to be most important as a first step towards a complete CIM strategy.

To develop maxims, and thereby the respective strategies, certain general attributes of maxims should be kept in sight. These include such generalities as:

- The range of validity of a maxim should be as wide as possible, independent of the particular CIM activity at hand. This of course does not rule out the fact that some maxims are more applicable to a particular CIM activity than to others. Thus certain maxims pertaining to numerical data processing would be more applicable in design activity than in production.

- In optimizing the performance of a system as complex as CIM, quite often, contradictory demands are made and a trade off between the contending factors becomes inevitable. This may sometimes be reflected in the apparently contradictory directions indicated by two maxims. For example, to increase the communication speed a maxim may advise the simplification of the network, whereas, to increase the reliability of communication, another maxim may suggest a more complex structure. Needless to say that under such circumstances a trade off, rather than a conflict is implied.
- The number of maxims, comprising a strategy, should be large enough to make the strategy comprehensive in its scope. On the other hand the number of maxims should not be so large as to render the strategy unwieldy.
- Several maxims may imply a certain redundancy. This redundancy can, for instance, be caused by the fact that a strategy needs to view a problem from different, yet not totally independent, vantage points. Thus, distribution of processing will have be considered from both points of view of local processing to reduce network load, and distributed processing to improve reliability and flexibility. An attempt has been made in this study to keep such redundancy to the minimum necessary.

In the next chapter maxims for data in CIM have been developed using the above methodology. Collectively these maxims form a data strategy for CIM.

Maxims, being statements of general validity, are somewhat like 'heuristic-rules' distilled from a substantial experience and appreciation of the field. To impart such an appreciation for processing and communication, we shall review the state of the art of these activities and their role in CIM. Based on this pool of knowledge and experience, maxims for these will be derived. These maxims will then collectively form the processing strategy and the communication strategy.

In chapter 11, the state of the art in processing and its role in CIM will be discussed. In CIM, processing plays a crucial and varied role, involving elements of both software and hardware at various levels of detail and sophistication. On the one hand there may be a need for general purpose number crunching hardware/software for finite element analysis in CAD. On the other hand one may need a highly dedicated servo processor for real time control of a robot arm. The processors for these two needs are of different design and possess different characteristics and the CIM designer will have to appreciate such differences. The idea here is to review the current knowledge and experience in processing and put it in the general CIM context. Here we shall introduce the major concepts and trends in processing and CIM and examine the interrelationship between them.

In chapter 12, the processing maxims will be developed. These maxims can be categorized into different groups, there are maxims about distribution of processing, maxims about real time aspects of processing, about fault management etc. Each maxim will be accompanied by a narrative to elucidate its role in CIM. The narrative accompanying each maxim will provide some background information and its relevance to the CIM activity. Now and then, an example may be offered to illustrate a point further.

In chapter 13, the communication state of the art and its role in CIM is discussed. In CIM, communication plays a crucial role. Many of the serious problems of CIM can be traced to communication bottlenecks. The communication needs in CIM are diverse and exacting and demand interfacing among different levels of activity. On the one hand an FMS cell should be able to communicate with an off-line CAD workstation for job design update, on the other hand it should be able to maintain synchronization among the various components of the cell, as also support a real-time dialogue with the sensors. Such diverse chores cannot be attended by an arbitrary communication network, and much attention should be paid to the optimization of communication. There are several internationally accepted standards. However, their implementation in a CIM environment leaves much room for arbitrariness and does not guarantee compatible interfacing of components. The relationship between the current communication knowledge and the CIM demands will be examined in this chapter. This should aid in the subsequent formulation of the maxims for communication.

In chapter 14 communication maxims, along with the narratives, are presented. These maxims refer to aspects of communication that play an important role in CIM. This includes such features as network security, reliability, protocols, hardware organization, maintenance etc. With these maxims serving as guideposts, an optimum configuration of communication for a given CIM environment can be arrived at.

To illustrate the utility of the strategies developed above we shall consider their application to some high level CIM activities. In principle one can choose the design of any CIM module and apply the strategies to it for optimization or streamlining. Within reasonable limits, the strategies developed here should perform this task for all CIM activities. To highlight the versatility, as also the limitations of the strategies, we have chosen sensors and graphics to which strategies will be applied. These two activities are both important to CIM and quite general in nature. At the same time they are very different in their functionality, so they offer a wide range over which the efficacy of the strategies can be tested.

In chapter 15, the processing strategy and communication strategy will be applied to sensor processing. Purely mechanical, or brute force, aspects of manual labour have already been largely replaced by machines. The key element of human labour contributing to the current industrial process is, not the human muscle power, but the keen sensory perception with which human beings are so well endowed. In view of this, one can safely say that the success of industrial automation will, to a large degree, depend on the success of sensor integration into CIM. In principle, in the industrial process wherever human intervention is needed for quality control, contingency management, risk avoidance, variability accommodation etc., an appropriate sensor with the necessary processor could replace the human operator. The maxims developed in the preceding chapters should help one in integrating such sensors and processors into CIM.

In chapter 16, after considering some salient features of graphics in CIM, the utility of maxims in optimizing graphics will be illustrated. Since the design of a product is the first step in the manufacturing process, due care must be exercised in this phase. Any errors or omissions allowed at this stage will propagate through the entire manufacturing process and could lead to undesirable results at the most unexpected places. In the design stage, heavy demands are made on human judgement, experience and knowledge. In the foreseeable future it does not appear likely that automation will fully eliminate the human element from the design process. However, much of the tedium associated with going-back-to-the-drawing-board can be eliminated by the powerful graphics techniques available now. A graphics facility, optimized by the strategies presented here, would allow the design engineer to exercise more design creativity as also to readily accept feedback from the shop-floor for design modification. The graphics techniques, available in current CAD, are varied and fast evolving. The assistance provided by the strategies as presented here, in optimizing graphics facilities, should prove valuable.

Chapter 10

Data strategy

Introduction

This strategy is designed to address a range of topics related to the management of one of the most important resources which a manufacturing organisation possesses - data.

The strategy is entirely concerned with the maxims which need to be observed in the management of data. It is not concerned with technological methodologies for storing and retrieving data - except in so far as identifying certain functional requirements which all CIM file or data base management systems must be designed to include provision for.

It is however necessary to explain why the proposals recommend a data strategy, rather than a data base strategy.

The detailed physical design of a computerised data (base) file comprises a very large number of different factors. These may include, but are by no means confined to, choice of basic access method; choice of indexing method; choice of method for the management of free space within the file; choice of different chaining methods for a variety of different purposes, choice of data formatting method, field lengths, field attributes (eg. may the field contain alphabetic characters, or may only numeric characters be used) - and many other factors for even a single elementary type of file. In addition, there are very many fundamentally different classes of file, which differ not merely in respect to matters of detail, but at a much higher and more philosophical level. Thus a Relational Data Base, (an ideal Data Base for information retrieval); a Knowledge Base, (required to support artificial intelligence based systems such as Expert Systems); a Codasyl type Data Base, (ideal for the maintenance of large volumes of data which are subject to volatile change and growth), represent but three of very many fundamentally different classes of data base which involve not just different decisions concerning a particular factor such as 'indexing' method - but entirely different factors.

For example, 'three value predicate logic with a single meaning for null' would apply to a Relational Data Base - but not to any of the other two types of data base mentioned; 'Frames, Scripts, Blackboards and Semantic Nets' would be appropriate to an Expert System Knowledge Base - but would be meaningless in the context of Relational or Codasyl type data bases.

The detailed design of the computer file required by a particular system represents a very major part of the system design activity for that system. File design is however even more important to the shape and nature of the overall system than it is to the design activity itself.

Two competing vendors may, and invariably will, wish to adopt very different approaches in providing a systems product to support a particular CIM activity - which in turn will condition the basic type of file they will need to employ.

For example, Vendor 1 may wish to develop a conventional 'algorithmic' based system, using traditional imperative computer programme coding techniques, to provide say a system for determining which machining processes are to be used in converting a steel billet into a finished part. In this case, the vendor would (from the three types already mentioned) choose a Codasyl type data base file.

A second vendor may wish to develop a competitive product to address the same activity, but to do so by the use of knowledge representation and inferencing, by means of declarative type computer programming. In this case the vendor would be obliged to use a 'Knowledge Base' type data base.

Remembering that any total CIM system will be a composite system, incorporating different system products from different IT vendors, it is inconceivable that there could be any single Data Base, or even any single collection of different types of data base, which could be introduced as an entity in itself - as a CIM data base.

Some of the world's largest and most technologically advanced IT vendors, have devoted huge resources and funds to the development of even a single data base management system - but there is still not a satisfactory distributed data base management system, even for distributed files of the same fundamental class. The problems involved in trying to construct a logical file (eg. to enable a single programme to retrieve items of data from many different files) from a multitude of files of many fundamentally different classes, are more difficult by many many orders of magnitude.

This refutation of a currently realisable CIM data base must not however be taken to imply that systems should not share data, and even physical files, wherever and whenever it is possible and optimal for them to do so. This will indeed be both possible and desirable in many cases. This is however very different from advocating that all CIM systems commonly share data and files, and that a strategy could be developed to enable this to be done.

Nothing is impossible of course, given infinite time and resources, and it may be that at some future time technology will have advanced to the point that a CIM data base becomes a realisable possibility. It will at that time be necessary to re-write CIM design rules.

The six basic CIM data strategy maxims described in this document are by no means comprehensive; many other maxims which are universally applicable across a very wide range of different areas outside manufacturing, apply equally to manufacturing.

Good provisions for the securing and recovery of computer files in respect to accidental damage or corruption, and good protection against unauthorised access to confidential data would represent two such universally applicable maxims.

It is not however the purpose of this document to teach basic Information Technology skills and expertise, but rather to highlight practices and disciplines which are essential to the integration of the many departments and disciplines which normally exist in any manufacturing organisation.

These maxims would not normally be found in conventional text books on data management - but are key to the realisation of CIM.

Many standard courses and text books are available for those wishing to obtain more general knowledge of the subject of data and computer file management and design.

Foreword to DS illustrations

In describing the various data strategy maxims considerable use has been made of sample applications, for the purpose of illustration and interpretation.

Wherever it has been appropriate to do so, a maxim has been illustrated by reference to a hypothetical 'stock recording' application.

Stock recording (as distinct from Inventory Management) has been selected to illustrate maxims principally because this subject, unlike many CIM topics, does not require special knowledge and training.

It must however be stressed that all maxims apply equally to all application topics - including all design and production engineering subjects and all administration and accounting subjects also.

Data strategy maxim 1

The principle of prime authorship

Maxim

It is a fundamental principle of Computer Integrated Manufacturing, that all departments and functions of any manufacturing organisation use that value for any item of data which is created and maintained by the single person or function which is officially authorised to supply that data.

Description

In a non-integrated operation, it is not unusual to find that a single item of data is separately maintained on several different computer files.

Thus, the product structure (or Bill of Material) used by say the Management Accounting Department in the calculation of Product Costs, differs from that used by say the Production Control department of the same factory, for the scheduling of production or external suppliers.

Although both of the above files may be maintained in reference to the same source document (issued by say the Product Engineering department) it is not unusual to find little correlation between the information contained on the two files. This renders cross-communication between the two departments difficult if not impossible - and leads to major inconsistency in the actions taken by the various departments of the company.

A second example will be considered in detail in order to illustrate some of the more important aspects of this maxim. The example to be studied will be that of Production Forecasts.

Within an imaginary company, the Production Engineering Director lays down production facilities to support a forecast production level of a); the Finance Director calculates a profit forecast for the company based upon a production forecast of b); the Personnel department authorises a labour force to support a production forecast of c); the Purchasing department negotiates purchase prices on a production forecast of d); the Material Control department schedules materials against a production forecast of e), etc., etc.

Each of these forecasts are made by the staff of the department concerned - and though these may be 'based' upon an official forecast produced by say the Sales and Marketing department, each department 'discounts' this official forecast in line with the 'view' of the department in question.

The consequence of these actions is the opposite of integration - and each department could better be described as an independent organisation, than as being a fully integrated part of a single organisation.

We might consider a possible alternative to this situation.

Imaginary company 2 maintains that a production forecast is nothing more than a probability value. The Sales and Marketing department of that company is accordingly instructed to issue production forecasts in the form of probability values - ie. there is an eighty percent probability that sales will not exceed two hundred units per week; it is ninety percent probable that sales will not exceed two hundred and fifty units; it is ninety five percent probable............etc., etc.

The company recognises that if inadequate production facilities are initially installed, it will be subsequently very costly to remedy this. The company accordingly rules that the Production Engineering department must initially facilitise the plant to support the ninetyve percent probability level of each product forecast.

Recognising that say labour deficits and material procurement programmes can be up-lifted without undue difficulty, the company further rules that the Personnel department and the Material Control department make initial decisions based upon say the eighty percent probability level.

Consistency and cohesiveness within the second company is accordingly better than in company 1.

There is however a second and equally important result which needs to be considered

Being deprived of the opportunity to create their own forecasts, and being obliged to base the work of their own departments on information which is created by an alternative department, departmental managers will place pressure upon this other department to provide accurate and timely information. This acts as a very positive incentive to improve the professional standards which each department operates.

In short, not only does the principle of prime authorship improve consistency and integration - it also tends to improve professionalism and accuracy.

The principle of prime authorship is particularly difficult to achieve in organisations which permit or encourage individual departments of the company to 'self-develop' their own applications, without reference to a single authority such as a Data Administrator or a central co-ordinating function.

Unless activities of this kind are most carefully controlled, CIM will be made entirely unrealisable.

The second DS maxim needs to be considered within the context of the principle of prime authorship.

Data strategy maxim 2

The principle of data effectivity

Maxim

It is a fundamental principle of Computer Integrated Manufacturing, that data of differing effectivity be derived from the most frequently updated occurrence of the data.

Description

This maxim represents a natural extension to Maxim 1, the Principle of Prime Authorship.

One item of data, that of "stock quantity" may be used to illustrate this maxim.

Many modern factories operate say an on-line, real-time stock recording system. Such systems enable the staff of the company to determine the stock of any given part on virtually a second-by-second basis. It is not unusual for such a system to be processed by means of a dedicated mini-computer.

In such a situation it is necessary to ask which departments and functions need, indeed wish, to be provided with the very latest and most up-to-date stock figure.

In producing period end accounts, the Accounts department may need to know what the stock position was at close of business on the last day of the previous financial accounting period. In this instance, the real-time stock position cannot and must not be used.

Other departments may need to be provided with the stock 'as at this morning', or 'as at last night'.

Few people will need, or in fact will want, to have access to the real-time value. Assuming that information is periodically transcribed, in this case from the stock recording mini-computer, to other computers within the network, this will influence the nature of the processing requirements. In this instance, it might reduce the need for the stock control computer to support a very large number of on-line, real-time enquiries, from a large number of different geographic locations - but might require batch 'file transfer' processing at regular pre-determined time intervals.

The **methods** which different vendors might employ in order to realise this principle are of considerably less concern - and many alternative methods will almost invariably be found to be necessary within any total CIM system.

One method which might be adopted would be, at a given time, to transcribe a value from the 'most current' field, to say the 'close of shift' field (which experience indicates as being one of the most useful 'cut-off' times for manufacturing data) ready for onward transmission to a central computer - at a time convenient to both computers.

In solutions of this kind however very special attention must be paid to transactions which are not immediately posted to computer files.

Most companies which operate on-line stock recording systems will post certain transactions immediately they happen, but not post other kinds of transactions until sometime after they have taken place.

Thus, the posting of receipts will often take place before the supplier's delivery vehicle is allowed to proceed past the main entrance to the site - whilst the posting of say a scrapped quantity might not take place for several minutes or even hours after the raising of a scrapping authorisation document.

Automatic transcription of the 'most current' field to the 'close of shift' field at a given instant in time, would, unless special provision is made, not include transactions which had occurred prior to this time, but had not been posted before this time.

Different solutions can be adopted to deal with problems of this kind - but it is important that the problem be recognised and addressed by any system.

Data strategy maxim 3

Generation of management information data

Maxim

Data files needed to support managerial decision making should wherever possible be derived from data files used to support mechanised procedures.

Description

CIM requires several different categories of computer system - the two most important being Management Information Systems and Mechanised Procedures.

Management Information Systems are defined as comprising all those systems which are designed to support managerial decision making, and include both 'reporting' and 'enquiry' systems. Management Information Systems may produce standard reports and enquiries, or may provide ad hoc enquiry or report generation facilities. In many instances such systems will include provision for both standard and ad reports and enquiries.

Mechanised procedures are defined as comprising all those systems which carry out routine company activities - ranging from the regular calculation and production of pay slips, to the physical control of machines and/or processes. Systems such as Computer Aided Design, which assist or enable an individual to carry out a specified business activity (as distinct from carrying out the activity without the need for human intervention) are, for the purposes of this maxim, also included in this category of system.

Simple or unsophisticated management information needs can often be supported on the same computer and from the same files which are created and maintained by mechanised procedures. This might, for example, be the case where just a restricted number of a limited range of standard reports or enquiries will meet all managerial decision making needs.

In many cases it will not however be possible to do this. There are many reasons for this, but two are sufficiently important as to warrant specific mention.

Computers which are introduced to process mechanised procedures are often sized (in storage, memory, cycle times, etc.) to meet the specified needs of the procedures in question. Large unpredictable numbers of randomly invoked enquiries or requests for reports would, in many instances, create unprocessable workloads for such machines - which could, in certain circumstances, lead to serious degradation in the performance of the primary tasks for which the computer was introduced. More often, large volumes of enquiries will delay response times needed to respond to such enquiries. Experience indicates that once enquiry facilities are provided, it is difficult or even impossible to control the numbers of enquiries which are made.

The second, and more significant reason, why management information needs might not be satisfied by the computers and files used to support mechanised procedures concerns sophistication of need.

Files and systems software used to support mechanised procedures are, in most cases, quite incapable of supporting sophisticated information manipulation and reporting needs. Files such as Relational Data Bases, which are necessary in such cases, would not, in most instances provide adequate support (either functionally or in terms of response time) for mechanised procedures.

For these and other reasons, it will often be necessary to transcribe data from the computers and files used for mechanised procedures, to other computers and files which are more appropriate to management information needs.

For reasons related to Maxim 1 the 'Principle of Prime Authorship', the data required to support Management Information needs should, wherever appropriate, be identical to that which is originated at the 'operational' level - ie, by a mechanised procedure.

Certain of the data needed to support management information needs, may not originate at an operational level - but where, as in most cases it does, it is essential that all management reports and enquiries use operationally created data.

Data Strategy Maxim 4 needs to be considered in relationship to Maxim 3.

Data strategy maxim 4

Management information data levels

Maxim

Data which is required to support managerial decision making, should be transcribed at the 'Transaction Summary Level'.

Description

In designing conventional systems, it is normally recommended that the system design activity commences by determining the 'outputs' and 'processes' which need to be supported. These determine which items of data will need to be maintained on computer files.

In the case of CIM Management Information Systems this is not possible, as the real need is to be able to answer ad hoc enquiries - which cannot entirely, or even largely, be specified in advance of the situations which occasion the need for the enquiry.

eg. Machine number X has broken down, and cannot be repaired for nine days; identify all outstanding jobs planned for this machine for which no alternative post-processed NC programme exists, which will become 'late' unless special action is taken, and which cannot be supplied from stock.

CIM management information systems are therefore 'management information systems' in the most literal meaning of the term - rather than 'information management' systems.

CIM management information support data bases are therefore not likely to be suitable support files for 'mechanised procedures' as described in Data Strategy Maxim 3. The principal purpose of any CIM management information system is to provide Managers with the information which they require in order to make a managerial decision. If the necessary files can eventually be used to support other types of requirement, this will be fortuitous and incidental to the design objective.

In order to determine which items of data must be maintained on CIM Management Information files, a different approach to data analysis is required. This approach is most easily explained by example.

Data levels

- Transaction Level

    The lowest 'level' at which items of data could be conceived as 'occurring' can be given the term 'transaction level'. Under this concept, each time a particular storeman made a particular issue of a quantity of a particular item, he would be deemed to have conducted a 'transaction'. Each item of data related to that particular transaction (eg. each item of information recorded on the issue document) would be deemed to be 'transaction level data'.

    It is probably correct to assume that if detailed information was stored in respect to every transaction that occurred throughout the company, most questions about the company could be answered. The amount of file space needed to store the information and the time required to process the information would however make the resultant system both extremely costly and technically impracticable.

- Global Summary Level

    A higher level at which data is frequently held could be defined as a 'global summary' level. Continuing the example of the hypothetical stock file, the 'total current stock of an item' would represent such a 'global summary level data item'. The data is 'summarised' because it represents the nett result of several occurrences of one type of transaction and is a 'global' summary because the value is determined by several different types of transaction. These might include issues, receipts, returns to supplier, scrapped quantities, etc.

    Each of these different types of transaction can be said to correspond to different factors or elements of the subject called 'stock'.

    The file space needed to store information at global summary level and the time required to process such data is well within the scope of data processing technology - and many systems are designed to maintain data at this level.

    The shortcoming of this approach within any management information system would, however, be quite serious, as such a system would be incapable of answering enquiries about receipts, issues, scrap, etc.

- Transaction Summary Level

    A suitable compromise therefore has to be determined, which is defined as the 'transaction summary' level. Items of information such as 'cumulative receipts to date'; 'cumulative returns to supplier'; 'cumulative scrap' would each under this approach represent transaction summary level data items. They are summary items because they represent the sum of a number of separate transactions; they are transaction summaries because they each represent a summarised total for a single type of transaction. Ideally, separate transaction summary values should be maintained for each financial or production planning period - and CIM management information data base files should be designed to accommodate several 'periodic' transaction summary values - eg. cumulative receipts in period 1; cumulative receipts in period 2, etc.

    Global Summary level data can, where required, be easily computed from Transaction Summary level information - eg. total stock equals cum receipts minus cum issues, cum scrap, cum returns to supplier. This means that it is not necessary to store Global Summary level information, in addition to Transaction Summary level data.

Data strategy maxim 5

Replication of data

Maxim

Data should be replicated and distributed to support processing needs.

Description

In practice, it will be necessary to replicate data on different physical data storage facilities - for both technical and performance reasons. It will not for example be either technically feasible or administratively desirable for all CIM systems to be required to obtain any one item of data, from a single data storage facility. This maxim negates such notional concepts as Centralised Data Bases, and even Distributed Data Bases - of the kind where any one item of data is stored only once (even though different items of data are stored on different files).

One example will serve to illustrate this maxim.

A numerical control programme for a machining centre needs to be replicated within several separate data storage facilities in order to facilitate satisfactory levels of operating efficiency. Where the machining centre forms part of a Flexible Manufacturing System cell, copies of the N.C. programme would normally be held in say the N.C. control unit of the machining centre and the cell supervisor computer - in addition to being held within the possibly remote computer used to initially develop and subsequently modify the programme. In particularly large and complex CIM operations, it may be necessary to hold further copies on other processors.

Few existing machine tool products offer sufficiently efficient and comprehensive communications and computer interfaces as to allow each separate machining instruction to be progressively down loaded from a geographically remote computer - as and when the machine tool requires each instruction. Any advantages which such an approach could offer would be insufficient to cause the majority of European machine tool and machine tool control unit manufacturers to modify their products to incorporate such a facility.

This example illustrates two of the factors which requires data to be severally repeated on different storage devices - operational efficiency and commercial acceptability. A further consideration concerns 'contention'.

Many items of data need to be accessed by a very large number of different CIM computer programmes. One extreme example of such a data item is the part or product number which is used to uniquely identify a part or product.

This item of data is necessary to almost every CIM activity, from design through to manufacture and distribution. Many administrative systems, from product costing to the sales ledger, require access to this item of data, as do most engineering and manufacturing systems.

In a large manufacturing company, as many as three thousand or more separate computer programmes may each need to access this same item of data.

No single data storage device could conceivably service the number of concurrent accesses which a large manufacturing organisation would therefore require - and this problem is compounded by the fact that in many instances concurrent file access is denied not only to single items of data, but to record segments, complete records, or even to entire data sets or files. In such extreme cases, only one programme may be allowed access to a complete section of a file, or even to a complete file, at any one moment in time.

In short, replication of individual items of data will be normal within any organisation which operates CIM systems. Replication of itself occasions few adverse consequences - other than less-than-minimum storage and processing costs. Replication of data does however pose considerable threats to data integrity - ie. of different occurrences of what should be the same item of data, having different values. This represents one of the most difficult obstacles in the realisation of the principle of prime authorship, discussed in Data Strategy 1 of this document.

The design of each computer application needs to address this problem in an effective manner - especially the design of computer programmes which transcribe data or update application files using data from alternative files.

Data strategy maxim 6

Distribution of data

Maxim

Data belonging to separate organisational entities should be held on physically discrete files.

Description

Centralised corporate data base philosophies are not appropriate to CIM systems. Many industrial organisations comprise several different factories or manufacturing sites with varying levels of autonomy. It is not unusual for different component parts or assemblies used in a company's products to be manufactured at different factories.

Very large factories are themselves quite frequently sub-divided into highly autonomous divisions or units - each with separately accountable management structures.

There are several reasons why these arrangements require separate files for separate organisational units, even if many of these units operate identical systems. Two of these reasons have special significance and are therefore worthy of description. The first of these concerns organisational accountability; the second concerns operational or processing logic.

Organisational accountability

Discrete and autonomous organisational units are, quite understandably, reluctant to make the fortunes of their unit entirely dependent upon the conduct or performance of other units over which they have no authority. The effects of errors or mistakes made within one unit, which destroys or severely corrupts a data base file, are much more serious if they are not contained or confined to the unit which occasioned the problem. The direct financial effect to the company of causing disruption to several factories is obviously much more serious than it would be if disruption were confined to a single factory or department of a factory. The complexity, and therefore the time and costs involved in recovering a corrupt corporate data base file would be greater by several magnitudes than the sum of the costs for recovering each 'unit' data base independently - even in the virtually impossible-to-conceive situation of every unit simultaneously corrupting or destroying its own data base.

The very advanced and sophisticated file security and recovery procedures which are provided within many modern data base management systems, do a great deal to minimise the effect of file loss or corruption - and the cost and effort needed to recover a lost or corrupted file.

No arrangements or facilities can however **completely** eliminate file corruption, or make recovery totally effortless. Non-containment of such effects within discrete organisational units would be extremely costly - and would quickly lose company-wide support and credibility for any CIM system which occasioned it.

Operational logic

This subject is considerably more complex than Organisational Accountability, and can best be illustrated by means of an example - from which a number of general principles can be derived.

A manufacturing company comprises say six manufacturing units. One of the products manufactured by this company comprises five major items or sub-assemblies (components).

Each of the five major components are machined and assembled by separate units (one to five) of the company. These components are all supplied to unit six, which assembles these together to create the finished product.

A system is required to 'explode' or analyse sales orders for finished products into manufacturing and purchase orders for component parts - including manufacturing orders for each of the parts and assemblies used on each of the five major components, together with purchase orders for the materials and purchased-out-parts used for each part and assembly.

Such a system might be designed in one of two ways - (a) as a 'centralised' single system - or (b) as a 'more-or-less-standardised' production planning and scheduling system, which is commonly but independently operated by each of the six manufacturing units.

Approach (a) would almost certainly use a 'centralised' data base, containing all necessary information about every assembly, part and raw material which is used in the production of the finished product. This data base would also need to contain machine capacity data, outstanding order data, etc., related to each of the six manufacturing units.

Approach (b) would best be supported by each manufacturing unit operating and maintaining separate versions of a 'standard design' data base, which contained information only in respect to those assemblies, parts and materials which were pertinent to their own operation.

Thus the 'product structure' information on the data base for unit six would include only six records: one record to represent the finished product, and one record for each of the five 'first-level-breakdown' components.

The data bases operated by units one to five would each contain one record for the component which they produce (which to them represent 'finished products') plus one record for each 'lower level' assembly, part or material used in the manufacture of their particular component.

Similarly, each unit data base would contain only such machine availability, and order status data, etc., which was pertinent to their own operation.

Under this approach each unit would receive 'sales' orders for the products which they made, and convert these to manufacturing orders - and 'purchase' orders, which would be despatched in some instances to independent external suppliers, and in other instances to other manufacturing units of the same company.

Under Maxim 6, CIM systems should adopt approach (b) ie. a 'unit' data base approach.

This approach is not only consistent with the proposals described under Organisational Accountability (the first of the two reasons for the maxim), but also offers other major advantages:

- eg Experience indicates that it is frequently not possible for all sites or units of a company to operate truly identical systems. This is true even where each unit carries out 'similar' operations by means of 'similar' production facilities. Where Unit 1 is say a pressing operation, Unit 2 is a foundry operation, Unit 3 is a machining operation, etc., absolute commonality of production planning and manufacturing systems is entirely infeasible. Centrally operated production planning and scheduling systems are, in practice therefore invariably 'composite' systems (containing different provisions for different units) which are more difficult to design, implement or administer than separate systems for each unit, developed from a common or base system.

- eg The 'unit' data base approach also offers greater flexibility in supporting changes in operating strategy - such as where new operating units are created or old ones closed. In a unit data base environment, changes of this kind require only the implementation of additional systems similar to the ones used in other units, or the discontinuance of systems. Operational changes do not, in this environment require changes to be made to operational systems.

Postscript to the data strategies

The data strategy maxims described in the preceding pages have been developed to be applicable to CIM system products, and have accordingly been largely addressed to vendors of CIM systems.

Clients for such products must however recognise that the successful adoption of these strategies is entirely dependent upon the steps which they themselves must take.

Thus, whilst CIM systems should, for example, include adequate provisions and support for the Principles of Prime Authorship as described under DS Maxim 1 - implementation of this principle can only be achieved by the client company itself.

If the senior management of a client company permits individual departmental managers to create and maintain whatever data they desire - without reference to any central authority, it would be quite impossible for that company to realise the Principle of Prime Authorship - or indeed any other data strategy maxim.

Client companies must recognise that company data is a most costly and valuable resource, which needs to be professionally managed and administered.

All except the very smallest manufacturing companies will need to recognise this within the organisational structures which they operate. One recommended approach is to appoint a CIM Data Administrator, to be responsible for all aspects of data management and control.

Typical responsibilities for a Data Administrator would include:

- Negotiating and determining prime authorship responsibility for each item of company data.
- Devising and negotiating rules relating to the privacy of each data item - including the identification of each department or individual that is permitted access to each data item.
- Maintaining a comprehensive and authoritative 'data dictionary' containing information about each item of data contained on computer files. This information would include an official description of each data item detailing any formula or algorithm used in its creation; details of the system which creates and maintains it; details of each system which uses it; details of the format, length and construction of the data item and the files upon which it is maintained, etc.
- Determining which items of data that are maintained by mechanised procedures (Maxim 3) are to be transcribed to the Management Information Systems Data Base for use in managerial decision support systems.

The Data Administrator may, in certain circumstances, also operate a company Information Centre, to advise management on the availability of data needed to support particular enquiries or requests - and to operate a service for writing special 'one-off', fourth generation programmes, to support ad hoc enquiries and reports for senior managers.

This latter accountability is dissimilar from the conventional Application Software development duties normally carried out by the Systems or Data Processing departments of a company.

The subject of Data Administration, like that of Data Base Design, is a major subject in its own right - about which many text books and papers have been written. It is not the intention of the present document to provide authoritative and comprehensive information on the subject - but to draw attention to the need for client organisations to pay proper attention to data management and administration. Companies intending to introduce CIM operating strategies will however need to obtain detailed training and advice on the subject of data administration.

Chapter 11

Processing: state of the art

Introduction

In this chapter an overview of the state of the art in processing is presented. A general description and analysis of the notion of processing is given, though some arguments will be specific for Computer Integrated Manufacturing.

Functionally, processing is the transformation of input to output. In many cases, the output will include a modification or extension of existing data. Processing is done by programs that are executed on a digital computer. To provide a general processing strategy it is necessary to take both hardware and software aspects of processing into account.

It should be noted, that there exists no strict logical boundary between hardware and software functions, although there exist a number of practical characteristics, that have to be taken into account. Eventually, basic processing functions have to be executed by hardware components, but almost any hardware function can be implemented in software (e.g. execution of floating point instructions without floating point hardware).

Processing systems vary in a number of aspects, the most important being:
- Reliability (the chance that errors occur and their impact);
- Extendibility (the possibility to increase the systems performance by adding or changing modules);
- Performance (speed of operation, functional scope);
- Only for hardware: environmental characteristics (size, cooling requirements, etc.).
- Only for software: portability (or hardware independency).

No attempt will be made to give exact definitions of each of these aspects. Reliability and extendibility will first be discussed in general. This will be followed by a discussion of current hardware technology and future trends. This chapter will be concluded by a discussion of the software aspects of processing.

11.1. Reliability

Reliability is an important issue for CIM. Reliability has two aspects:
- the chance that a particular unit breaks down
- the impact of the breakdown of a particular unit on the entire system.

The break-down of a particular unit can be caused by both hardware and software failure. Possible hardware failures are:
- Breakdown of power supply.
- Breakdown of cooling.
- Malfunctioning mechanical components (in disks, tape units, etc.).
- Malfunctioning Integrated Circuit (IC).
- Electrical noise (through air or through power supply).
- Design error.

To a certain degree hardware errors can be avoided by careful design. Especially the chance that a chip breaks down is very small as long as the power supply and the cooling system function correctly. Intrusion of electrical noise can be stopped by careful shielding and special transformers to prevent intrusion via the power supply. It is nevertheless impossible to guarantee the correct functioning of a processing unit. When it is absolutely necessary that a particular unit remains functioning it is necessary to provide for two copies of that unit that can check each other's behaviour and take over if necessary. When a breakdown may occur but quick repair is essential it is sufficient to have a set of spare components available. In that case there needs to be sufficient checking and diagnostics to ensure that an error will be found in time.

For software failures, there is basically just one cause, and that is a design error. This design error can be located in the part that breaks down, but can also be located elsewhere, and sometimes has its origin in the specification phase of a system.

Although progress has been made in software design methodology during the past decades, complex software still contains (many) errors. This is unlikely to change in the near future. Propagation of errors can be minimized if:
- functional interfaces between modules are specified in a clear and unambiguous way;
- input data formats are well structured;
- careful input data checking is performed. If this is not done then errors generated in one module are likely to cause errors in other modules as well.

11.2. Extendibility

Extendibility of computer systems can be realized in two ways:
- the possibility to upgrade an existing system to increase capacity and performance;
- the possibility to upgrade an existing system *without* having to take the system down for some time.

Extendibility in the first sense is currently realised in many systems for the addition of extra primary memory. Extending the system is easy, only an other memory board has to be plugged in, but to do this it is necessary that the system stops operating during the time the modification is made.

Extending the processing power is usually much more complicated. If possible at all, it currently requires to buy and install a larger version of the computer already in use. This may change with the advent of multiprocessor systems. With such a system, extending processing power is just as simple as adding memory, just plug in another processor board.

Extending a processing unit without taking the system down for a certain amount of time is currently rarely achieved. Generally speaking, a module can be added to a processing system during operation if its power supply can be switched off separately and if plugging in and out the connectors does not generate unacceptable levels of internal noise. With current technology this is usually only the case if the connection between the module and the processing system is serial (as is the case with network links and the interface between disk and disk controller).

The use of networks can increase extendibility considerably. Adding processing power will imply adding a new processing server machine to the network. Even with current technology adding nodes can be done without disturbing other activities on the network.

Extensions to software products take place in the so called "maintenance phase". This "maintenance phase" is currently considered to be the most troublesome phase of the life cycle of a software product. Analysis performed by the United States Department of Defence has demonstrated that the cost of software maintenance can be as high as 80 % of the total software costs.

Much research in Software Engineering is currently devoted to design methodologies to produce software that will be easier to maintain. Although such methodologies seem promising, experience outside academic institutions has demonstrated that these methodologies are no guarantee for success. It is likely that progress in this field will be made but the problem of software maintenance will probably remain for a long time.

11.3. Processing Hardware

Computers are available in a wide range of performance. Performance of a computer can not be characterized by a single number. Different computers are often optimised for different tasks and consequently the performance ratio of two systems depends on the application. A supercomputer may outperform a mini by several orders of magnitude in floating point operations, but may turn out to be less efficient than the mini for real-time control operations.

The currently available computer systems are often characterized as either a micro, a mini or a mainframe. This characterization is certainly not independent of the current technology. In fact, new developments lead to new categories, like 'supercomputer' for high performance number crunchers, and 'workstation' for advanced graphical micros or minis. Rather than to try to work out this characterization in detail and to try to classify every particular system, it is necessary to analyze the basic differences in capabilities that can be found.

Technological Development

To analyse the ongoing technological developments a taxonomy of processing equipment is needed. The best known taxonomy for classification of general purpose computer systems is due to Flynn. [Flynn1972] He classified machines on the simple view that there could be more than one data stream and more than one instruction stream in any computer, with the possibility that a single instruction could process either a single stream of data or more. Flynn denoted three main types of machines by the acronyms SISD, SIMD, MIMD.

a   Single Instruction Single Data (SISD) (Classical von Neumann architecture)

b   Single Instruction Multiple Data (SIMD) (Array processor)

c   Multiple Instruction Multiple Data (MIMD) (Multi processor system)

Classification of a particular machine in one of these three categories can be problematical. For example, Flynn puts pipeline machines into the second category because a pipeline is considered as a time-multiplexed version of a processor array (with each processor element in lock-step). But others regard pipelining just as an internal way to speed up certain operations and consequently classify a pipelined machine as SISD. We will here take the programmer's point of view and consider a machine only SIMD when the instruction set contains instructions which operate on multiple data.

Apart from general purpose processors there are also special purpose processing systems. Such systems have been built for Lisp (special architectures for support of a symbolic language), for graphics workstations, for picture analysis (special architectures for neighbourhood calculations) and are under design for Prolog (special architectures for inference machines).

Furthermore, general purpose architectures, that differ radically from the classical von Neumann architecture have been designed (data flow machines). This last category, however, is very much in a research phase and it is not to be expected, that any results achieved in this field will become important for commercial applications in CIM within the next decade.

Characterization of performance is often done in Mips (Million instructions per second) or in Megaflops (Million floating point operations per second). The first one can only be used when two SISD processing elements are compared. Megaflops can be used more generally, although one should realise that especially for vector processors the theoretical maximum and the value achieved in ordinary programs can differ widely. Consequently the number of Megaflops is only meaningful for a combination of a computer architecture and an algorithm.

The processing power of a computer system can be characterized by:

1   The type of its processing (SISD, SIMD, MIMD).
2   The speed of its processing (combination of actual CPU speed and primary memory speed).
3   The size of its primary memory
4   The possibilities for transfer between primary and secondary memory and the speed of such transfers (to disks and tapes).

Each of these aspects will be discussed below in detail.

1   The classification of computer systems as SISD, SIMD and MIMD has been discussed above.

2   SISD machines used to give a wide range in processing capabilities. Due to the rapid development of micro processor technology, and the fact that the speed of the fastest SISD machines has hardly increased in the past decade, this range of processing power is quickly getting smaller. Currently the difference in speed between the fastest SISD processors (e.g. the CRAY-1 [Cray1980] or CDC 205 [CDC1980] in SISD mode) and advanced microprocessors (e.g. the MC 68000 [Motorola1982]) is one order of magnitude for integer operations †. It is expected that this difference will diminish even further during the next five years and that the floating point possibilities of the microprocessors will match their integer capability.

SIMD machines are currently the fastest processors available. They are based on very sophisticated pipelining techniques. The gain in speed which can be offered by an SIMD architecture over an SISD architecture is currently of the order of 40, measured for useful programs. A serious problem in reaching this gain is that a drastic amount of reprogramming is necessary and that new algorithms are often needed. Additionally it is quite often not possible to use the fastest algorithms available (e.g. for matrix inversion) since those algorithms are only suited for SISD architectures.

---

† A 32 bit integer add operation on the 68000 takes 1 to 2 microseconds, depending on the location of the operands; on the Cray-1 instruction timing is complex and depends on possibilities for pipelining and vector processing; if no vectorization is possible, the Cray-1 will execute roughly at a rate of 25 Mflops.

MIMD machines are quite often based on microprocessors and consequently their current performance is not comparable to SIMD machines. Initial experience with MIMD machines has demonstrated, that there is a class of problems, which can be easily divided over different processors on a high level. If this is not possible, the problems in actually using the available parallelism are certainly as great as they are for SIMD machines.

3   Memory size used to be an important difference between micros, minis and mainframes. Current technology has resulted in a situation where even personal computers can have of the order of one Megabyte primary memory Furthermore, modern multi-user micro systems can have their memory extended to of the order of ten Megabyte. When this is compared with the 40 Megabyte some large mainframes possess then it is clear that memory size is not a useful aspect for distinguishing between processing systems. The only exception is for real-time control systems where physical size limitations are important and there is consequently a low limit for the number of boards which can be plugged in additionally.

4   This aspect is still an important point of distinction between micros, minis and mainframes. The capabilities of mainframes for transfer between primary and secondary memory are still, and will probably remain, much better then those of micros or minis. This is a very important point for databases where large amounts of data have to be transferred between primary and secondary memory during a search. It is possible that microsystems will even on this point become a threat to mainframes, but only if a large number of them can be combined in a network. Such a network would then have to implement a single distributed data base.

Main processor categories

As noted before, different computer systems are often optimised for different applications. In this respect, the following main classes of applications can be distinguished, all of them occurring in Computer Integrated Manufacturing:

-   real-time control
-   number crunching
-   data base management
-   graphical interaction
-   text-processing, administrative operations, etc.

This results in five main types of computer systems with different functionality.

I   "The real-time controller"
    This is a SISD or MIMD architecture, with probably no secondary memory and constraints on the primary memory. These constraints are rather loose when the memory is limited to the on-board available (of the order of 1 Megabyte) but become very strict when the primary memory is limited to on-chip available RAM and ROM (of the order of 1 to 10 Kilobyte).

II "The number cruncher"
    This is a SIMD machine, functioning as a server on a network for intensive number crunching calculations, with little local secondary storage, primarily for the implementation of virtual memory. This machine can also be attached directly to a personal workstation. In the future also MIMD architectures may implement this function, once the individual processors of the MIMD machine have become fast enough.

III "The data base machine"
    This is a SISD machine with large amounts of secondary storage and an architecture optimised for fast transfers between primary and secondary memory. This machine will resemble current mainframes in many aspects, but it will be used in a much more

specialized way.

IV   "The personal workstation"
This is a SISD machine, preferably connected by means of a local area network to remote file servers and data bases, with little or no local secondary storage but good interaction with the user, probably with graphics.

V   Computer systems used for text processing and administrative operations differ widely in size and performance, ranging from stand-alone simple personal computers to large mainframes. The general characteristic of these systems is not that they are optimized for this application but rather the lack of optimization for one specific item alone.

Apart from the above mentioned computer systems it is likely that for some CIM applications special purpose computing devices will become available. These machines will be used where the four previously mentioned categories are not capable to do the necessary processing in the given time. Among applications where this is likely to happen are: analysis of pictures in real-time (e.g. vision systems for robot control) and inference machines for logic languages, such as Prolog. At the present time there also exists a tendency to implement traditional software functions into hardware using VLSI techniques (e.g. Fourier transformations, matrix-multiplications, heavily used and/or time critical operating system functions).

Configurations of processing units

Mostly, a processing unit is not used stand alone but in a combination with other processing systems. To do so it is necessary that the processing systems can communicate with each other. An analysis of the different means and technologies for communication is given in chapter 13.

The currently most popular configuration of processing units is a central computer centre with one or more mainframes with connections to locally placed mini computers. In the future this may well change to configurations where the bulk of the processing power will be available locally and not in a central computer centre.

One of the most crucial issues in designing and setting up processing system in CIM is the configuration of the processing systems, the distribution of tasks over these processing systems and the control relations between them.

11.4. Processing Software

In general, a distinction is made between system software and application software. The system software includes editors, compilers, debuggers, operating system software, network protocols, etc. In most cases, the system software will be delivered by the computer manufacturer and will generally not be specific for CIM. Application software is often purchased from software vendors that are not necessarily computer manufacturers and is generally specific for a particular CIM-module, although for some areas, like bookkeeping and some other administrative tasks, the software may have a more general applicability. Examples of application software are: NC programs, packages for finite element analysis and packages for material requirements planning.

The boundary between system- and application software, however, is not clearly defined, but is general indicated by practical considerations as mentioned above; furthermore, a technological trend can be observed, that application functions, when their usage becomes more wide-spread, tend to move towards the direction of system functions (e.g. basic graphics functions).

System Software

As has been demonstrated above, different computer systems will give optimal performance for different applications. This is also reflected in the system software, especially in the operating system. Operating systems are of a special importance because the operating system determines largely the type of work that the computer system will be suited for. The requirements for the operating system depend not only on the type of application, but also on the computer system. For example, when a mainframe is used, important aspects will be resource sharing, scheduling and accounting, whereas when a cluster of micros is used the transparency of the network for the user is the primary concern. The application influences the requirements for the operating system as well. The requirements for an operating system supporting real-time control differ from those for an operating system supporting geometric modelling as application. If the application is real-time control, then important aspects of the operating system will be: the time needed by a process switch in response to an external event and a guaranteed response time for these events. If the application is geometric modelling, then important aspects are: a good user interface by means of a graphical workstation and good database facilities.

Since the requirements of CIM modules differ in many respects, many different computers from different vendors with different software packages are inevitable, at present. This variety of requirements and, as a consequence, of computer systems to be used, imposes many interfacing problems, since in CIM many different systems must interact with each other and/or humans (operators, programmers, etc.).

Therefore, it is highly desirable, that these systems, as far as possible, form a family of mutually compatible systems such that:

i) these systems can, when needed, exchange information in a simple and straightforward way;
ii) individuals, who are acquainted with a particular member of the family (e.g. with a CAD system) can as easily as possible learn to manage another (e.g. a robot system). One necessary, but not sufficient, condition to achieve this is, that these different systems with different functions should use the same interface (e.g. command language) to those functions, that they have in common. Another condition is, that sets of supported functions should be as orthogonal as possible: a module should not incorporate functions, that are already supported by others.

Examples of the technological trend towards families of mutual compatible systems are:

i) the advent of the UNIX † operating systems, which already are also used in industries for a great variety of application tasks;
ii) the development of Ada ‡ , planned to eliminate the uncontrolled growth of the number of different and often totally incompatible computer languages, that have been developed during the past decades, by eliminating the need for them.

To achieve a higher level of reliability and flexibility in software products it is important that this trend will be carried through.

Application Software

The application software for CIM can be divided in several categories according to the general applicability.

---

† UNIX is a registered trademark of Bell Laboratories
‡ Ada is registered trademark of the Unites States Department of Defence

- Standard packages that can be used by a large number of different companies. Some of these will not be specific for CIM, like text processing and other administrative tasks, and others will be specific for CIM like CAD systems for the design of mechanical components. These packages can be used directly by the end user.
- Company specific software that is either a general product to be used during a longer period of time, or for example a NC program that will only be used as long as a particular product is made. For this software two types of processing resources are needed. One is the processing needed to *develop* this software (editing, compiling, testing), the other is the processing power needed during *execution* of this software.
- In addition to the above mentioned categories also basic software components for CIM can be distinguished. Characteristic for such basic software components is that they handle basic CIM tasks which occur in many different applications in a company and in many different companies alike. Examples are:
  - Graphics modules
  - Modellers.
  - Software support for sensors, robots, NC machines, etc.
  - Finite element packages.

These basic packages can be used to build branch specific packages that can function as complete systems. For example, a graphics package and a 2D modeller can be used as the basic building blocks for a CAD system for Printed Circuit Boards (PCB's). The importance of the above mentioned basic components for CIM is that they greatly facilitate the design of specific packages, for use in a particular branch of manufacturing. This increased ease of design will also lead to an increase in reliability and flexibility of the software product.

The analysis of the processing and communication requirements of the basic software components is important to determine the processing requirements of standard packages, because the processing requirements of standard packages will be largely determined by the constituting basic software components.

The processing for basic software components can differ widely, both in the type of processing needed and in the overall processing power needed. In some cases these components make use of a dedicated processing unit to handle their processing. Currently this is already done in graphics systems. It is very well possible that in the future this will be done more often.

11.5. Conclusions

The current trend in computer technology seems quite clear. Decrease of price for the same performance, rather than increase of performance for the same price. This implies that a large number of computer systems will have to be used to support CIM. Distributing CIM processing tasks over many computer systems is not only necessary, but also desirable to increase flexibility and reliability. Networks will be needed for the communication between different computer systems. A general overview of networks is given in chapter 13.

In software products more and more use will be made of basic components that will be used as building blocks for either standard packages or company specific software. This will decrease the development costs and ensure greater reliability and flexibility. The emergence of mutually compatible families of software products will make it easier for users to learn handling a new product (e.g. due to a similar command structure). Despite these positive trends it seems likely that software maintenance will remain a major problem.

In the next chapter, a carefully selected set of important maxims will be presented, which when properly applied, will aid in fulfilling the intrinsic difficult processing requirements of Computer Integrated Manufacturing.

References

CDC1980. CDC, *CYBER 200/Model 205 Technical Description,* publ. CDC Inc. (Nov. 1980).

Cray1980. Cray, *CRAY-1 Hardware Reference Manual No. 2240004, Rev. E.* 1980.

Flynn1972. Flynn, M.J. , "Some Computer Organizations and their Effectiveness," *IEEE Trans. Comput.* **C-21**(9) pp. 948-960 (Sept. 1972).

Motorola1982. Motorola, *MC 68000 16-Bit Microprocessor User's Manual,* Prentice-Hall, Inc., Englewood Cliffs, N.J. 07632 (1982).

Chapter 12

Processing strategy

Introduction.

In this chapter the processing strategy is presented. As already mentioned in chapter 9, such a processing strategy consists of a set of maxims or guidelines of general validity, which will aid in optimizing processing for various processing tasks throughout Computer Integrated Manufacturing systems.

Each maxim takes the following shape: title of the maxim, a short description (consisting of one to three lines) and thereupon a narrative explaining the notion behind the maxim, attended with examples when applicable. In the narratives much of the material from chapter 11, concerning the state of the art in processing, will be used.

Processing strategy maxim 1.

The principle of intelligent local processing.

In Computer Integrated Manufacturing, many processes have to exist locally, in order to control or exchange information with persons and devices external to the computer. It is essential that these processes be assigned a complete, well-defined task which interfaces to the rest of the CIM system at a sufficiently high level, so as to ensure that no details of the process irrelevant to the rest of the system, will penetrate outside the local process.

Local processing is enforced by physical, environmental requirements such as data which is only available locally (e.g. the scene for a vision system) or the control information which has to be exchanged with a physical device (e.g. a crane controller). Local processing may exist on a small scale, for instance an axis control system for a manipulator, or on a large scale, such as local control for a manufacturing cell.

Local processing is always concerned with an external medium with which data has to be communicated. The reasons for providing local processing can be very diverse:
- the frequency of communication or the amount of data communicated would make remote processing impossible or very expensive.
- the process can be too specific for a remote, more general purpose computing system.
- the local processing system is an extension of the computing hardware built in the device.
- the local process is provided with the device and cannot be replaced because the specific knowledge about that device is not available.

The maxim is not primarily concerned with the reasons to distribute processing. Given the need for distributed processing, it addresses the scope of the local process, that is, which parts of the processing tasks must be localized. Local processing requirements constitute a minimal set of constraints for a specific task. Given task and local processing requirements there must be an interface to the rest of the system.

This implies that any local processing task deals with two interfaces. The local, interface to the external world and the higher level interface to the rest of the system. The disparity between those two interfaces determines the amount of processing which is situated locally. This is sometimes referred to as the amount of intelligence assigned to and realized by the local process.

Builders of Computer Integrated Manufacturing systems have to make any local device fit in with the overall system architecture. Providing local processes with sufficient intelligence for the purpose of interfacing it to the rest of the system is one of the basic means to achieve this harmony.

Manufacturers of devices such as NC-controllers, graphical workstations and robots must allow these devices to be embedded in a local process, because no two systems for Computer Integrated Manufacturing are alike, and the local processing requirements are to be determined by the specific manufacturing tasks.

The creation of a practical task-inventory is an intermediate step towards standardization in Computer Integrated Manufacturing. Successful standardization can only be achieved through definition of complete task sets: only those are suitable for general purpose functional or communication interfaces.

On the other hand, standards must accommodate situations where enhanced local intelligence is exploited. Such standards will be widely applicable and have the flexibility to serve future developments.

The creation of a local processing margin between external interface and CIM system interface, is a practical solution for using existing standards as well as encouraging factory-wide standardization.

Vision systems are pre-eminently coupled to local processes. The image analysis of the scene on the lowest level is very much vision hardware (camera) dependent. The colour or black and white raster representation of the raw image is too bulky for it to be transmitted elsewhere. This is why vision systems have built-in processors, programmed for the initial elementary analysis, such as noise filtering, thresholding and image enhancement. Also the sampling of depth information (e.g. strip-lighting) is done locally. This incomplete process now has to be extended by the image analysis task required for the application.

The maxim states that this preferably has to be done by a local extension, until a suitable interface is obtained. Scene analysis tasks can be very well supported by special purpose hardware, such as pipeline processors or parallel processors.

A second example is robot-control. The first robots had one central processor to control the entire robot. Movements were very difficult to model, as the processor could manipulate only one joint at a time. Management of sensor-data was likewise interleaved and intricate.

Modern robots use local processors. A model that comes to mind is one processing unit per joint, where one processor controls the actuator and another handles data-reduction and data-transmission. This architecture would leave room for a global processor (the robot-controller) that is solely concerned with specification of tasks to the joint-controllers such as trajectory-calculations, sensor-processing at a global level and the like. This example illustrates that by local processing a better architecture can be obtained which does not only offer more functionality but is also easier to understand by its creators and maintainers.

A third example is drawn from the area of transaction-management in computer aided design. Contrary to business-databases where transactions span a fairly short period of time and concern a relatively small amount of data, transactions in an engineering design-environment (corresponding to the design of a part) can take days, weeks or even months; the amount of data involved is incomparably much larger than the amount of data involved in business-database transactions.

Implementation of these transactions on a centralized processor system would lead to very heavy loads on the processor and the network linking the processor with its storage and communication devices. An engineering design environment would be implemented much more satisfactorily in a distributed environment where transactions are done locally on workstations connected by a network. Once a transaction is finished, its result (the data representing the design) is transmitted to a central location on the network where the overall design is kept. In this way data-communication can be reduced drastically if workstations are capable of storing the data that represent the design and allow the designer to operate on those data by graphical or other means. Moreover, as the network becomes large and can serve more users than a centralized system, significant resource-sharing becomes possible.

Processing strategy maxim 2.

Systems design should aim for the distribution of processing.

CIM subsystems should be designed and implemented in such a way that they will be able to execute on distributed computer architectures. They should anticipate an increasing degree of distribution.

Computer Integrated Manufacturing systems are, by the nature and diversity of the tasks to be carried out, very complex, not only from the point of view of systems design but also due to the computational complexity of many of the identifiable tasks, even at a relatively low and non-integrated level. It is therefore inconceivable that any serious system for Computer Integrated Manufacturing could be built around a single processor.

A system for Computer Integrated Manufacturing configured around a single centralized processor is not practical. It is much more effective to bring processing power at the places where it is needed rather than transport the data to a central site and perhaps move the output of the operation back to the original place after processing has been done. In a centralized configuration these tasks will have to be carried out by one processor. Disadvantages of a centralized approach would be that the processor will have to be very powerful if the total system's performance is to match the performance of a distributed one; the network around the centralized processor will easily get overloaded, not to mention the fragility of such a system: when the central processor breaks down, the entire system stops functioning.

On the other hand there are practical bounds to the amount of decentralization: a balance has to be found for the tendency to decentralize for better performance and the amount of hardware available to realize the concurrency implied by the system.

The advantages of distributed processing have been recognized for Computer Integrated Manufacturing as they have been for other application areas. A substantial part of the severe processing requirements for Computer Integrated Manufacturing such as high reliability and reconfigurability can be adequately dealt with by applying distributed processing architectures. Moreover, the organizational prerequisites for distributing tasks and processes are existent if not traditional in most plants.

Currently there is a clear tendency towards replacing the predominantly mainframe-oriented installations by distributed configurations. For existing Computer Integrated Manufacturing sub-systems this change may require extensive reimplementation of large modules to make them suitable for the new situation. Needless to say that this difficult and expensive process would not have been necessary if the systems that are to be modified would have been designed for a distributed environment.

The key to systems design is the identification of a logical structure such that the realized system will behave as specified. Identification of a logical structure entails identification of the different tasks a system has to perform: these tasks play a role comparable to the list of parts in mechanical design. The logical structure should be such that the tasks implied are mutually as independent as possible in order to achieve realization on different configurations of processors.

One of the ways to master the inherent complexity of a processing task is to decompose it into a number of sub-tasks whose interdependence is minimal. Distribution of processing is only one aspect of modularization. Given the attraction to distribute processing it is natural to decompose a task into modules capable of concurrent execution. Evidently, distributed processing implies the distribution of data for which the reader is referred to the relevant maxims in the data strategy.

As we cannot expect that large integrated processing systems for Computer Integrated Manufacturing will emerge as immutable black boxes there will always be room for adaptation of these systems. It is almost a necessity that these systems are adaptable and extensible, as it will enable their users to *incrementally* enhance or modify their equipment. Another aspect is that sub-systems for Computer Integrated Manufacturing themselves are subject to enhancement and modification and, for evident economic reasons, it must be as easy as possible to replace old systems by new ones.

The maxim states that one of the aims of processing systems design is to leave as much room as possible for those adaptations. Examples of those adaptations are:

- Better possibilities for incremental modification of a given configuration of processors and processes;
- General purpose processors must be replaceable by (configurations of) special purpose processors or different general purpose processors;
- On the fly reconfigurability *without shutdown* is in many situations necessary or even essential;
- Better possibilities for reallocation of processes to processors (e.g. for tuning purposes).

To realize this kind of flexibility all sub-systems for Computer Integrated Manufacturing must be equipped with standard interfaces to other sub-systems (*cf.* processing strategy maxim 9).

In a system for Computer Integrated Manufacturing the logical processing tasks to be performed are distributed over several processors placed in a network. The frequency and intensity of communication in the network depends on the relation between the tasks performed. Non-communicating processes are independent and can thus be run on different processors that may not even be connected at all. An integrated system consists of processes that do communicate data and thus one can ask how to allocate different processors to the processes such that the load of the network and the load of the different processors is in some reasonable balance. Normally, one tends to combine two or more processes on one processor if their communication needs would strain the network and one tends to distribute processes over two or more processors when their data-exchanges are more moderate. Note, however, that once it is decided that two or more processes should be run on the same processor the combined process should *still* be easily distributable, at least in principle, if only to allow the flexibility mentioned above.

Design of distributed systems may not yet be a well-developed practical art that many people are familiar with; nevertheless, for the development of computer integrated manufacture it is essential that systems are designed for distributed processing.

Processing strategy maxim 3.

Systems design should aim for deadlock-freedom.

There should be no situation where a process will block indefinitely because of deadlock.

A process is said to be deadlocked if it is waiting for an event that will never occur. So, effectively, the process stops functioning. Deadlock can occur everywhere in a system and at each level. Classical examples can be found in the area of operating systems. Two jobs, for instance, are deadlocked if one cannot proceed without having access to a resource that is allocated to the second job, while the second job cannot proceed without the resources currently allocated to the first one. Distributed systems can also be plagued by deadlock when one process cannot proceed until another process does something and that other process cannot proceed because it is waiting for the first one to do something. Both processes have effectively stopped and there is no way out of this situation except external intervention. In a factory, similar situations may occur. Consider a robot which is waiting for another robot to get out of its way. Deadlock occurs when the other robot is also waiting for the first one to clear the way. The example presupposes, of course, that these robots have the means to detect each other, otherwise the result of this deadlock situation could have been more spectacular.

A very appealing solution to the deadlock problem is to design a system such that deadlock cannot occur. However, in many situations such a solution can lead to poor resource utilization, because in this case deadlock situations can only be prevented by knowing all resource requirements of the processes in advance of scheduling, so that all resources are at the disposal of the process during the the period it is active, irrespective of the actual use that is made of these resources.

Another approach, automatic deadlock recovery, is to design a system such that although deadlock-situations may occur, these situations are side-stepped as soon as they approach: the system is able to predict when a deadlock will occur. Processes in such a system must still request their resources in a disciplined way, either by requesting them all at once, or by relinquishing all the held resources when making a new request, or a linear order must be imposed on the resources and processes are only allowed to request a resource i if and only if they do not hold a resource j such that $i<j$. A similar situation holds for networks. Environments with those disciplines are restricted and, for many areas of Computer Integrated Manufacturing, unreasonable.

If there is no guarantee that deadlock will not occur, the maxim states that a system must be equipped with adequate means to detect a deadlock-situation (i.e. has deadlock occurred and which resources are involved?) and for recovery of the situation once it has been detected. In these systems deadlock can only be resolved by explicit intervention of an operator. In this case, the operator must know the cause of the deadlock, the time it will take to recover from the deadlock and how often deadlock occurs in the system.

Some systems are not deadlock-free, but the probability of deadlock is negligible. Reducing deadlock-probability is a valid and useful technique to realize this maxim.

It should perhaps be stressed that in the design of real-time systems special care should be taken with respect to deadlock-situations as their occurrence as well as their recovery may easily jeopardize real-time behaviour. Moreover, systems where deadlock is excluded can use the important and effective techniques for detecting irregularities based on time-outs, because in these systems time-outs are not caused by deadlock.

Processing strategy maxim 4.

Every (sub)system for computer integrated manufacturing must provide diagnostics information.

Every sub-system of a Computer Integrated Manufacturing system will be able to provide diagnostics about the occurrence, type, severity and location of faults.

Diagnostic information is information about a process made available to the outside world (supervisory system), in order to determine the occurrence, type, severity and exact location of an error. Any subsystem is supervised by a control-system. A control-system is a part of the outside world of the subsystem and could be local or global, machine or human. The maxim states that any Computer Integrated Manufacturing (sub)system must be able to provide its control-system with diagnostic information, either on request, on a regular basis without explicit request, or such that the supervisor has access to all (diagnostical) information present in the processes it controls. The request from the control system for diagnostic information can be triggered by an error signal coming from an error detection system. The control system may, however, also check the subsystem by every now and then requesting diagnostic information, without being triggered by an error detection system.

Both error detection and provision of diagnostic information are of the utmost importance to ensure the reliability (i.e. the continuity and integrity) of a process. The reliability can be degraded by frequent occurrence of (small) errors or by errors which take a long time to correct. Since no system can be built in such a way that nothing goes wrong, error detection and provision of diagnostic information is always necessary. So, if an error does occur, a good error detection system is needed to detect that error as early and as close to the place where it originated as possible (*cf.* processing strategy maxim 5). This means that error detection must be done preferably at all strategic places and as frequently as needed, without interfering too much with the actual process.

Note, however, that there are situations where it can be cheaper to allow small errors to occur, which can be solved easily and quickly, as opposed to a system where the chances for errors are minimal, but if an error does occur, its influence can be so severe that it stops the system from operating for a long time.

With respect to control there are several types of control-systems. One (rather extreme) type is a real time surveillance system which is completely aware of the actual situation, so that on error occurrence the type, location, severity and cause of the error are known. Thereupon the error can either be corrected immediately or the information can be passed on to a higher level of control.

Except for critical real time processes, where survivability demands immediate error recovery, such a control-system does not appear to be realistic: if realizable, it will probably cost too much effort.

Another, perhaps more realistic type is a control-system that only checks for error-occurrence. It collects just enough information to determine whether a process is functioning correctly or not. If the error cannot be identified by the control-system, external action is needed, thus the process(es) which it controls can best be stopped until the corrective action can take place. If a process is stopped, it must be done in such a way that it is possible to restart it from the point of stopping without introducing new errors.

If an error is identifiable, state evaluation must follow (processing for diagnostics). Special care must be taken to ensure that this state evaluation does not introduce new errors, otherwise the system might end up in an endless loop of error detection/creation.

It will always be necessary to have an error detection system and provision of diagnostic information in one way or another, since error occurrence is inevitable. So, the most logical set up is to develop a system such that at each step of the design, the required diagnostic information is specified, as part of the system specification. This can be achieved by fixing the pre- and post-conditions for each process, implying a modular set up of the Computer Integrated Manufacturing system. Unfortunately, this situation is not common practice at the present time.

However, there are a few existing strategies that deal with error-occurrences. The first that will be discussed here is the so-called skipping strategy. Skipping strategies are based on the inclusion of so-called skipping points, or check-points in a data-stream. These data can represent a process (program-text) or they can be data to be used by a process that is active. Examples of skipping points in a program-text are brackets of different types (**begin .. end**, **if .. then .. else**, goto's and their labels, (..), [..], etc.). Examples of skipping points in what is more ordinarily thought of as data are end-of-file markers in files, header- and trailer-labels surrounding a data-set, source and destination of messages in a network, to name a few. We do not mean the operator-initiated checkpoint-restart facilities available in some systems.

These skipping points can be used as labels to jump to if something is wrong in a process. Skipping points can also be grouped: a header-tag could initiate a process, intermediate skipping points could be supplied for sub-processes and at the end of a process a trailer-tag must be present. In this way a system can match its internal state with the skipping points supplied in the data, so that detection and diagnostics of errors or irregularities can be more accurate. Skipping points can be given types, so that a process can anticipate on the type of skipping point it will have to process - if the expected type mismatches with the type that has actually been found, then adequate diagnostical information can easily be generated. When skipping points are used, it must be ascertained, sometimes by means of special preprocessors, that all the necessary skipping points are indeed present: does each operation have a begin and an end, is each left parenthesis accompanied by a right parenthesis, etc. On error occurrence, the process can skip to the next available skipping point, using the state information to determine the kind of skipping point (e.g. level, type). The necessary resynchronization of different processes running in parallel can also be taken care of at these skipping points, thus preventing fault propagation (*cf.* processing strategy maxim 5). At a later stage, further state evaluation might follow in order to find the cause, type, etc. of the error.

As an example of the skipping strategy, consider an automated factory where some product is made: each product is transported on one or more conveyors along several workstations, beginning as a collection of parts, ending as a complete product. During its fabrication a log can be kept, so that at the entering and leaving of a workstation, the actual log can be compared with what it should be. On the occurrence of a mismatch of logs, the complete product as it is at that moment can be placed on a side-track, where at a later stage state evaluation might follow. With this method, each workstation can continue working on the next product.

Processing strategy maxim 5.

Faults should not propagate.

In computer integrated systems the environment of modules that are in an error-state should be shielded from the consequences of that error.

As production-losses due to system-crashes in a Computer Integrated Manufacturing plant will usually be non-recoverable and therefore can be very substantial, an important question is how to reduce probabilities of these disasters and their consequences to a minimum.

Errors occurring in modules for Computer Integrated Manufacturing should not cause the failure of modules other than the one in which they occur. Moreover, techniques should be used which ensure, depending on the nature of the error, the survival of modules in which errors occur. Protecting the environment from the consequences of errors in a local module implies that these modules must be capable of handling their own error-recovery. Attaining this may appear to be an ambitious goal, but it will be the basis for the evolution to the future construction of fault-tolerant systems.

If a system would allow faults to propagate, either to different processors or by way of incorrect data that will impair further processing then situations can easily arise where, due to a single error at an identifiable point processing in an entire factory becomes afflicted by this error. Even worse, when there is more than one error propagating in a system, these errors can interact, thereby making diagnosis and recovery very much more complicated.

Preventing errors to propagate is the basis for both building systems that are capable reliable and systems for which it is easy to localize the cause of the failure. The localization of errors is a major problem in large computing systems, and therefore an equally important issue in processing systems for Computer Integrated Manufacturing: the speed at which recovery from an error is possible depends on it. So the non-propagation of errors contributes to the prevention of large production-losses whenever they occur.

The ability to make reliable sub-systems (by using redundant hard- and software) presupposes that a failing unit does not corrupt other units: making a unit inactive when it has damaged its environment does not undo the damage done and, in many cases, halting a process will not stop further damage.

In machining for instance, a faulty part can be produced due to the use of worn-out tools or the use of low-quality material for its production. Both types of errors should be checked upon, in order to prevent the propagation of such errors through the Computer Integrated Manufacturing system. This can be done by inspection of the tools used and parts made at appropriate points in the machining process. When an error is detected during such an inspection, the subsystem can recover from the error by moving to the next so-called skipping point, where the faulty part can be removed and the error can be corrected. Subsequently, a new part can be produced, or, when the faulty part can be saved, it can be re-machined. The necessary re-synchronization of different processes running in parallel is also taken care of at the skipping points. Another example is a server that erroneously advertises the fact that it can process any request without delay: suddenly all requests for its service are directed to that server with net the result that it will no longer be available on the network.

The early detection of errors is essential for cost-effective production. A fine-grained maze of skipping points in a system for Computer Integrated Manufacturing can help considerably in avoiding errors to propagate.

For the next example, suppose that a process in a network wants to start up a process elsewhere in the network. If the request is not handled quickly enough - which could, for example, be found out on the basis of time-outs - the request to run a process is directed to a different node of the network. If, due to latencies, the time-out technique is not adequate, it is possible that two (or more) requests are granted. These processes in their turn may start up other processes, so that eventually the system gets choked without doing much useful work at all.

The factors responsible for this undesirable phenomenon may not simply be attributable to the apparently incorrect use of a time-out in combination with reposting the request to a different processor. Reposting a request elsewhere can be a reasonable strategy by itself, if load-balance of processors is important, or when real-time issues come into play. These phenomena could even occur as a consequence of quite subtle mismatches between protocols and services requested by the processes running in the system.

A following example of these phenomena is a local process in a network that is able to select a server elsewhere on the net by inspecting the queues for the different processors. By having access to these global data, this process that was designed for its local effects, suddenly becomes a window to the entire system and the processing taking place, with all the possibilities for disasters this situation may create either due to errors within the local process or in the network-manager. Needless to say that errors of this kind could impair the operation of a whole system immediately, and diagnostics, let alone recovery-measures, are very hard to do, if at all possible.

Given the fact that faults do not propagate in a system, the problem of reliability of the processes themselves remains. One important issue here is the *availability* of computer systems, especially in environments with real-time demands. We can distinguish two types of reliability.

The first type, *fail-safeness*, is concerned with designing systems such that their availability is as high as can practically be achieved. Usually, these systems are designed such that there are at least two of all hardware components, so that in the event of component failure a backup component is immediately pressed into service. In this kind of system, one design criterion is to ascertain that total system capability is available at all times. The system programmer or system controller is mainly concerned with error detection and immediate component substitution so that in case of error the failing component can be localized as quickly as possible and its substitute can be immediately utilized.

The second type of reliability, *fail-softness*, does not stress availability at all cost. Backup units need not be available for all critical components, and, in case of a breakdown of one of the parts, the system is allowed to run in a state where the performance may be degraded, but it must remain functionally complete.

Software faults can be divided into two categories: those that cause a system to stop working, and those that are latent (in which case the system seems to work properly, but the results it produces are erroneous). The first category of errors can be processed by exception handling mechanisms. These mechanisms enable a system to regain control in case of errors such as overflow, so that the propagation of such errors through the system can be prevented. The second category of errors is more difficult to handle, because it requires some kind of checking mechanism in order to detect such errors. Of course, checking must take place before the start of a new process and at the end of a finished process. In general, one could not make an a priori statement about the desirability of checking in between those points. Much will depend on the trade off between the cost of checking and the consequences of undetected errors.

Processing strategy maxim 6.

Real-time processing tasks should be well-specified.

All real-time processing tasks for a Computer Integrated Manufacturing system should be identified and their time-requirements should be specified.

Three types of processing can be identified in Computer Integrated Manufacturing: off-line processing, on-line processing and real-time processing. Especially real-time processing can present some critical problems.

A processing system is *off-line* if

i) the input is terminated by an end-of-input marker;
ii) all output resulting from the entire input needs to be produced only after the end-of-input marker has been read.

A typical example of off-line processing is the batch-processing of a job.

A processing system is *on-line* if

i) there is no *a priori* bound on the length of the input;
ii) after reading an input item the system must first process that item and produce output before it reads the next input item.

An example of on-line processing is interactive computation.

A processing system is *real-time* if

i) it is an on-line processing system;
ii) there is an *a priori constant* bound on the *waiting time* for output.

Examples of real-time processing can be found in process-control, flexible manufacturing systems, robot-control, etc.

It should be stressed that the real-time aspects of a system do not have to apply to every accepted input item. When some but not all input items generate real-time behaviour in a processing system, the processing system is said to operate in real-time with respect to just those items.

As to the bound on the waiting time before output is produced, it must be constant and known in advance. Needless to say that the constant bound on the waiting time is independent of any factor: the constant must for example be independent of any previous input.

By waiting time we mean the time that passes between reading the input and producing the output as measured by an external agent independent of the real-time processing system. The exact bound on the waiting time for a real-time system is of course dependent on the particular application the system is designed for. There are tasks that require very sharp bounds (e.g robot-sensor complexes), tasks such as controlling flexible manufacturing systems that allow medium waiting-times and tasks that have longer completion-times but which must nevertheless operate in real-time (storage-control systems, payroll-processing).

The maxim states that

- per application all real-time tasks are to be identified;
- per real-time task the bound on the waiting time for output is to be specified.

It may appear that verification of the real-time properties of a system would not be difficult: take a stopwatch and observe the system in action. However, if the system is poorly designed there will be situations where the real-time properties are disturbed. These situations may arise infrequently. Examples are internal housekeeping tasks of the system such as deadlock-recovery or rehashing. As the internal organization of a system is not easily observable from the outside and real-time tasks are very crucial to the operation of a factory, the need is stressed to specify the real-time tasks long before actual systems implementation.

Following a proper specification of the real time requirements, the necessary conditions are fulfilled for concentrating the control of the complete real time environment to one process. This greatly simplifies guaranteeing the requested reaction times, the major reason being that only one scheduler is needed to meet all of the requirements.

Processing strategy maxim 7.

Software design should aim for time-critical parts to be small and few.

The parts of programs the performance of which are critical with respect to timing constraints should be small and few.

The time-critical parts of a program are those parts where the larger part of its running time is spent. Usually 90% of the running time of a program is spent in only 10% of its code. The time-critical parts of a program play an essential role in the design of real-time systems: they are those parts on which its real-time behaviour is based.

The maxim states that given the specification of a real-time system the software design for that system should be such that the code to realize the time-critical parts must not be large and that the time-critical code may not be dispersed throughout the system.

The larger the program for a certain task, the less is its understandability. When programs are large, it will be more difficult to make a system behave according to its specifications and, once realized, it will be harder to maintain as the code for it becomes larger. This holds *a fortiori* for the time-critical parts of a program.

If the time-critical code is dispersed throughout the system then the system is ill-structured with respect to its time-critical parts: the system, if realized at all, will then be prey to all the disadvantages associated with ill-structuredness - it will be totally unmanageable.

As time-critical systems play a crucial role in many areas of Computer Integrated Manufacturing and the malfunctioning of those systems can be detrimental to the work in a factory, it is essential that they are behaving as specified and that they are accessible to maintenance. Moreover, if the time-critical parts are large or difficult to identify, then it is impossible to reprogram those parts in microcode.

Processing strategy maxim 8.

To achieve simplicity, extendibility, portability, flexibility and reliability high-level programming languages should be used.

Programs designed to be used in a Computer Integrated Manufacturing environment shall be written in high-level languages.

High-level languages in general have the following features:
- facilities for modular construction of processing tasks;
- more orientation towards humans than low-level programming languages; it is easier to express abstract concepts in a high-level programming language than in a low-level programming language;
- machine-independence.

High-level languages with these properties offer major advantages when used both for the construction of modules for Computer Integrated Manufacturing systems and for the definition of how the different modules interact. It is stated explicitly that this maxim applies to both subjects.

Using a well-chosen set of abstractions facilitates understanding complex tasks and problems. In programming this will not only lead to shorter development times, it also leads to more reliable programs that are easier to understand for others. In systems for Computer Integrated Manufacturing this has an analogy in the field of parts programming, where parts are designed and manufactured using a parameterized abstract model of the part. This abstract model enables designers to create reliable parts in a simple way.

Extendibility and flexibility are improved by the modularity offered by high-level languages, that is processing (sub)tasks can be extended or modified easily. Moreover, software modules for similar tasks will have similar structures, so that already existing software modules can easily be adapted for the construction of new processing tasks.
Portability is improved by the fact that high-level languages are machine independent; thus, migration of processing tasks from one processor to another is easy.

Of course, code generated by high-level language compilers might be less efficient than code generated by hand. However, due to improved compiler construction technology this difference is decreasing steadily. Moreover, cases are known where rewriting complex systems from assembly language to a higher level language actually improved run-time behaviour.

The advantages of using high-level languages apply to real-time programming as well. Due to decreasing hardware costs in combination with increasing performance, the use of high-level languages in this field is growing rapidly. Nevertheless, in some cases time constraints may be so severe that certain time-critical parts require low-level language coding. However, in most cases time-critical parts of real-time processing tasks can be kept small and few (*cf.* processing strategy maxim 7).

Processing strategy maxim 9.

Basic components should be used where feasible.

Systems designed to be used in a Computer Integrated Manufacturing environment should be built on the basis of basic components.

In Computer Integrated Manufacturing systems basic hardware and software components should be used as much as possible. A Computer Integrated Manufacturing system, although complex and of an advanced nature, is not special in this respect: many modules can be constructed from basic components (for example, all graphical processing in Computer Integrated Manufacturing can be done using standard graphical software). In constructing systems for Computer Integrated Manufacturing, a major part of the effort will be spent in making a correct interface between the large variety of different basic components.

The use of such standard basic components offers considerable advantages, some of which are:

- modules can be built with low cost: standard basic components have to be developed once, and can be used many times afterwards;
- the construction of modules for Computer Integrated Manufacturing requires less time: standard basic components can be taken off the shelf, and can be applied quickly, since their properties are already known to system designers;
- basic components make it possible to introduce and extend Computer Integrated Manufacturing in a given company in smaller steps;
- basic components make it easier to customize plants with respect to Computer Integrated Manufacturing: components from different vendors have characteristics that are more attractive for a specific type of plant; they also will probably have less attractive features for this plant, and so the user will be able to assemble a system for Computer Integrated Manufacturing that will best suit his needs.

Two categories of standard basic components can be distinguished: in-house developed components as opposed to commercially available components. The maintenance of commercially available hardware and software is usually done by the supplier. In that case, no (or little) in-house knowledge about the internal functioning of the components is necessary. Of course, one should know all about the functionality of such components. Besides, commercially available hardware and software components generally are much cheaper than components developed in-house, partly because they obviously take no time for development from the part of the user.

Moreover, system development will be less risky, since standard components obtained from outside suppliers can be judged and tested before they are bought and used. This will affect the reliability of systems for Computer Integrated Manufacturing considerably. Also, spare parts of frequently used standard hardware components can be kept in stock without high costs, thus enabling quick repair of malfunctioning components.

Using standard basic components will decrease the total cost and time of development and maintenance for Computer Integrated Manufacturing systems. The use of basic components as building blocks for such a system has an important implication: if basic components are available and constructed as replaceable systems, there must also be standard interfaces between those components.

Processing strategy maxim 10.

Systems design should aim for programmer-independence, device-independence and machine-independence.

Systems designed to be used in a Computer Integrated Manufacturing environment should not be dependent on the programmers that made them, nor on the devices and computing machinery that were present in the plant when the system was developed.

Information technology, and the information technology for Computer Integrated Manufacturing in particular, lacks the traditions that characterize many older fields of science and engineering. Consequently, systems once installed will usually need several cycles of enhancement and modification during their operational period.

In a field as young and as rapidly evolving as information technology it is inevitable that the users of this technology will want to renew their systems for more effective or more efficient ones. Companies using Computer Integrated Manufacturing technology will want to expand the applications of these systems.

So, to be as effective as possible, systems for Computer Integrated Manufacturing and their components should be very flexible (*cf.* processing strategy maxim 2). Three objectives have to be met in order to achieve this flexibility.

First, programs should be *portable*, so that systems are not dependent on one particular brand of computers. It would be undesirable to be unable to change an old computer for a new one of the same type. When changing from a general purpose machine to special purpose hardware however, portability problems will arise when one wants to exploit the special purpose hardware to its full extent - this is partly due to the fact that we are consciously willing to pay a price for increased performance.

Second, programs should be *device-independent*. At a particular installation, a system must be capable to operate with a choice of different devices having a similar function. In other words, if a system would not be device independent, then the devices it controls would not be replaceable, and therefore the system's modifiability would be impaired.

Third, and equally important, systems design should aim for *programmer-portability*. It is very undesirable that programmers, as they move from one application or package to the next, be forced to learn and understand new conventions, jargons and concepts at every move. On the other hand, if programmers do not move to different projects, they may form their own concepts which can be difficult to learn for new programmers. Systems must be designed with a clear and understandable conceptual structure, so that programmers are able to express their programs in terms of that conceptual structure.

It should be stressed that computer-oriented standardization efforts develop with precisely those objectives in mind: program-portability (i.e. computer independence), device independence and programmer portability.

It is therefore important that systems be portable, modifiable and have a clear and general conceptual structure not only to achieve device- and machine-independence, but also, and equally importantly, to achieve programmer-independence.

Characteristics pertaining to a specific device make its programming equally specific for that particular device: desirable features such as portability and device-independence are lost. To achieve device-independence all device-specific features, such as instructions, typical ways to establish a particular state, etc, must be restricted to a level as close as possible to the level that drives the operation of the device.

The portability of systems has been a long-standing issue in their design. Indeed, it is very undesirable if systems are not portable from one machine to another, both for systems designers and systems users.

The main vehicle to achieve machine independence used so far has been the application of high-level programming languages. Programs written in a high-level language must be translated to machine-specific code and this is enough to achieve machine independence for many (simple) applications.

There are applications that require more than a simple translation phase to be machine-independent. Portability problems can - for more sophisticated application areas - demand another layer of abstractions to achieve a practical level of machine-independence. An example that comes to mind is the recently accepted ISO-standard for graphical computer applications GKS, which addresses a whole application-area. GKS is implemented as an extra layer between the programming level and the operating system. An application program merely calls procedures the meaning of which is specified in the standard.

Another aspect of the portability problem is the replacement of general-purpose computers by special-purpose machines such as array-processors. Special-purpose machines may have language-translators to enable them to run most programs originally designed for general-purpose machines. It is usually not guaranteed that the systems as realized on these special-purpose machines will perform optimally, due to the special architecture that these machines have. It is not simple for a compiler to detect opportunities for use of the special hardware from high-level language program constructs that originate from general-purpose architectures. It is, for example, well-known that programs that were written with general-purpose machines in mind, need significant modification before they can exploit the hardware of an array-processor satisfactorily.

It can be expected that when systems are built on widely used application-oriented subsystems these problems are less severe, since the application-oriented sub-systems may have a special implementation exploiting the new hardware and most of the processing-time of the application will be spent in the specially coded sub-systems.

The conventions and methods used in the design and construction of software should be uniform to prevent programmers from being forced to learn and understand new conventions, concepts and methods as they move from one application to another.

For the design-phase this implies the use of (a standard convention and) a standard framework for the formulation of concepts.

For the coding-phase this implies the use of a standard (high-level) programming language well-suited for the application. Moreover, a uniform programming style, standard lay-out of programs and their documentation should be used in order to obtain standard programming constructs.

For the overall control of software development a standard, sophisticated programming environment should be used. Such a programming environment should contain tools for source- and object library management, or automatic recompilation of software modules depending on modules which have been changed.

Processing strategy maxim 11.

Programming systems and control systems should be clearly separated.

Systems designed to be used in a Computer Integrated Manufacturing environment should allow programming systems and control systems to run independently.

In Computer Integrated Manufacturing, supervisory tasks have to be supported on different levels of the system. Typical examples of these supervisory tasks are: data acquisition, equipment control, quality assurance, material flow control, etc. A system which supports these supervisory tasks is known as *control system*. The control level is characterized by the fact that it is associated with a device operating in the real world. Therefore, a control system always operates in real time (*cf.* processing strategy maxim 6).

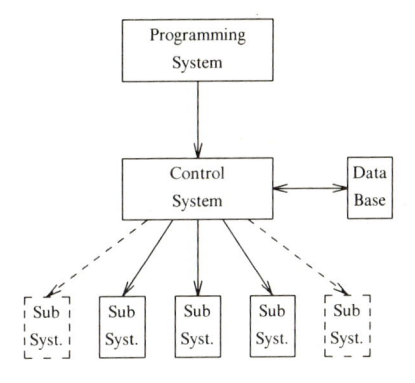

Figure 12.1 : A control system supervising several sub-systems.

In order to perform these supervisory tasks, the control system has to be supplied with a specification of these tasks. The system that supports the specification of tasks is the *programming system*. In figure 12.1, the structure around a control system having access to a database and supervising several subsystems is shown.
A programming system can be the subsystem of a higher level control system, in which case a hierarchical control structure is formed (see figure 12.2). A control system in a hierarchical structure, can communicate with control systems at the same level, and it can communicate with lower level control systems via the programming system of that low level control system.

Generating information that specifies a task as done by a programming system is of a different nature than seeing to it these tasks are actually performed as done by a control system. This different nature and the hierarchical relation that exists between the programming system and the control system, suggests a separation of the two systems. There are more well

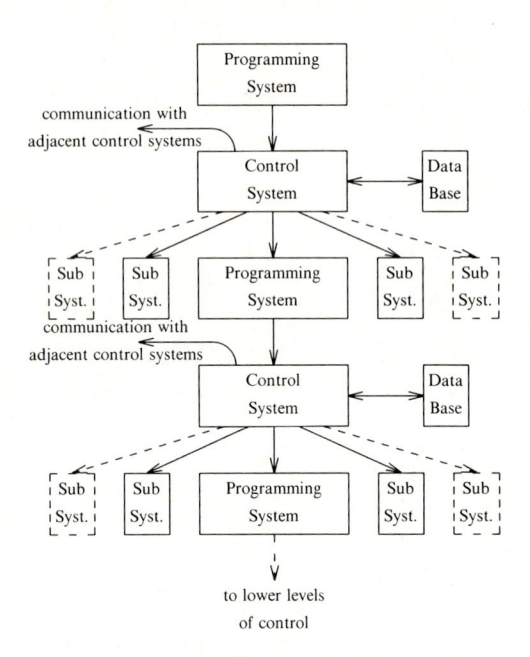

Figure 12.2 : Recursive structure: A programming system accepts a certain complex task and subdivides it in subtasks, that have to be performed by subsystems, controlled by a controller. These subsystems in their turn contain one or more subprogramming systems, which subdivide the subtasks in sub-subtasks etc. Eventually, the complex task will be split up in numerous elementary operations.

founded reasons to clearly separate these two systems as we shall see later on.

In Computer Integrated Manufacturing, one could think, on a high level of abstraction, of a design system, where the specifications of objects are determined. The accompanying control system then would be the manufacturing system that actually manufactures these objects.
As example of a low level programming system and control system, one could think of the programming system (or programmer in a non-automated system) of a NC-machine and the controller of that NC-machine.

It is not hard to imagine the different nature of a design system and a manufacturing system, that for instance manifests itself in the way object representations are handled. In a design system, the primary goal is that object representations can be easily manipulated and altered. Object representations in a design system are dynamic entities. In a manufacturing system on the other hand, the representation of the objects to be manufactured is static. The main concern here is to assure the constant quality of the objects, flexibility comes second.
Also the real time demands of the two systems differ. Although there are some real time demands in interactive design systems, real time demands in a manufacturing system are of a much more pressing nature.
Due to these differences, it is not surprising to find different structures and programming languages in both systems.
In practice, one will find a clear separation between design systems and manufacturing systems. This separation is not a result of well thought of reasons, but it is more a result of the difficulty to integrate both systems.

This is different for low level programming and control systems, where nowadays integrated systems *can* be found. Some CNC-machines for instance, have an integrated programming feature, that allows shop floor programming. This feature is very useful indeed, not only because it allows a gradual introduction of NC-technology without great organizational provisions. Shop floor programming also allows stand alone operation of integrated systems. This provides an escape to local programming when the centralized programming system fails and thus makes the central facility fail-soft.

On the other hand one has to assure never to mix the programming system and the control system. When both systems are sharing one processor, this means among other things they are not allowed to share variables (*cf.* processing strategy maxim 2).

The same is true for an integrated design and manufacturing system. There it is evident that an object representation, upon which designers are still is working, should never be used by the manufacturing system as long as it remains under development.

Reasons to separate programming- and control systems are the following.

Separation will raise the efficiency of the controlled system (for instance a NC-machine), because this machine can continue operation while the programming takes place.

Furthermore, programming itself can be done more efficiently, because the program first can be checked (for instance by a simulation system), it is easier to alter and it is reusable.

The separation makes it also possible to generate machine independent programs, that just have to pass a functional interface to become applicable for the particular machine. In this way, one programming system is able to program several (different) machines.

The separation furthermore enables the programming system to be distributed to other parts of the system.

Taking into account the directions in which manufacturing systems develop, integration of programming systems and control systems in the future will also be possible for the higher levels of the system, so that also at these levels one will have to assure that a sufficient separation between programming systems and control systems is maintained.

A clear trend in the design of flexible manufacturing systems is processing in small batches. Currently, flexible manufacturing systems process large batches of one version of a product alternated by large batches of a different version of that product. One way to increase the flexibility of a manufacturing system is to reduce the size of those batches. A feasible way to achieve this is to design manufacturing systems that are operator-controlled, where the operators do a limited amount of programming to the effect that they bring the flexible manufacturing system into the right state to make a different version of the product. This would imply a larger amount of programmability in the control-system for the system, while at the same time the logical separation between programming and control for that manufacturing system must be maintained.

Processing strategy maxim 12.

The principle of lazy evaluation.

Systems designed to be used in a Computer Integrated Manufacturing environment should begin processing only when all data for their processing task are ready and available to be processed.

The principle of lazy evaluation is a maxim for deciding where and when computer processing should take place. Many computing processes need to evaluate part of the environment before they can actually start their specific processing task. The principle of lazy evaluation states that the evaluation of the external state is done only when and as far as it is certain that the result of the partial evaluation will be used immediately following.

The alternative would be to immediately evaluate the external state each time one of the components changes value, so that any process using the state could run immediately. The latter will become very expensive when the state vector changes more often before any process needs to run. This is bound to be the case when state information originates from different sources.

When a state changes component wise, intermediate combinations of component values may be invalid or erroneous. The state evaluation mechanism would have to detect this and signal an error. In a more tolerant environment however, one would like to only signal an error when the state is actually applied. In the other cases (i.e., where the state is not yet used) further change of state variables might remove the error condition. If the principle of lazy evaluation were applied the system would never be in doubt about whether or not to react to error conditions because it is certain that the erroneous elements are about to be used. In between real use of the state, the system would be very tolerant.

The lazy evaluation option is relevant in many situations occurring in Computer Integrated Manufacturing environments, as exemplified by a robot that needs to place a part on an conveyor for transportation to the next cell. To do this the following conditions must be true:

- the part must be ready for transport (e.g., the treatment of the part at the robots cell is completed).
- there must be a place available on the conveyor to put the part.

Now the lazy evaluation principle states that the process (e.g. the vision system) which looks for a place on the conveyor system will only become active for this function when a part is actually ready and available for placement. The alternative would be to recalculate placement options each time the state of the conveyor system changes. Associated with a lazy evaluation mechanism is a more complex state administration which keeps trace of the state components that need to be updated before use.

The consequences of the lazy evaluation pronciple are twofold: In situations where at first sight lazy evaluation would lead to unacceptable delays (because data are not ready when needed), the processing task must be broken down in refined tasks in such a way that the amount of data to be kept up to date in advance, is reduced to an acceptable level (considering resources) and that the place where processing has to stop because further data need to be evaluated is well defined first, and does not cause unacceptable delays.

The other consequence is that the lazy evaluation mechanism itself requires that for each process which needs to become active, it is explicitly known which pre-processing need to be checked or triggered. Conversely, for each data item it must be specified which process or data item use this particular item.

Processing strategy maxim 13.

The principle of equal resource utilization.

A processing task must utilize approximately equal amounts of resources for successive executions with approximately equal amounts of data.

This maxim addresses the predictability of runtime behaviour of processes running in a system for Computer Integrated Manufacturing. Tasks designed for Computer Integrated Manufacturing systems must have stable and reliable runtime characteristics. These characteristics include the utilization of resources like processor-time, memory, disks and tapes, communication lines, the number of processes started up and completion times.

Runtime behaviour may be unpredictable for several reasons. Large databases exist that people use normally and satisfactorily during the week. The computer spends its weekends collecting garbage for that database. Working with the database will probably not lead to serious problems as long as the amount of secondary storage is large enough so that garbage collection actually does take place in the weekends. However, when the garbage collector is called during the week, perhaps because somebody needed much disk-storage, the database will probably not be available for the rest of the whole week.

A second example that illustrates a possible reason for unpredictable runtime behaviour can be found in time-sharing environments. In time-sharing environments users can usually start up a number of processes that will have diverse resource claims. When one user suddenly starts up many demanding processes this will influence the completion-times of every other process in the system: suddenly all the users get very low response-times and it is hard to find the cause as it can only be discovered by means of the very time-sharing system that appears to stop working! So the principle of equal resource utilization applies as well to the users of a computing system.

Both examples illustrate that causes of unpredictable behaviour can come from global non-transparent side-effects that influence the performance of processes. In the first example the architecture of the database was such that it could endanger not only its performance, but even its availability. The reason for this lies in the fact that transactions on the database generate new data and may invalidate old data. One approach to the management of invalidated data is to mark them as such so that the system will correctly skip these records, but these data are not removed from the system and thus they continue to occupy storage. When storage is nearly used to completion, all these invalidated records and tables have to be removed so that storage is 'cleaned up' and work can continue. This garbage collection process can be very time-consuming, so if many entities in the database are changed or if many new relations are constructed, the system will, due to the side-effects generated by these operations, quickly grow to its maximum size and consequently the garbage collector must be called into action. Note that the end-users of the database may be entirely unaware of the architecture of the system and so for them the system simply appears to be inoperative for mysterious reasons, certainly when they are not aware of having done much more or much different work than in other periods. The second example illustrates that due to the behaviour of one user (or, equivalently, one process suddenly starting up many other processes) the performance of an entire sub-system may be severely degraded. This phenomenon is also caused by side-effects of the operations of one process.

There are a number of reasons for requiring this predictability.

- Process planning. A process in a factory that unexpectedly uses much more resources than it normally does, is by itself undesirable, but it usually has a cascading effect on the rest of the work in the factory: it can virtually block the traffic in a network if it suddenly starts communicating very much data or it can lead to very long idle times for following processes if the completion time is suddenly very long.
- Human factors. In a Computer Integrated Manufacturing environment people will apply automated processes. They must observe and react to these processes, which can become very hard if those processes behave irregularly: confidence in the correct functioning is undermined and, on the other hand, real malfunctioning may be harder to discover. Irritation and fatigue will increase especially when processes have to be watched intensively. Many skills humans have will become useless, especially the skills of learning and anticipation.
- Planning and programming. Both task planning and programming are based on predictable quantitative aspects of the behaviour of tasks. Predictability is served by a small variance of those quantitative aspects. Optimization and scheduling will fail if there are too many exceptions with respect to the aspects on the basis of which optimization or scheduling is done.

Processing strategy maxim 14.

The principle of distributed capacity.

Overcapacity should be distributed evenly over systems.

Much work in computing environments is characterized by relatively long periods of modest resource-claims, alternated by shorter periods where sudden bursts of activity take place. In environments for program development, for example, users are normally editing or debugging their programs, but sometimes they must recompile an entire system, which would result in such an activity-burst. A system must have a certain amount of overcapacity to accommodate for these sudden activity-peaks. As these peaks usually relate to one or more different resources, the overcapacity of a system should be balanced with respect to its resources.

A similar situation holds for networks. The nodes of a network can be thought of as processors, file-servers, printers, workstations, etc. Multiple copies of these entities can exist in a network. The nodes can send and receive messages. Several different types of processors, workstations etc. can be present in a network. A network inherits the set of resources of a single computing system augmented by the explicit availability of communication links between nodes and the presence of a finer structure given by the different capacities of the resources; a network can accommodate mainframes, microprocessors and processors for special applications, to name an example.

If the overcapacity of those resources is not evenly distributed over the system, there will be one or more subnets on the network that have no capacity left and will therefore not be able to cope with activity bursts, with all the consequences and possibly interacting effects this may have in for instance a factory.

The maxim does not only apply to systems globally, but it also applies to subsystems. When there are multiple resources having different capacities on a system, then it will not only be so that their basic load tends to be higher in absolute terms, but also the load-peaks tend to be higher to the extent that the capacities of these resources are larger. For instance, a mainframe in a network will not only have a higher load factor than a microprocessor in the same network, the peaks will also tend to be higher. Overcapacity will have to be distributed evenly over systems relative to their respective capacities.

A similar point holds with respect to time-sharing systems: one tends to mix I/O-bound and CPU-bound processes in order to avoid these load-peaks. Similar considerations hold for networks. It is a good rule to do processing in the neighbourhood of the resources (*cf.* processing strategy maxims 1 and 2). This will reduce the claims on the communications links and consequently improve the completion times of these processes and global network preformance. Another good approach in avoiding load-peaks in network communications is to move the resources to the processes that need them most often if the resources allow such operations (e.g. files). In this way near-optimal placements of (movable) resources can be obtained dynamically, similar to self-organizing data-structures in other areas of computation.

Processing strategy maxim 15.

The principle of itemized processing.

Parts to be processed by systems designed to be used in a Computer Integrated Manufacturing should be input in an integrated way.

This maxim is concerned with reducing the complexity of systems for Computer Integrated Manufacturing especially with respect to control.
The maxim states that processing for a certain task is only started when all information required for this task is available. Moreover, the particular service for this task will receive this information as one unit.

The most important effect of this maxim will be that no process is burdened with collecting data (place-wise) prior to running a task or, even worse, will start processing until halfway through the task the process has to be suspended because the remaining information is not yet available.

An example would be a machining line which should not start machining a certain part until all the tools and materials needed are installed or available along the line.
If this principle is not strictly followed, unfinished tasks will come into existence and proliferate through the system. Processes which have progressed halfway can only be suspended and restarted at great expense. Errors or failures will have more widespread negative effects because rescheduling will be much more restricted. This principle also constitutes one of the basic guidelines for standardization of functional interfaces. Standards need to support the concept of itemized processing. In this way pre- and/or post processors can be built which will provide processing tasks with uniform commands based on the itemization. For each task assignment either on the side of the originator or on the side of the receiver the itemization will take place. The buffering necessary will be in the pre or post-processor.

Conclusion.

This concludes the maxims concerning processing in Computer Integrated Manufacturing. Some of the maxims from this chapter will be put into practice in two areas which are specific for Computer Integrated Manufacturing: sensors (discussed in chapter 15) and graphics (discussed in chapter 16).

Chapter 13

Communication: state of the art

Introduction

The rise of computer networking during the last years made it possible to connect computer systems in a highly reliable way. As a consequence of this, computer networks can now be used in Computer Integrated Manufacturing. Clearly, the concept of Computer Integrated Manufacturing would be inconceivable without computer networks.

The major task of the network is to take care of the integration of the various computer systems. By this, computer networks will contribute to high reliability, cost reduction and high flexibility.

In Computer Integrated Manufacturing reliability is considered to be a most important design issue. The cost of a failure of the computer system can be prohibitively high, for example owing to production losses. The consequence of production losses should not be underestimated: their cost can be higher than the cost of a complete, integrated computer system. Reliability can be improved by using computer networks. To achieve this, the system should be built in such a way that when one computer breaks down its task will automatically be performed by another one. Clearly, an improvement of reliability implies cost reduction, even when extra investments are needed.

The use of computer networks contributes to cost reduction in more ways. Trends in hardware development lead to a better price performance ratio for small computers than for large ones. As the development time for small computers is short, later (and therefore cheaper) technology can be used. This results in networks of mini- and microcomputers more and more becoming economically justifiable.

The advantages of highly flexible manufacturing systems are well known. The most important one is that a system of high flexibility makes it easier to realize wishes of customers, wishes that are of a miscellaneous nature. As the production process can easily be influenced with the aid of computer networks, it will be easy to tune the production to the demand.

13.1. Communication facilities

In the following sections an overview of the state of the art in computer networking is given, with the emphasis on aspects that are of major importance to Computer Integrated Manufacturing.

Network architecture

The architecture of modern computer networks is a seven-layer structure (see Figure 13.1).

Layer	Protocol	Layer
Application	.....	Application
Presentation	.....	Presentation
Session	.....	Session
Transport	.....	Transport
Network	.....	Network
Data Link	.....	Data Link
Physical	.....	Physical
Transmission medium		

Host A — Host B

Figure 13.1: Open Systems Interconnection between Host A and Host B.

This structure has been agreed upon by the International Organization of Standardization, [ISO/TC97/SC16/N2271981] and is commonly known as the ISO-OSI model (OSI = Open Systems Interconnection). Although many existing networks do not follow this standard completely, the concept is extremely attractive. The purpose of each layer in the OSI model is to offer a set of functions to its 'upper neighbour' and to conceal the problem of implementing these primitives on top of lower layers. The higher layer has to use the functions of the next lower layer only. As a matter of course, the higher the layer, the more elaborate the functions it offers. The inter-layer interface has been designed carefully, so that changes that enable a better implementation of some layer (e.g. technological improvements) do not affect the implementation of higher layers.

A short description of each layer will be given now. More details can be found in Tanenbaum. [Tanenbaum1981] Layer one, the *physical* layer, is concerned with the transmission of bits over a transmission channel. The meaning of the bits is of absolutely no concern to this layer. Layer two, the *data link* layer, adds the possibility of sending data frames over the network. Error checking is done, so that transmission errors cannot be seen by the next layer, called *network* layer. Its task is to divide data packets into (level two) data frames and to prevent too many data packets from using the same transmission line, causing congestion. The next layer is called *transport* layer. It accepts data from its upper neighbour layer, splits it into packets, and ensures that these packets arrive at their destination in the correct order. The transport layer is a real source-to-destination layer, as opposed to lower ones. In layer five, the *session* layer, connection establishment as well as connection management is performed. In this way the functions offered by the transport layer can be made more application oriented.

The sixth layer, called *presentation* layer, has as main function the reduction of the amount of data sent over the net. Data compression usually will be done in this layer. Another function of the presentation layer is data encryption, a topic that is extremely important in long haul networks. The top layer is called *application* layer. It is up to the user to define protocols in this layer.

13.2. Transmission - analogue versus digital

The world of data transmission can be divided into two groups: analogue and digital.

Historically, analogue transmission, in which some physical quantity continuously varying in time is sent over a transmission line, has been most important. As communication between computers becomes more and more important, digital transmission is taking over this leading position.

The difference between analogue and digital transmission is that in analogue transmission the precise value of the signal is used, whereas in digital transmission the value of this signal will be viewed as being zero or one. Digital transmission has several advantages over analogue transmission. [Tanenbaum1981] As the set of possible values to be send comprises only two elements the error rate is potentially low. A weakened signal can almost always be recognised, restored to its original value and repeated accordingly. Apart from this, several types of data can easily be multiplexed over one transmission line.

It should be weighed carefully in what form data will become available and in what form it should be transmitted. When digital data is transmitted in an analogue form, it has to be converted by a modem (modulator-demodulator). Also, when analogue signals are transmitted digitally, they should be converted by a codec (coder-decoder).

13.3. Transmission media and their characteristics

The purpose of computer networks is to transmit data over a communication channel. The hardware chosen largely influences important system parameters like performance, reliability and maintainability. Several transmission media are used; a characterization of each of them will be given in the following sections.

13.3.1. Coaxial cable

A very popular transmission medium is coaxial cable. Among its advantages are a low error rate, high data rates and low cost. The error rate varies from one bit in 10^7 bits to one bit in 10^{11} bits. In existing local networks, the transmission medium is essentially error free.

As always in the real world, there is no advantage without an associated disadvantage. Here it is the sensitivity to electrical noise that takes care of a large increase of the error rate in environments in which such electrical noise is generated. It usually will be hard to isolate the cable sufficiently.

Coaxial cable can be used to support two sorts of transmission channels: baseband and broadband.

Baseband technology

When a coaxial cable provides a baseband channel, digital data is directly put onto the cable; the cable is used as a single channel. Digital information departs from the transmitting node in both directions. The transmission rate can be up to 50 Mbps, depending on the distance that has to be covered. The great advantage of the coaxial cable is the ease of installing or removing nodes from the network, often without disturbing current network traffic. As distinct from optical fibres, coaxial cable can be used in many network configurations, not only in point-to-point networks.

Broadband technology

Broadband-based networks use a technology that is widely used for cable television (CATV). The physical channel is divided into several separate logical channels by use of Frequency Division Multiplexing (FDM). Broadband cable television uses directional broadcasting, putting some constraints on networks using this technology. A solution to the problem of having directional transmission only, whereas bidirectional is needed is to let one node act as a repeater: it has to repeat the incoming data stream in a different frequency range (to preclude interference). The consequence of this policy is that a message generally has to go farther than with bidirectional transmission: first to the repeater node, then back to the destination which means that transmission delays are increased. The advantage of broadband networks is that multiple logical channels can be supported, each with a different bandwidth. This means that all kinds of data (voice, video, digital data) can easily be multiplexed on one physical channel.

13.3.2. Twisted pair cable

Twisted pair cable technology is frequently used by telephone companies. Depending on wire length it can handle frequencies up to 10 KHz. Its main advantage is its low cost, its most important disadvantage is its susceptibility to external interference. This means that twisted pair cables work well for the transmission purposes they historically were intended for: telephone calls and other low frequency applications. However, they are totally inadequate as transmission medium in a high performance, high reliability network.

13.3.3. Optical fibres

Optical fibres have several desirable properties. First, the transmission rate can be very high (up to 10^9 bits per second). Secondly, the error rate is low: about one error in 10^9 bits. Thirdly, this error rate does not increase in hostile environments: it is not influenced by radio frequency interference, electromagnetic interference or electromagnetic pulses. In contrast to conventional communication media crosstalk is negligible, even when many fibres are cabled together. Another characteristic of optical fibres is that they are best suited to digital transmission. Additionally, fibre optic cables are thin and have a low weight, which makes them easy to install.

But there are also some disadvantages. The most important one is the difficulty to retransmit an incoming signal on multiple output lines. This implies that optical fibres are best suited to point-to-point networks, that is networks in which each node has exactly two neighbours. Furthermore, it is difficult (or almost impossible) to add additional nodes to the network, implying that the topology of the network should remain constant during its operation, providing for less flexibility. Other disadvantages are the lack of standardization of components and, given the current state of the art, the fact that fibre optic technology is expensive. However, as soon as fibre optic technology will be used more, prices will rapidly fall and this will make the new technology cost effective.

Now that the specific advantages and disadvantages of fibre optic technology are collected, it is time to become more specific and make conclusions from this information. Fibre optic technology is superior to other transmission media in several important ways: principally in networks where noise will be generated the reliability of the network can be increased when fibres are used. Since the cost are still high, it should be weighed up carefully whether the gain in reliability is worth the extra cost. As long as it is impossible to install or remove nodes from the network without disturbing network traffic, fibre optic technology can be used best in networks where the topology is stable for relatively long periods.

For these kind of networks, optical fibres will soon be the best available transmission medium.

13.3.4. Satellites

Communication satellites can be seen as repeaters. A signal that is sent to it will be broadcast in another frequency range, so that other earth stations can receive it. Satellites that are used for transmission purposes are usually at an altitude of approximately 36,000 kilometers, so the time that elapses before the transmitted data can be received will be substantial: 270 msec. Apart from this, the error rate of a satellite channel is dependent on meteorological conditions. Protocol designers should take error rates of one in 10^4 bits into account.

Satellites can be of great importance in long haul networks, especially when much of the data should be broadcast. Often the only alternative way of communication is to use the existing (low-bandwidth) telephone system. When large amounts of data have to be sent satellites can perform better. However, they are not suited well to local networks mainly because of the large propagation delay.

13.3.5. Radio

Radio channels can be used in two different ways, one being broadcast, the other point-to-point. The data rate that can be achieved depends on the frequency of the radio signal. For reasonable data rates this implies frequencies in the MHz or GHz range. The error rate of radio channels is higher than with other transmission media. This means that radio channels are not particularly suited to communications in Computer Integrated Manufacturing. A possible use of radio channels, however, can be the connection of two cable based networks, for instance because the distance between them is large and it presents difficulties to use another transmission medium.

13.4. Local Area Networks

Some characteristics of local area networks are: [Tanenbaum1981]

1. They are small, the size of the of the network does not exceed more than a few tens of kilometers; usually a local area network is contained in a complex of buildings.
2. They are usually owned by a single organization.
3. They have a data rate of more than 1Mbps (Million bits per second).

As these three conditions will generally be satisfied in Computer Integrated Manufacturing, the importance of local area networks is elucidated.

13.4.1. Ring Networks

Ring networks are called this, because their topology is shaped like a ring. Every node of the ring has a point-to-point connection to two neighbours, except for the trivial case of a network of two nodes. The connections can be made by using coaxial cable, twisted pair cable or optical fibres. Since a ring consists of point-to-point connections, one can choose the most appropriate transmission medium for each connection separately.

Token rings

The most important type of ring is the token ring. When the ring is idle a bit pattern, called the token, circulates around the ring, from node to node. The token performs the duties of a permission to send, which means that a node can send only when the token has just passed by. To prevent other nodes from transmitting, before sending is actually started the last bit of the token is inverted. The hereby generated bit pattern is called a connector. To prevent token and connector to appear in the data, bit stuffing will be used. The concept of bit stuffing will be explained by an example. If the token consists of eight 1 bits, the sending node will generate a 0 bit after seven succeeding 1 bits in the data, whereas the receiving node will remove this 0 bit. By this, it is precluded that tokens appear in the data on the ring even when they appear in the original user data. The data transmission will begin with the transmission of a connector, after which the actual data is sent, followed by a new token.

Contention rings

A disadvantage of a token ring is the delay introduced by waiting for a token under conditions of low load. Because of this a new ring design was suggested, the contention ring. On a token ring a token circulates even when there is no traffic, but a contention ring is quiet in these circumstances. Eventually, a node decides to send, it puts its data on the ring, followed by a token. Other nodes can monitor the traffic on the ring and wait for the token before they transmit. This means that when the contention ring is full of packets it acts like a token ring. The only problem is that two nodes might both decide to start data transmission because they both think that the channel is idle. In that case a collision occurs and both have to wait a random time before their new attempt.

Slotted ring

Another kind of ring is the slotted ring or empty packet ring. In this ring a number of fixed size packet slots circulates around the ring, each with a leading bit to indicate whether the slot is empty or not. A node that want to transmit waits until an empty packet slot arrives, marks it as full and puts its data in it.

A disadvantage of this system can be that each packet has the same size. When this size is small, messages have to be divided into several packets, if it is large, bandwidth is waisted because packets will not be filled completely. Furthermore, sometimes artificial delays are needed in the repeaters, because otherwise the ring cannot contain all packets simultaneously. This can be caused by a large number of packet slots, a large packet size or a small number of nodes.

A well known slotted ring is the Cambridge ring. [Wilkes1979] Packets consist of two bytes of data. The first implementation of the ring had several twisted pair connections as well as one fibre optic link. Nowadays they are often used in many network areas.

Problems in the design of a ring

This section is based on an article by Saltzer et al. [Saltzer1983] The main problems in ring design are the reliability of the repeater string, distributed initialization and closed loop clock coordination.

In the basic token ring a failure of one node can cause the failure of the entire network, since all nodes are active repeaters. A possible solution is to let every inter node transmission line go through a central point, the wire center. At this center, bypassing schemes can be used to disconnect failing nodes or lines. The resulting configuration, a star shaped ring, is commercially available.

The problem of distributed initialization is also solved. [Saltzer1981] As we have seen in the previous section the ring is in trouble when the token is lost. Although the probability of transmission errors in Local Area Networks is usually low, it is not zero and therefore the aforementioned problem can occur.

To handle this problem a strategy has been developed. It consists of two parts, one that detects the trouble, the other solves it. Discovery of token loss can be done by a simple timer. When token loss is detected, a special, so-called jamming signal is put on the ring to ensure that every node knows about the current situation. Then, every node tries to reinitialize the ring, one after the other, based on each node's station number. If an attempt fails, for example owing to collisions or transmission errors, the same node is prohibited to try again. This simple scheme guarantees that, with a high probability the ring will be functioning soon after token loss.

13.4.2. Carrier sense networks

The transmission media most suited to carrier sense networks are coaxial cable and twisted pair cables. In contrast to the token ring, the topology of the net is not prescribed in a carrier sense network. It can be linear, tree, segmented etc. The basic point of carrier sense networks is their access protocol. It is based on the fact that the propagation delay (i.e. the time necessary to reach the farthest node of the network and return) is small compared to the packet transmission time. This means that a node can listen whether other nodes are transmitting, and will try to make access with the channel only, when it does not hear any transmissions. Although, because of the propagation delay collisions cannot be prevented completely, this protocol performs better than a protocol without the listen-before-transmit-feature.

The most widely known carrier sense network is Ethernet †. [Metcalfe1976] † Ethernet is a trademark of Xerox Corporation Its transmission medium is coaxial cable. The access is based on the above-mentioned listen-before-transmit-feature. Whenever a collision is detected, the stations involved stop their transmission, wait some time, different for each station and depending on the number of failed transmissions, and later try to access the net as usual. This protocol is called CSMA/CD for Carrier Sense Multiple Access with Collision Detection. Ethernet is a most commonly used local area networks.

13.4.3. Dual rooted tree

A local network design that may become important in the future is the dual rooted tree, developed at the University of Toronto. [Boulton1983] It uses glass fibres as its transmission medium and, as its name suggests, its topology is a dual rooted tree. The access protocol is relatively simple but requires that a node can listen to all its incoming transmission lines simultaneously and handle accordingly. Under these conditions the performance can be very high. However, the dual rooted tree is merely a research project, not involved in standardization efforts of the IEEE 802 committee. [Graube1982, Clancy1982]

13.4.4. A comparison of Ethernet and token ring

The most commonly used local area networks nowadays are carrier sense networks and token rings. Because of this, many comparisons of the two have been made. [Chanson1982, Pogran1983] We will try to summarise the results, emphasising arguments that can be important in Computer Integrated Manufacturing. Several characteristics of local area networks are of interest: performance, reliability, maintainability and fairness of the access protocol.

Performance

Intuitively, a carrier sense network performs well, when it is lightly loaded because a node can send immediately when it wants to, and the probability of collisions is low. On the other hand, in a highly loaded carrier sense network the likelihood of collisions is high, which means that the packet delay can be high. In comparison to a carrier sense network with CSMA/CD a token ring will perform worse under conditions of low load, because a node has to wait for the arrival of the token, but performs better under conditions of high load, because it is essentially collision free.

In a carrier sense network the performance is highly dependent on the ratio propagation delay versus packet transmission time. When this ratio is high, for example because of the large size of the network, or because of the high data rate of the transmission medium, the advantage of the collision detect mechanism disappears almost completely. This implies that when better transmission media become available, the performance of a token ring will improve more than the performance of a carrier sense network.

A more formal approach to performance can be found in, [Bux1981] and. [Stuck1983] The performance of an Ethernet is described in. [Shoch1980] Tobagi and Hunt analyse the performance of an CSMA/CD network. [Tobagi1980]

We need to emphasise that the policy of using highly loaded networks is not considered to be a good one, especially not in networks with real time constraints.

Reliability

Reliability is considered to be one of the most important issues for comparison. Essential is that a failure of one node of the network will not cause damage to the rest of the network. In the earliest token ring designs, such a failure would disconnect the network, a highly undesirable situation. Star-shaped rings however, [Salwen1983] do overcome this problem, carrier sense networks do not have it.

Another point of interest is the failure rate of hardware and software. As Graube puts it [Graube1982] there should be only one undetected transmission error per year. In Computer Integrated Manufacturing there can be situations in which even this error rate can be too high. But, in a hostile environment like in some factories, the probability of transmission errors can be high due to electromagnetic induction, unless special precautions have been taken. The most elegant solution to the problem is to make use of optical fibre technology. An adventitious advantage is the possibility of high transmission rates. At this moment the problem of turning a broadcast network into a fibre optics network is not satisfactorily solved. [Saltzer1983] Token rings can easily be adapted to optical fibres.

A point worth noting is the desired upper bound on the packet transmission delay. Unfortunately, given the present state of the art, there is no guaranteed upper bound. In both networks designs transmission errors can ruin the data, leading to retransmissions. Moreover in an Ethernet there is a (low) probability of repeated collisions, whereas in a token ring token loss, owing to transmission errors, and resulting reinitialization procedures can theoretically increase the transmission delay considerably.

Maintainability

It should be easy to repair failing connections and nodes. It should be easy to add new nodes to the network and to remove unused nodes. When the topology of the network is changed this should not influence the normal operation of the network. In both the star shaped ring and an Ethernet it is easy to remove a failing node from the network. Detection of the failing node is easier in the ring. [Saltzer1983]

Fairness of the access protocol

In the token ring a fair access protocol is provided, that is a protocol in which no particular node can sabotage the entire network by sending packets continuously. In a broadcast network with CSMA/CD the situation is different. Because in most systems the time a node waits before retrying after a collision is positively correlated to the number of collisions a packet has suffered from, a packet that has been involved in several collisions is discriminated against.

In both network designs, it is not easy to have high priority messages. In Computer Integrated Manufacturing this is an undesirable situation, because there are excellent applications of these. One example is the use of alarm messages in a factory. However, as transmission media become faster, the need of high priority messages will probably become less important.

13.4.5. Local networks and standardization

When the International Organization of Standardization started its network standardization efforts, leading to the model of Open Systems Interconnection, this model was designed with a view to long haul networks. Local networks differ from long haul networks in many ways and this finds expression in the way the physical and data link layer of local networks are designed. Although local networks do not fit in the OSI model completely, some standardization was clearly needed. The IEEE 802 committee has the intention to provide standards for local networks. Layers above the data link layer of OSI are beyond the scope of these efforts, which means that local networks and long haul networks adhere to the same standards from the network layer.

As indicated already, several incompatible schemes for local networks exist. Therefore the draft of the IEEE 802 standard distinguishes three different types: contention networks with CSMA/CD, token rings and token busses. For each of these types a standard will be created.

13.5. Wide Area Networks

In Computer Integrated Manufacturing wide area networks are not as important as local area networks. However, for some applications they play an important role, for instance the research and development department of an industrial organization will benefit largely from the use of wide area networks, because their use gives the opportunity to react adequately to developments in the 'outside world'. It is evident that the use of wide area networks in this way gives enormous advantages, but, on the other hand, this implies serious threats to the security of the integrated computer system. The most secure way to ensure that researchers of other organizations will not copy (or modify) private information is to let a special computer serve as an 'information-conveyor'. Information can be brought to this special purpose computer from other in house computers (for instance via a local network) and from the outside world, but it should not be possible to reach the in house computers via this computer. Still, special precautions have to be taken to ensure that information can be used only by the people it is intended for.

Another field where wide area networks are used is as connection between several geographically dispersed local area networks. By this, integration of the complete computer system can be accomplished. Since often the telephone system is used to achieve interconnection, protection and security problem need to be considered carefully.

13.6. Communication types associated with CIM-modules

Now that the state of the art in computer networking is described sufficiently, it is time to become more specific and draw the attention to communication requirements in Computer Integrated Manufacturing. In the next sections communication in CIM and its influence to hard- and software is described.

For each CIM module the communication requirements need to be classified. The first question, a very important one, is what is the type of the required communication, for example does it need high data rates, is the communication bursty, what are the reliability constraints etc. The answers to these questions will sometimes make the use of certain types of networks impossible, for instance because they do not support sufficiently high effective data rates.

Next, environmental issues need to be considered and other issues related to the circumstances in which the communication takes place. To give some examples:

- The environment of the communication network can influence the choice of the transmission medium, especially when electrical noise is generated frequently
- The real time constraints of the communication will have influence on the choice of the hardware
- The size of the network will have immediate consequences on performance

These characteristics all have their own relevance. The implications for networking of the most relevant characteristics should be analyzed first. If these implications do not uniquely imply certain design decisions, less relevant characteristics will help in finding good and efficient network solutions. These solutions need not necessarily be unique; when there are no special requirements for a specific CIM module, many existing networks will satisfy the conditions, but, on the other hand, sometimes the requirements completely prescribe the network.

If the communication type and the 'environmental circumstances' are known it will be easy to identify which combination of basic communication elements will suit the application best. This leads to a deliberate choice of these elements. Ultimately, the communication strategy will lead to an adequate support of communication in Computer Integrated Manufacturing.

It needs to be emphasised that the implications of the requirements of global CIM modules will be principally in the area of the hardware that can be used. When the submodules of CIM (e.g. a geometric modeller) are considered into more detail this will certainly influence the used communication protocols, especially those of the higher layers of the OSI model.

13.7. Characteristics of communication in Computer Integrated Manufacturing

In this section a description of communication in Computer Integrated Manufacturing is given. The purpose of this classification is to simplify the recognition of important constraints that are imposed on computer networks used in this field.

A global classification of communication in Computer Integrated Manufacturing consists of three parts: the first about communication during the design phase, the second about communication in manufacture and the third about communication occurring in both areas.

First we give a definition of a logical network.

A *logical network* is a distributed system, consisting of processors, software and physical network, designed to perform a specific task.
For each node of the network it is specified to which of the other nodes transmission can take place.

13.7.1. Design

Communication during the design phase, including computer aided design (CAD), computer aided engineering (CAE) and computer aided testing (CAT) is of a miscellaneous nature. Three important types of communication can be distinguished:

1. The transmission of data that enables an eminent interaction between designers, engineers and their work stations.
2. The transmission of data related to the use of distributed data bases.
3. The transmission of data related to analysis, the finite element method, the result of the design, etc. They are taken together because they can be characterized by the large amount of data involved.

The first type does not have real time constraints: it is not necessary to design the communication network in a way that precludes late arrival of messages. However, the speed with which results are obtained is important, because engineers and designers will be more inspired when the response time of the system is short, leading to a higher productivity. As a matter of course, this is not only related to the communication strategy, but also to the processing strategy, because when a good response time is required, it is necessary that information is produced fast, and, when this information becomes available, it is transported fast to its destination. As a consequence the network should have a high capacity.

The ever increasing integration in Computer Integrated Manufacturing results in an increase of communication in the design phase, especially the transport of data out of huge data bases is needed. The use of distributed data bases has an enormous influence on the communication strategy. As long as these data bases do not change it can be advantageous to have several copies of them because this largely decreases the volume of communication required. The consequence of having several copies is, however, that updates of the data base require a lot of communication because all copies need to be updated as well. Apart from this, the problem of having multiple updates on multiple copies at the same time is not solved yet, and needs further investigation.

If every node of the network knows exactly where data can be found it can decide where a data base query has to be processed, based on this information as well as on information about the network load. If it is not known in advance where data bases are maintained, broadcast messages to investigate where particular data is stored are needed. Problems arising from this design should be investigated further.

The communication between a CAD system and a CAE system consists of the transmission of large amounts of data, for instance solid models and sculptured surfaces. There are no real time constraints but it is necessary that the transmission delay is as small as possible. Given the amounts of data this requires a high performance network (one of which the effective data rate is high). The reliability of this network is not as important as of the shop floor network. Because of the gargantuan amounts of transactions the overall performance of the network should be good. If the network fails the failing part should be detected quickly and replaced without causing the network to fail completely.

13.7.2. Manufacture

In manufacturing three types of logical networks can be distinguished:

1 The control network
2 The monitoring network
3 The management network

The control network is responsible for the driving of machines, robots etc. In general, messages will be short but very important. As an example a description of the communication between robots and vision systems will be given. If a robot is used to acquire parts, it needs information about the scene that the vision system can see. More to the point the robot has to know the position of the part to be acquired exactly. Although this can be handled straightforward, currently no standard is available for communication protocols between robots and vision system. This situation is undesirable because it prevents integration of systems of different manufacturers. The introduction of a standardized communication protocol is made difficult by the existence of many already commercially established, incompatible robot languages and languages for vision systems. Often it is easiest to instruct a robot in its 'own' language. If a standardized communication protocol is used, this would not necessarily contain the special instructions for a particular robot, implying that the messages should be translated or interpreted, which can slow down the effective communication rate largely. One might expect, nevertheless, that when a standard has been developed, manufacturers of robots as well as vision systems will adapt the languages they use, to make them better suited to the communication standard. It can be concluded that standardization is necessary, but made difficult by the number of languages already available.

It is evident that the arrival of wrong data in manufacturing can lead to disastrous consequences. For instance, when a moving crane is not slowed down in time, it can devastate a great part of the machinery on the work floor. Reliability thus will be a key issue in this type of communication. However, it is not possible to separate reliability and performance completely, because they are closely related. The reliability of a real time system is dependent on timely delivery of exact data. The timely delivery depends on the performance of the network under various loads as well as on its sensitivity to transmission errors, whereas the correctness of the data is related to the protocols used.

It is important to note that a gain in flexibility of the manufacturing increases the need for communication; sometimes this will influence the choice of the physical network. To conclude, this type of communication requires a high performance, reliable local network.

The monitoring network guards the correct functioning of every subsystem. To achieve this, it collects information from the control network. Several conditions should be satisfied.

1. Input material should always be available. Often this material will be produced by other subsystems, which implies that these subsystems should produce sufficient quantities to keep their 'successor' going. The monitoring network should detect when raw material is needed and act accordingly.
2. Machines and instruments should be in good repair. When it is detected that instruments do not function as required, the network should monitor this and act appropriately.
3. The monitoring network should establish that every subsystem meets its production demands. If production losses are inevitable this should be logged. In this way it will be easy to design a global system that takes care of the total production line.

In general, communication in a monitoring network will be local. It follows from the requirements that a real time network is needed.

The third type of network in manufacturing is the management network. Its main goal is to optimise the functioning of the production line as a whole. To achieve this, the output of the monitoring network of each subsystem should be inspected in detail, for instance to detect which subsystem is a bottleneck of the production line. Moreover, it is necessary to know what percentage of the production is lost in a particular subsystem. The management network collects this information from the monitoring network and acts appropriately when certain production units are performing badly. Production planning can be done easier when conscientious information concerning the current production is available. When the production is carefully analyzed it can be seen easily whether the requested production has been achieved or not and how the production planning has to be adapted to become reliable.

In contrast to the earlier mentioned logical networks, the management network will have to operate correctly during long periods of unattendance, namely as long as no information is needed. Of course it should always collect statistical and monitoring data and provide that this data is available immediately when needed. If the network is not working properly, this will, in general, not immediately be noticed. This implies that the network ought to be 'self-checking' that is it should have facilities to establish that failures of parts of the network can be detected and corrected. In this way, no human intervention will be needed to ensure that the network is functioning correctly. Moreover the network should have facilities for automatic rerouting etc.

13.7.3. General

A type of communication needed in both the design phase and the manufacturing phase is the communication related to the maintenance of the network. Even when networks are built in an extremely reliable way, it is unavoidable that errors in hard- and software occur.

These errors need to be detected quickly so that appropriate action can be taken. Often it will be easier to keep the network in good repair when it is possible to plug terminals into the network, for instance to detect which node is malfunctioning. These possibilities can be used best in networks without real time constraints.

13.7.4. Conclusions

It can be concluded that each type of logical network needed has its own requirements and responsibilities. There is also a clear need for communication between logical networks.

If the requirements of a logical network should not be violated against, it is necessary to minimize the mutual communication between logical networks.

Primarily in networks with real time constraints it will be easier to meet the obligations when the performance of the network is minimally affected by inter network communication. But speaking more generally and in terms of other networks it seems unnecessary and even unwise to let the inter network communication become large.

To give an example how the communication between networks can be minimized, consider the data flow between monitoring network and management network. Starting from the principle that normally the production system will perform well, the need for data transmission can be decreased largely when the monitoring network only reports production losses to the management network, so that when all is going well no communication is needed.

Furthermore, the requirements of each logical network are different. The differences are so large that it is unlikely that all requirements can be fulfilled in one existing (physical) network. At least nowadays it not possible to meet all requirements in one physical network and it is not likely that this situation will change in the near future.

Therefore communication in Computer Integrated Manufacturing should make use of mixed technology (i.e. various physical networks).

Even for each individual logical network the technology is not necessarily prescribed. Sometimes more feasible solutions to the communications problem exist, and the choice will depend on other characteristics. When these characteristics are considered into detail one can choose the network that will suit the application best, depending on reliability constraints, performance constraints, etc.

References

Boulton1983. Boulton, P.I.P. and Lee, E.S., "Bus, Ring, Star and Tree Local Area Networks," *Computer Communications Review* **13**(3) pp. 19-24 (July 1983).

Bux1981. Bux, W., "Local Area Subnetworks: A performance comparison," *IEEE transactions on communications* **COM 29**(10) pp. 1465-1473 (October 1981).

Chanson1982. Chanson, S.T., Kamur, A., and Nadkarni, A.V., "Performance of some local area network technologies," TR 82-8, University of British Columbia, Department of Computer Science (1982).

Clancy1982. Clancy, G.J. and Harrison, T.J., "Local Area Network Standards: A status report," pp. 149-161 in *Distributed Computer Control Systems 1981*, ed. W.E. Miller, (1982).

Graube1982. Graube, M., "Local Area Nets: a pair of standards," *IEEE Spectrum* **19**(6) pp. 60-64 (June 1982).

ISO/TC97/SC16/N2271981. ISO/TC97/SC16/N227, "Reference Model of Open Systems Interconnection," *Computer Communications Review* **11**(2) pp. 15-65 (April 1981).

Metcalfe1976. Metcalfe, R.M. and Boggs, D.R., "Ethernet: Distributed Packet Switching for Local Computer Networks," *Comm. ACM* **19** pp. 395-404 (1976).

Pogran1983. Pogran, K.T., "Selection of a local area network for the Cronus distributed operating system," *Computer Communications Review* **13**(3) pp. 25-35 (July 1983).

Saltzer1981. Saltzer, J.H., "Communication ring initialization without central control," MIT/LCS/TM202, Massachusetts Institute of Technology (December 1981).

Saltzer1983. Saltzer, J.H., Pogran, K.T., and Clark, D.D., "Why a ring?," *Computer Networks* **7**(4) pp. 223-231 (1983).

Salwen1983. Salwen, H.C., "In praise of ring architecture for local area networks," *Computer Design*, pp. 183-192 (March 1983).

Shoch1980. Shoch, J.F. and Hupp, J.A., "Measured Performance of an Ethernet Local Network," *Communications of the ACM* **23**(12) pp. 711-721 (December 1980).

Stuck1983. Stuck, B.W., "Calculating the maximum mean data rate in local area networks," *Computer* **16**(5) pp. 72-76 (May 1983).

Tanenbaum1981. Tanenbaum, A.S., *Computer Networks*, Prentice Hall, Englewood Cliffs (1981).

Tobagi1980. Tobagi, F.A. and Hunt, V.B., "Performance analysis of Carrier Sense Multiple Access with Collision Detect," *Computer Networks* **4** pp. 245-259 (1980).

Wilkes1979. Wilkes, M.V. and Wheeler, D.J., "The Cambridge Digital Communication Ring." pp. 47-61 in *Proceedings of the Mitre NBS Local Area Communications Network Symposium held in Boston*, (1979).

Chapter 14

Communication strategy

Introduction

The communication strategy is designed to address a most important issue in Computer Integrated Manufacturing: communication between different computers. It consists of a number of communication strategy maxims, each one stating a different property of communication in CIM.

Some of the communication strategy maxims will be applicable in Computer Integrated Manufacturing only. Other maxims have a more general value for communication. However, it is considered necessary to name these maxims also, because of two reasons.

First, the ideas that underlie these maxims are not only important for communication in general, but they will often have an even more significant value for communication in CIM, because of the stringent requirements that are found in this application area. The narratives that will be given for each maxim, will clarify this further.
Secondly, it is inconceivable that a communication strategy for Computer Integrated Manufacturing can do without ideas that have proven their value for communication in general.

In the following sections the communication strategy maxims will be given. Each maxim is attended with a short narrative that explains why one should adhere to it. For many of the communication maxims it is also described how they can be realized in existing CIM-systems.

Communication strategy maxim 1.

Network modification should not lead to shut down.

In a complex structure, like that of a computer system and its associated network in Computer Integrated Manufacturing, there will frequently be a need for modification of the network. Sometimes the network needs to be extended (i.e., the size of the network is increased), sometimes it has to be expanded (i.e., the number of nodes is increased). Also it is possible that components of the network are to be replaced by new, often more modern substitutes, that are better able to fulfil the high requirements that are prescribed by the users. In Computer Integrated Manufacturing typically a network will be present, that has to be adapted piece by piece, because it is not economically justifiable to replace the network as a whole. Therefore it is clear that the network very often is not a static, but instead a dynamic entity. The consequence of a dynamic approach is that flexibility of the concern can be achieved easily.

If the network is built in a way that precludes modification of it 'on the fly' this leads to an undesirable situation. If, for example, the network responsible for the production process is built that way modification of the network would lead to production losses. Because of the cost of these, the consequence could be that replacement of components will be postponed and therefore the network will not resemble the current state of the art in computer networks.

Also in other parts of Computer Integrated Manufacturing modification of the network should not affect the normal use of the network. In this connection we may think of the design process where designers cannot do their job properly when the network is down due to changes in the topology of the network.

Now that the need for modification of the network is made clear, and the importance of the fact that network should work properly as much as possible, it is time to draw conclusions from these data and combine them into a maxim. The conclusion is straightforward: if the network is to be modified, network traffic should be affected minimally.

It is worth noting that network modification without shut down is technically realisable. For an elaborate illustration of this, the reader is referred to chapter 13. Suffice it to say here that many of the currently available local area networks have provisions for modification without shut down. The costs of these provisions are not high.

Another form of modification occurs in the network when new protocols are installed. In general these protocols will be high layer protocols, because low layer protocols are often part of the network interface unit. As long as the interface the new protocols expect corresponds to the one provided for by existing hard- and software, this type of modification will cause no trouble.

Communication strategy maxim 2.

De facto standard protocols should be used in CIM as much as possible.

Historically, protocols were designed in an ad hoc manner, depending on the application, the transmission medium etc. This is customary for a developing technology, because only in this way one can learn what is really important and what is not. During the last few years several standards for communication protocols have been developed by several independent standardization committees. Gradually, these standards are going to be accepted in most application areas, including industrial applications. Internationally accepted standardized protocols are principally low level protocols, because most standardization committees follow a bottom up approach. Still, not much is known about standardized high level protocols, because few experience with these protocols has been gained.

The range of network applications in Computer Integrated Manufacturing is very large. Apart from applications that can also be found in other application areas, a lot of special CIM applications, each one needing its own protocol, can be found in CIM.

Since the application range is large it cannot be expected that only one application protocol will be sufficient for CIM. Instead, it can be stated that different applications need different protocols but this difference need to be concealed as much as possible. If we consider the standardization efforts of the International Organization of Standardization, this means that application dependency needs to be seen in the application protocols only, so that the lower layer protocols can be the same for each different application.

On the other hand, the low layer protocols (e.g., the physical layer protocol as ISO calls it) are dependent on the transmission medium, which means that they will differ when different transmission media are used.

It may seem that standardized protocols are an illusion. However, appearances are deceptive; it is possible that protocols are designed in such a way that from a certain layer on, the functionality satisfies the standard conditions. In other words, the interface that presented by the protocols is the same, although the way they are actually implemented may differ largely.

The main reason why this maxim is important to CIM is the dramatic reduction of the cost of software development. Experience, gained during several years, becomes available for CIM and it would be unreasonable not to use it. Moreover, if non standard, low level protocols developed by a company itself are used in CIM, it will be difficult to test them thoroughly. Current protocol verification techniques cannot guarantee that a high level protocol is designed in a way that precludes errors. This means that these errors will be found during tests. The influence of these errors however, can be severe, especially in a factory environment, and in some cases even disastrous. Principally in these circumstances, it is important that protocol design techniques, developed for other application areas, will be used as much as possible.

There are more advantages of the use of standardized protocols. The consequence of this design concept is that it will be easy to extend the network with components of (possibly) different manufacturers, which, because they use the same interface, can be coupled fairly simple. Another advantage is that network interconnection, an issue that tends to become more important in the future when more integration will be realized, will be made easier by this approach. Even multinational manufacturing organizations will benefit from these developments.

Communication strategy maxim 3.

The specification of a network should clarify which timing requirements can be handled by the network.

A timing requirement will be defined as the, by the application prescribed, maximum time that is allowed to elapse before a message has to reach its destination. In some parts of Computer Integrated Manufacturing these requirements are strict and explicitly known, other sections of CIM have less stringent timing constraints.

In order to be able to meet timing requirements, they must be taken care of at *specification time*, that is, they must be part of the network specification. The next consequent requirement is to implement the network according to the specifications.

The specification of most networks (especially Local Area Networks) that are currently available does not accommodate for timing requirements. Current specifications often do not comprise more than the maximum throughput of the network, information that can be very useful, but that is not exactly the need of Computer Integrated Manufacturing. If the specification of the network does not say which timing requirements can be handled by it, this means that, given a network, one cannot estimate the probability that data do arrive within a prescribed time interval. Hence the choice of the network cannot be done in a responsible way.

Now consider the situation that the specification of the network *does* accommodate for timing requirements. Since it is possible to quantify the timing requirements of network traffic rather precisely, (think e.g. of a control network) one can use this quantification and see whether a specific physical network can fulfil the prerequisites.

It is evident, that having the possibility to accommodate for timing requirements alone, is not sufficient. A mechanism should be provided for, that checks if different combinations of traffic, all with different time requirements, can be handled by the network. If not, the 'network manager' should be notified of this. The mechanism ensures that the network does not become overloaded (i.e., it cannot cope with all the prescribed timing requirements). Also, it guarantees that the influence of network extension or the effect of an increase of network traffic can be quantified, which gives the possibility to see how it affects the performance of the network as a whole.

Communication strategy maxim 4.

A network should allow for a mixture of different hardware technologies.

Different transmission media are used in communication networks. For convenience we repeat the most important: optical fibres, coaxial cable, twisted pair cable, twin pair cable, radio and satellite. We refer to chapter 13 for a thorough description of these types of hardware. From the description there, it follows that all technologies have their own significance for CIM.

At first sight the use of different hardware technologies does not seem necessary. It seems that one can choose the best technology available and use it in all networks. It needs to be accentuated, however, that a 'best' technology does not exist. Different applications need different hardware.

During the last few years, the importance of fibre optics technology for computer communication became apparent. This technology indeed is important and its use in Computer Integrated Manufacturing can solve many problems, but it cannot be used for every application. To give an example, laser communication will in some cases be better suited to the task of sending instructions to a moving crane.

Another area where different hardware technologies are used is the interconnection of networks. Even when a network uses only one type of hardware, another network might use another technology. Consider, for example, an Ethernet, a public data network and a network in a hostile environment. The connection between these networks, the gateway, should hide the differences and make sure that the communication will not be disturbed by them.

Given that the use of different technologies is inevitable, one might still consider various, uncoupled networks, each having its 'own' hardware technology. This 'solution' however, is contrary to the main reason for computer networks in Computer Integrated Manufacturing, namely the integration of computer systems.

If the network does allow for the mixture of different hardware, the network can be replaced gradually. The performance of existing networks can be improved piece by piece as needed, which means that the expenses of these improvements can be spread over a longer period. Also, when new hardware is developed, it can be incorporated into existing networks.

Communication strategy maxim 5.

Special precautions
should take care of protection against unauthorized access
and provide for data security.

From the maxims that form the data strategy for Computer Integrated Manufacturing, the importance of the correctness of data can be noted. These maxims, for example, ensure that only that value of any data item is used that is created and maintained by the single person or function that is officially authorized to supply that data (principle of prime authorship).

Assuming that these rules are fulfilled, still some problems have to be envisaged. These problems are related to the way in which data is transmitted over the communication links. It has to be guaranteed that data is received exactly as it was sent. Especially when public data networks are used, it is not easy to fulfil this requirement, and, as stated before, it can be useful to use them in Computer Integrated Manufacturing. The communication between several management departments of one business concern is a good illustration of this. Data that is transmitted for these purposes shall often be confidential or even secret.

The part of the system that takes care of the security should guarantee that even if the data sent over the communication links is compromised, it is of no use to anyone else than the intended receiver.

The general requirements for data communication are:
- Try to ensure that data will be received as it was sent; if this is impossible, make sure that this is logged.
- Ensure that intruders cannot decode the data.

A system, that will fulfil these requirements, should have [Voydock1983] mechanisms to
- prevent release of message contents
- prevent traffic analysis
- detect message stream modification
- detect denial of message service
- detect spurious association initiation

The protocols of the presentation layer take care of data security. Part of their job is to encrypt the data to ensure that no one but the intended receiver can decrypt it. The processing time required to do these operations need not be high because special encryption chips are available. Some encryption algorithms do increase the size of the data, which means that more bandwidth is needed. On the other hand, the presentation layer takes also care of data compression implying a contrary effect.

It is worth noting that encryption techniques, because they are situated in the presentation layer, are not depending on the transmission media used.

Communication strategy maxim 6.

A network should be designed in such a way that it can cope with environmental noise.

Communication in Computer Integrated Manufacturing is in many respects similar to communication in other application areas like, for example, office automation. That is why much of the work presented before has a 'general value', that is it is not restricted to Computer Integrated Manufacturing only.

In some ways, however, communication in CIM is different from that in other application areas. One of the aspects involved is the fact that real time requirements are important. Another significant aspect is that the manufacturing process, by its nature, produces a significant quantity of noise. This noise will generally be neglectable in other application areas.

Environmental noise is probably the most important factor that makes communication is CIM special. In the next sections the influence of machinery on environmental conditions is described.

We want to stress here that environmental circumstances are not unalterable. Even in highly flexible machining lines that are used nowadays, eventually the production line has to be changed, probably because a new product is announced. This means that unexpected sources of noise can come into existence. The consequence of this all is that a network that is used close to a production line should take noise into account, even when this is not generated currently.

Hostile environments

Hostile environments can be classified according to different aspects of hostility: those that affect computer systems and communication links, and those that affect human beings. The importance of the first group of aspects will be clear, the second is significant for a different reason, namely that computer maintenance in environments that are unfriendly to people will be difficult.

Many machines that can be found in Computer Integrated Manufacturing require high currents. Electrical welding and switching of high current machines generate noise in the form of sources of interference and electromagnetic pulses. This noise will influence the communication if no special precautions are taken.

Furthermore, the manufacturing process can be attended by high temperatures. This is not only unpleasant to men, but can also influence the performance of temperature sensitive electronic components. Special equipment is necessary to keep the temperature below specific limits.

Another aspect, that affects the repairability of the system, is the distance between different machines of a manufacturing system. From the point of view of the transport system, it will be advantageous to keep this distance small; by doing so the time to move a workpiece from one machine to the next can be decreased, without the need for an increase of velocity. From the point of view of repairability of computer systems and communication links however, there is a lower limit on the distance between the various systems. If the distance is decreased below this limit it will be impossible to replace failing components.

The quality of the air is also important. Many manufacturing processes can influence it. A paint spraying process causes the presence of small portions of paint in the air. Grinding processes cause the presence of small pieces of metal. All this does not contribute to good air quality and precautions are necessary to ensure that the effect on maintenance does not become too large.

Now that a brief classification of hostile influences on the environment is given, it is time to become more specific and see how these influences can be quantified.

The state of the art of communication does not solve questions about tolerance limits. If an elaborate analysis of communication in Computer Integrated Manufacturing is to be made, more attention has to be paid to this subject. Only then it will be possible to solve questions like: "What is the minimum distance of this type of transmission medium and that type of electrical welder."

If solutions to these types of problems are known, the communication requirements of CIM can be quantified easier. Much more research is needed here and future projects that deal with Computer Integrated Manufacturing will have to investigate the problems into more detail.

Communication strategy maxim 7.

A failure of one component of the network should not affect the traffic on the rest of the network.

The importance of this maxim will be illustrated by combining it with several processing maxims. One of the programming maxims states that system design should aim for the distribution of processing. The consequence of this maxim is that processes that are logically independent can be physically separated as well. In other words, these processes should be self-contained as much as possible. Because of this, it would not be right if a failure of one node of the network would cause to stop other processes, that are independent of processes at the failing node. Instead, one should endeavour to keep them functioning.

This does not only apply to failing nodes (i.e., computers) but also to failing communication links. If a communication link stops functioning this will certainly affect network traffic. The performance of the network will degrade, but it should not occur that nodes that need to communicate are unable to do so.
Fortunately, it is not difficult to achieve this requirement, as will be shown later.

Diagnostics is another area that will illustrate the need for the above-mentioned maxim. When a process that is attached to the network stops working, the diagnostic system will have to discover this and act accordingly. Some of these actions cause network traffic, for instance to a diagnostics server. Apparently, it would make no sense to develop a highly sophisticated diagnostic system, if the network is not able to support it.

We will discuss now how this maxim can be adhered to by existing network technologies. First two of the most common Local Area Network technologies are compared: carrier sense networks and token rings.

Reliability of these networks is a important aspect. If they were extremely reliable, the need for this maxim would be less apparent, because then failures of network components are so uncommon that the extra design effort to meet the maxim would be a waste of time and money. Although the constantly advancing technology makes networks more reliable, still the high requirements of Computer Integrated Manufacturing cannot be met (*cf.* communication strategy maxim 6 about hostile environments).

It was impossible to meet this communication maxim in the early implementations of a token ring. Every node of the ring acted as an active repeater, which meant that a failure of one of them effectively caused the network to halt. Nowadays, the star shaped ring is developed and failing parts of the network are simply by-passed.

In a carrier sense network there are difficulties if one node, due to a failure, constantly tries to send messages over the network. This implies repeated collisions, which means that the effective communication halts, if this situation is not detected.

With the help of ideas originated in graph theory, this maxim can be explained further. We will illustrate that the above reasoning is a special case of a general theory.

Define the arc connectivity of two nodes as the minimum number of arcs that must be removed to disconnect them, and define the arc connectivity of a graph as the minimum of the arc connectivities of all pairs of nodes in that graph. Analogous, the node connectivity of two nodes is defined as the minimum number of nodes whose removal will disconnect these two nodes, whereas the node connectivity of the graph is the minimum of all node connectivities for all pairs of nodes.

Some graphs and their connectivities are shown in figure 14.1.

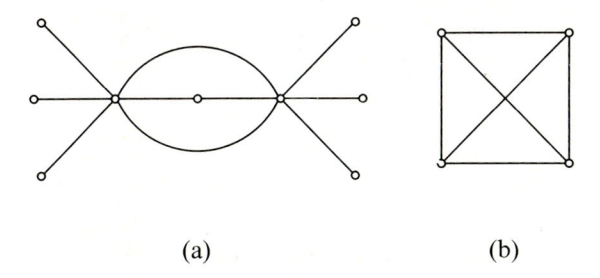

(a)   (b)

Figure 14.1 : (a) A graph with arc connectivity 1 and node connectivity 1.
(b) A graph with arc connectivity 3 and node connectivity 3. Note that there is no way to disconnect two nodes by removing other nodes if there is an arc between them. In this case the node connectivity is *defined* as the number of nodes in the graph minus one.

It is required that the arc connectivity of the network is at least two, if the above maxim is adhered to. Also the node connectivity should be at least two. If the connectivities of the network are increased, this means that the reliability of the network as a whole is increased too.

Communication strategy maxim 8.

A trade off should be made between network performance and network reliability depending on the application.

Reliability of the networks is, as stated before, a key issue in Computer Integrated Manufacturing. On the lowest layers of the OSI-model almost all transmission errors are corrected by a cyclic redundancy code, for which hardware is available. To satisfy the extremely high demands on reliability in Computer Integrated Manufacturing more warrants are needed. These demands can be met by software in the higher level protocols.

It should be noted that, especially when the number of different orders to a machine is small, transmission errors that are not corrected by the low layer hard- or software can easily be detected, because an arbitrary error is very unlikely to produce a valid command to the machine. Therefore we can conclude that:

The command interface to machines ought to be kept as small as possible.

This requirement only is not sufficient. To illustrate this, we will give a simple example, namely an example of a machine with only two valid commands, say 0 and 1. If, due to a transmission error a bit is inverted, this produces another valid command. Therefore we may conclude that it is not only important to have a small command interface, but also that the distance between valid commands (i.e., the number of bits that, given a command, have to be changed to produce another valid command) has to be large.

This is an easy way to increase reliability without affecting the performance too much. When more guarantees are built in in higher layer protocols this will undoubtedly affect the performance of these protocols. A lot of processing has to be done, to check all possible error conditions. Especially when local area networks are used (with their high transmission rates), the time required to perform this checking can easily become larger than the time needed for real data transmission. By this, the effective transmission rate of the network can be much smaller than the transmission rate that is theoretically possible. If the performance of the network decreases too much, it will be impossible to cope with real time constraints. Here, reliability of the network is not worth much: timely arrival of data will still be impossible because of the degraded performance.

Therefore, the trade off between network performance and network reliability has to be made carefully, depending on the application. This trade off should be made and fixed, *prior* to any other operations like scheduling and programming.

Communication strategy maxim 9.

The network should allow the use of messages with different priorities.

There is a fundamental need for a priority mechanism for messages in Computer Integrated Manufacturing. Certain circumstances can cause a situation that asks for fast and efficient actions. As an example, consider the shut down of a specific machine. It is conceivable, that its malfunction will influence the rest of the production process. Other machines may have to be stopped, because otherwise severe material damage will be caused.

Not only situations that cause an alarm need high priority messages. When a central laser is used that has to switch operations rapidly high priority traffic can take care of the switch messages. More generally, when scarce or expensive resources are used, a high utilization can be realized if the network is able to support fast switching of operations. The messages required to enable this are more important than 'normal' messages.

In general, to prevent undesirable situations, messages can be sent with appropriate instructions. But if these messages are not handled in a special way, it is possible that earlier generated instructions, which are not valuable any more, will be carried out first. Therefore there should be a way to indicate the special meaning of important messages, to ensure that they will be send before 'normal traffic'.

The realization of this maxim can cause some difficulties. For example, in many of the local area networks that are currently available, the need for different priority traffic was not recognised. Both in the token ring, where important messages can locally be put in front of the queue, but still have to wait until the token passes by, and in a contention network, where the protocols used preclude that one message is considered to be more important than others, different priorities are also difficult to implement. This means that more research is needed here.

One way to solve the problems is to have a separate network for important messages. Every node of the network should know that this network contains the messages that are really important. However, as this solution involves the purchase of extra hardware, it will be costly; therefore it should be considered whether this solution is economically justifiable. It is almost certain that hardware prices are continuing to drop rapidly in the near future, and, on the other hand, failures in the production process will become more and more costly. This means that the solution involving separate networks will, as time goes by, become more justifiable.

Communication strategy maxim 10.

The network should provide for diagnostics and error reporting.

This communication maxim is closely related to the processing maxim about diagnostics. Both in processing and in communication a diagnostic system is vital. The need for a diagnostic system in network systems will be illustrated with examples.

In case of errors, first it should be detected which parts of the network have ceased functioning. In general, the transport protocol will detect errors, because this protocols tries to maintain a reliable connection between source and destination. Sometimes the transport protocol will try to search for alternative routes to reach the destination. This is possible only when it has almost complete information about the topology of the network. Whether alternatives are searched for or not does not make difference to the error reporting. As soon as an error is detected a message has to be send to the 'network manager' (this can either be a computer program or an operator). The combination of these diagnostic messages enables an easy identification of malfunctioning parts of the network. Once these parts are identified, appropriate actions can be started.

The transport protocol can support error reporting in another way. It can send information about the number of retransmissions needed, subdivided according to destination. If detailed information about this is available, it is possible to see which communication links are noisy, etc. If the performance of links degrades too much, actions are needed, possibly including the replacement of network components. It is almost needless to say that the transport protocol supplies information about retransmissions only when retransmissions are actually needed; it would not contribute to a good performance of the system if the transport protocol was always busy, giving information about the fact that a particular destination is reachable.

Communication strategy maxim 7 states that a failure of one component of the network should not affect traffic on the rest of the network. This means that if a communication link stops functioning, the network will try to find an alternative route from source to destination. At the moment the need for an alternative is clear, a diagnostic message has to be send to the 'network manager'. By this it can be seen which components of the network fail, and whether appropriate actions are required immediately or not.

Communication strategy maxim 11.

The number of special application protocols needs to be as small as possible.

In current standardization efforts by the International Organization of Standardization and other organizations a bottom-up approach is followed. Consequently, the high level protocols, like the application protocol, are least well developed, and it is not known yet in what direction the standardization will go.

Computer Integrated Manufacturing needs application protocols that are different from those found in other application areas (like office automation). However, the framework of these protocols can be made in an uniform way. This frees the implementor of these protocols from designing the same thing several times.

One way to achieve this is the minimalization of overlap of related protocols. An example can be given from computer graphics. Special protocols are developed for graphics and geometry. Much of the work done by the graphics protocol can be used in geometry protocols as well. Therefore these protocols need to be developed collectively. If they are developed separately and by different standardization committees, it seems almost impossible that they are designed in a way that minimizes their overlap.

It seems that it is not possible to retain the differences of the user's view if the number of application protocols is kept as small as possible, and if they are all made following the same approach. However, the retainment is considered useful, because it cannot be expected that the user changes his view of the data (i.e., the meaning that he gives to it) just because the application protocol does not provide for the facilities to use that view. Instead, it should be possible that the user preserves his own way of manipulating data, the consequence being that the user does not have to make changes to the protocols. This can be achieved by having separate programs that define and maintain a user's view on top of the application layer protocols.

In this way it becomes easy to establish a precise separation between the user on the one hand and the implementor of the programs on the other. As it cannot be expected that every user knows the well established rules of Information Technology the need for a strict separation is apparent.

Communication strategy maxim 12.

A network should be able to carry different types of data.

Computer networks have primarily emerged for use in carrying information among terminals, hosts and network servers. They support a wide range of applications: file transfer, electronic mail, terminal access to time sharing environments etc.

In all these applications, data is transmitted over the network in its 'natural' form (i.e., digitally). Not all data needed by the manufacturing process is available digitally, however. The spoken word still represents an effective mode of communication and it is likely that this will not change in the near future (the unmanned factory is certainly not the only CIM goal).

Until recently, the telephone system was primarily used for voice traffic. Telephone lines are not well suited to the transmission of digital data as is produced by computers, since the need for communication is bursty. Most of the time no traffic is required, but sometimes high data rates are needed and the telephone system is not able to support them appropriately. This means that the telephone system cannot be used for all data transmission, so if a uniform way of communication is required, telephone lines are not adequate.

On the other hand, it *is* possible to use a computer network for voice traffic. We think that such a network should be available in Computer Integrated Manufacturing, having as consequence that a *mechanism* for sending voice over the network is provided for. It might depend on economical factors, which *policy* is chosen to transmit the spoken word, telephone system or communication network.

The communication network should not be able to transmit data of the above type only. In CIM also video is significant. It is important that during the design process rapid exchange of information about the design is possible. In one case this may involve the transmission of a formal specification of the design, in another case it might be sufficient to just send a picture.

Communication strategy maxim 13.

A uniform and flexible addressing scheme is needed.

To achieve high flexibility in Computer Integrated Manufacturing it is required that the computer population is highly dynamic, which means that computers can be plugged into the network or removed from it as needed.

Some of the problems to realize this flexibility can be solved by using a uniform and flexible addressing scheme.

A major requirement of an addressing scheme is that an address has to be location independent. As such, an address will be more than a name in the real world than like an address. By this we mean that an address (as used here) is indissoluble connected to an object, like a name is connected to a person in the real world.

If an address is not location independent it will be difficult to move services around. Moving services around is useful, partly because it can reduce network traffic. To see this, consider a file server, used by the manufacturing process. Due to an extension of the network, caused by changes in the manufacturing process, the place of the file server, at first chosen to minimize network load, gradually becomes more and more inefficient. Reconfigurability is therefore necessary, and this is made easier if the address of the file server remains constant, because processes that are using the file server do not have to bother about changing addresses.

Furthermore, the addressing scheme should be rich enough to enable both traffic to nodes in distant interconnected networks and to nodes that are located in the same (local) network. A mechanism has to be provided for, that makes it possible not to include the 'global information' in local addresses. If such a mechanism does not exist the processing time to establish a correct global address can be so large that the high transmission speed of a local network cannot be used.

Communication strategy maxim 14.

Protocols on fast, reliable links may not become inefficient because of the presence of slow, unreliable links.

In Computer Integrated Manufacturing a wide variety of transmission links is used, because of the wide variety of applications and the large differences in environmental conditions. Some transmission links are chosen because of their resistance against environmental noise, others because the public data network prescribes them.

This maxim states a property of low-layer protocols: up to the transport layer that provides a reliable, secure service to its users.

In Computer Integrated Manufacturing, it will sometimes be difficult to achieve a reliable service; communication in hostile environments will illustrate that. A way out of this problem is to develop one transport protocol (and some associated lower layer protocols) that is extremely reliable even if the transmission medium is unreliable. This causes overhead in both processing and communication but for unreliable links it may well be worth the effort.

The abovementioned maxim states that this transport protocol is not an adequate solution for fast reliable links. During the last few years, an enormous gain in transmission speed (and reliability) of local area networks has been achieved. In these networks no longer the transmission speed is the limiting factor but instead the processing power needed to supply the data fast enough. When the transport protocol is built in a way that guarantees a reliable connection even when the transmission link is very unreliable, and the transmission medium *is* reliable itself, the consequence would be that a lot of processing time is wasted. According as the transmission speed increases, this disadvantage increases as well.

References

Voydock1983. Voydock, V.L. and Kent, S.T., "Security Mechanisms in High Level Protocols," *Computing Surveys* **15**(2) pp. 135-171 (June 1983).

Chapter 15

Sensor Systems and Computer Integrated Manufacturing

Introduction

In this chapter, the applicability of the processing maxims from chapter 12 and the communication maxims from chapter 14 upon sensor systems will be shown.
Sensor systems have been chosen because sensors play an important role in Computer Integrated Manufacturing (CIM) as will be elucidated in the next section.

15.1. Motivation of Choice

Some of the specific tasks, that appear in the manufacturing process, are control of material flow, quality control, checking of tool wear, part handling etc. These tasks, in the near past, were carried out by people, who, in order to be able to perform these tasks, made extensive use of their sensing capabilities.

Recently, with the growing level of automation, these tasks more and more are taken over by dedicated systems, which -due to the improved technology of the last decades- can do these jobs better and cheaper. In order to be able to perform their task, these dedicated systems need, like people, to be equipped with sensing capabilities.
Although the requirements of these sensing capabilities can vary greatly, most of these sensor systems show a remarkable resemblance in global setup. This uniform basis and the intensive and numerous use of sensor systems in CIM, are the reasons that sensing can be considered as one of the key features toward CIM.

15.2. About Sensors

Sensor systems are able to give dedicated systems their independent sensing capabilities. Sensors are able to provide information about the physical state of an external system, independent of the controlling mechanism of that system. The fact that sensors are independent subsystems, exploiting their own way of collecting information, is the basis for both their usefulness and the method of embedding them in a control system. The independence is why a sensor system can give feedback about the effect of controlling actions, can trigger control mechanisms or provide error checking and diagnostics.
The independence of the sensor subsystem implies that the sensor signal needs to be communicated to the control system. Hence, there is always a (communication) interface between sensor and controller. This interface may also pass sensor control signals for resetting or requesting a certain sensor task from the control system to the sensor system.

The sensor itself is a transducer which produces an electrical signal related to a physical parameter of interest. This is based on the conversion of one form of energy into electrical energy. If the first form of energy is modulated, carrying information, the electrical output signal will be modulated, carrying the same information. A well known example of this phenomenon is a microphone.
Sensors are designed to react dominantly to a specific property. However, often other physical parameters unintendedly contribute to this reaction as well. In that case, the impure signal, as coming from the sensor, has to be compensated for or filtered to yield a pure sensor signal. For instance, due to the intrinsic properties of sensor material, very often sensors show an almost unavoidable temperature dependence.
Depending on the nature of the sensor and its task, this clean sensor signal has some external properties such as cycle time, resolution, whether it delivers absolute or relative values etc.
In the next sections a description will be given of the sensor classification and of sensor complexes. In appendix 4, one can find some more discussions about sensor classifications and a discussion about sensor applications.

15.2.1. Sensor Classification

The following eight aspects of sensors are frequently chosen as the basis of a classification (see fig.15.1): converted energy; transduction principle; influence on the environment; adaptability; connection with object; relation of signal and parameter; relation to manipulator; property of output signal. A further explanation about this classification is given in appendix 4, section 1.
Depending on the point of interest, the importance of these classifications can be quite different.

Aspect	Classification
Converted Energy	Radiant Mechanical Thermal Electrical Magnetical Chemical
Transduction Principle	Self-generating Modulating
Influence on Environment	Passive Active
Adaptability	Non-adaptive Adaptive
Connection with Object	Contact Non-contact
Relation of Signal/Parameter	Absolute Relative Linear Logarithmic Inverse .....
Relation to Manipulator	Internal External
Output Signal	Analogue Digital

Figure 15.1 : The different classifications of sensors based on eight different aspects.

Classification can be helpful in order to determine the processing requirements of the sensor, as shown by the following examples.
Take for instance the output signal property. If the sensor supplies an analogue signal, noise reduction is most likely to be required. Noise reduction, on the other hand, normally will not be necessary for digital signals.
Furthermore, a relative sensor may require an initialization process and a process that keeps up the absolute value.
An adaptive sensor, on the other hand, will need a controller that takes care of the adaptation.

A thorough classification of sensors, based on their processing and communication requirements, is considered to be an important step that would stimulate the automation process.

15.2.2. Sensor Complexes

There are several ways in which sensors can be implemented in the system.
The most primitive way is that they operate as individual sensors, completely independent of other sensors and other parts of the system that have a connection to the outside world.
In a less primitive way, not only the pure sensor signal as coming from the individual sensor, but also its context is relevant for the interpretation of the signal. This is true for the following examples of non primitive sensors:

Sensor array, which is a combination of individual sensors of one kind, grouped in a specific set-up in order to acquire a broad view of the parameter of interest. Examples of this kind are artificial skin (tactile sensor array), vision systems (CCD array's), temperature sensing array's etc.

Directed sensor, in which case not only the pure sensor signal, but also its momentarily (if flexible) orientation is of interest. For example, a force sensor, mounted on a gripper finger in order to be able to decide whether an object can be lifted or not. The information as supplied by the sensor will only be the magnitude of the force on the gripper finger. However, due to combination with the knowledge of the position of that gripper finger, also the direction of that force can be derived.

Active sensor, where it is vital for the interpretation of the detected information to combine it with certain parameters of the source. Examples are computer vision with structured lighting (such as strip lighting), triangulation range finders etc.

In most cases however, there is a need to combine information from different kinds of sensors, to obtain a sufficiently detailed model of the physical state of the outside world (workspace). Sensor complexes are combinations of individual and/or non primitive sensors that work together in order to obtain this sufficiently detailed model. For this reason, interaction between the different sensors is desirable. This interaction will have to be coordinated by a specific controller, which must be capable of:

- requesting the at the time most relevant information.
- reconciling conflicting information.
- recognising concordant information.

A sensor system is understood to be a sensor complex and its sensor controller. The sensor controller is not to be confused with the external (to the sensor system) control system to which the sensor data will be sent.

The data representation of the sensor information has to be chosen in such a way that mapping of the relevant information obtained onto the data representation is straightforward. This representation has to suppress the peculiarities of the sensor that is actually used. Depending on the type of sensor, the representation can be in the form of boolean, integer, real, point (x,y) or array (x,y) and combinations thereof.

15.3. Sensors in CIM

In order to elucidate the use of sensors in CIM, a brief discussion of how manufacturing tasks are accomplished will be given.

15.3.1. Realization of Manufacturing Tasks

In a non-automated manufacturing system, a manufacturing task is executed by a worker, who due to his skill is able to determine the sequence of operations he actually has to perform to accomplish that manufacturing task.

Fig.15.2 shows the connection mechanism between a manufacturing task and a production system. This principle is valid independent of the degree of automation of the system.

The manufacturing task is accepted by a programming system. This programming system is the system that determines the sequence of operations that will have to be performed in order to accomplish the manufacturing task. The result of this is the manufacturing program, which contains all the information a controlling system needs to know to be able to actually control the corresponding manufacturing system.
The manufacturing system is the physical device that performs the actual manufacturing.

With this outline in mind it will be possible to identify the different aspects of the programming and controlling system. Next, the processing requirements for these two systems and in particular the way they differ can be explained.

```
                    ┌──────────────┐
                    │ Manufacturing│
                    │     Task     │
                    └──────┬───────┘
                           ▼
                 ┌───────────────────┐
                 │ Programming System│
                 └─────────┬─────────┘
                           ▼
                    ┌──────────────┐
                    │ Manufacturing│
                    │   Program    │
                    └──────┬───────┘
                           ▼
                 ┌───────────────────┐
                 │  Control System   │
                 └─────────┬─────────┘
                           ▼
  ┌──────────┐    ┌───────────────────┐    ┌──────────────┐
  │  Blanks  │───▶│ Production System │───▶│ End Products │
  └──────────┘    └───────────────────┘    └──────────────┘
```

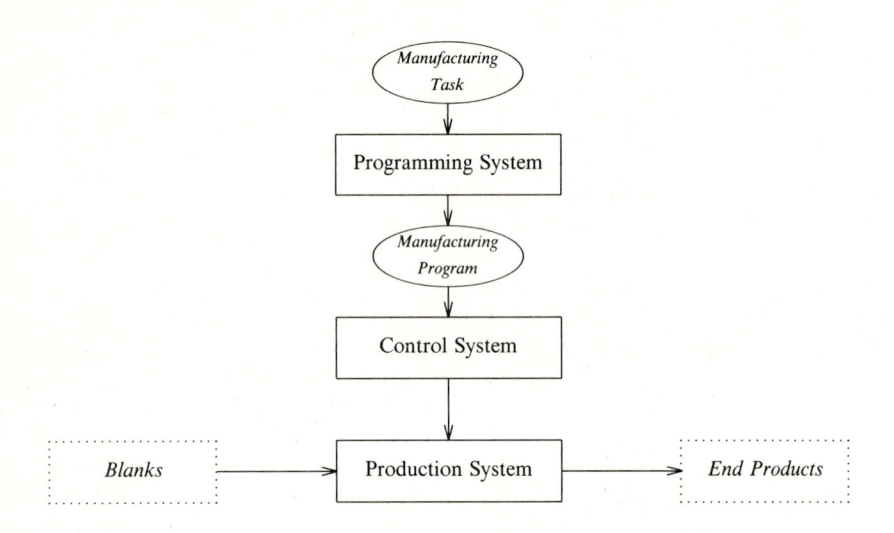

Figure 15.2 : Connection mechanism between manufacturing task and production system.

15.3.1.1. Programming systems for Manufacturing

Given the manufacturing task, the programming system has to compile the manufacturing program, that is the sequence of operations (referring to the particular class of manufacturing system) for the manufacturing system.
In general a programming system will not have strict real time requirements, although it may be interactive.

The manufacturing program generated, preferably has to be machine independent and adapted to the relevant machine type of the accessory class, by a post processor.
There should be no dependence of the kind of information carrier that is used (punched tape, magnetic tape or communication network) for transferring the program to the device.
Part of the task description parameters are provided by the CIM environment in which the programming system is embedded. Hence, programming systems should have a facility which allows easy linking to a higher level in the CIM structure.
They preferably have to make use of high level task-oriented programming languages.

For NC-machines and robots, programming systems are dominantly geometry oriented.
They need a data base for storage of technical and geometrical data.
Graphics techniques for program checking and simulation can be a very useful programming aid.

15.3.1.2. Control Systems for Manufacturing

Control systems inherently have to meet strict real time requirements. The set of real time requirements may have varying characteristics. This leads to control systems having a hierarchical structure, where each level in the hierarchy deals with a subset, having common real-time characteristics. Hierarchical control systems enable the distribution of control tasks and local intelligence.
In order to be able to take the necessary decisions, control systems need information about the state of the manufacturing system and its environment. This is the point where sensor systems enter the picture. More about the relation between sensor systems and control systems will be given in section 15.3.2.
Control systems should not use movement oriented, but task or world model oriented

programming languages. This will enable the use of a world model and thus simplifies the connection to an advanced programming system and the implementation of sensor control and monitoring.

15.3.1.3. Integration of Programming and Control Systems

The development of control systems using microprocessors made it possible to integrate the programming system and control system. Nevertheless, due to the different nature of the two systems, the programming and controlling tasks must always be strictly separated. This does not necessarily mean that these systems have to operate on separate processors, but it does mean that they have to operate independent of each other.
As a result, this division between programming and control system enables simultaneous programming and machine operation, thus reducing the down time of the system. It also enables simple modification, checking generated programs, the use of higher programming languages and coupling with a higher level in the CIM structure.
This separation allows shifting of the programming system, to higher levels of the CIM structure. This feature becomes more important with an increasing level of automation.
Preferably, shop floor programming should be possible to ensure that the introduction of NC-technology can be done gradually.

15.3.2. Sensor Systems and Control Systems

The general way a control system operates is that it compares a desired state of a physical device with the current state of that device. It uses this comparison to take the adequate correction measures. The control system obtains the information about the desired state and the current state via the information system (see fig.15.3).

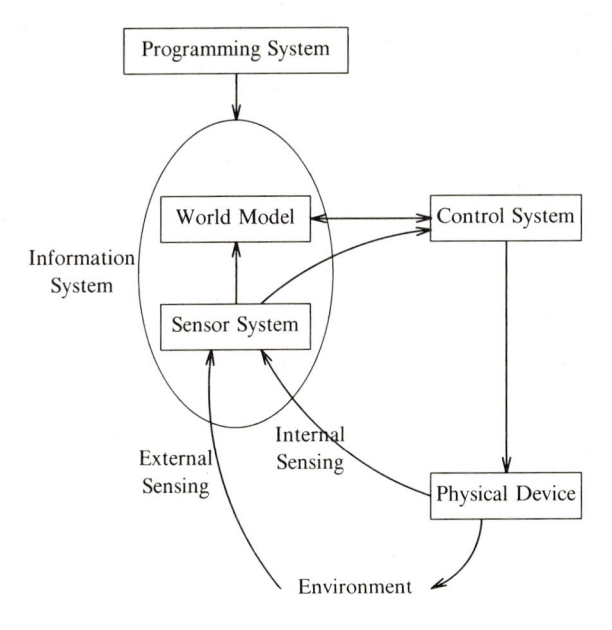

Figure 15.3 : The general structure around a control system.

The desired state is provided by the programming system, while information about the current

state is supplied by the sensor system and the world model.
This world model contains information about parts such as shape, weight, their dynamic position etc., but it may also contain information about tools. The dynamic part of the world model can be adjusted by a priori known results of control actions and by information supplied by the sensor system.
The more sophisticated sensor systems have internal and external sensors. Internal sensors supply the information about the state of the physical device, while external sensors supply the information about the relevant environment. Both world model and sensor system are a part of the information system.

As stated in section 15.3.1.2 control systems usually have real time requirements. In order to be able to control its matching manufacturing system in an interactive way, the information of the state of the system and its environment also has to be supplied in real time. In complex situations there are varying timing requirements for varying information requirements. This fact will very often lead to a hierarchical structure of both the control system and the sensor system. In that case the information exchange will have to occur at several levels of the hierarchies as shown in fig.15.4.
At each level the response time and information a sensor system level can provide must match the response time and information the corresponding control system level requires. For instance, at low levels, exchange of hardly processed information, such as servo feedback signals will take place with a typical response time of 10^{-3} seconds. Examples of servo feedback signals are joint position, velocity control and limit switch signals.
At higher levels, however, more highly processed information, such as part identity will be exchanged with a typical response time of several seconds.

In order to be able to feed information to higher levels of the sensor system, several processing tasks have to be accomplished. For instance, for a vision system these tasks may be: coordinate transformation and scaling at the lowest levels, edge detection and position determination at medium levels and identification and quality determination at the highest levels.
At each level the result of the processing of the level below is used.

In the control system the information flow is in the opposite direction. Here the result of the processing of one level is used at the level below. Typical processing tasks, in order of appearance while descending the control hierarchy, are: complex task decomposition, elemental move generation, primitive action generation and coordinate transformation.

The justifications of a hierarchical structure of both the sensor and the control system are: simplicity, modularisation, easy expansibility, easy evolution etc. Another structure can be justifiable as well if there are well founded reasons to do so, such as timing constraints etc. One can however notice, that the gradual increase in processing investment, that goes with higher levels of information exchange between a hierarchical sensor and control system, generally is nicely coupled with the gradual reduction of the required response time.

Sensor Systems and Computer Integrated Manufacturing

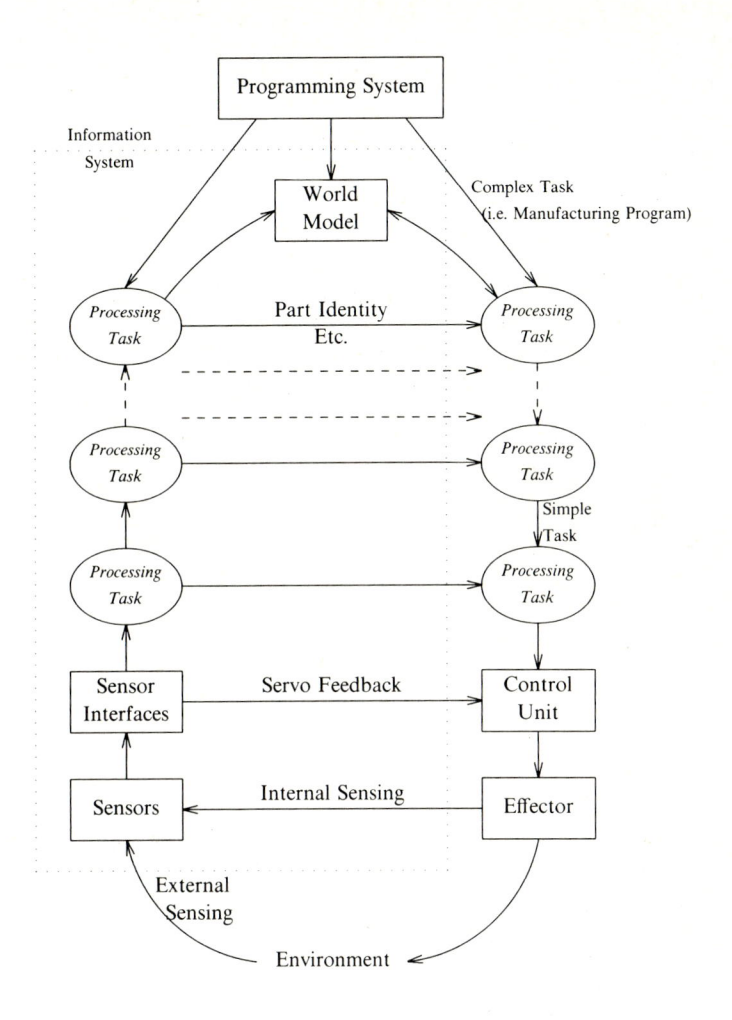

Figure 15.4: The hierarchical structure of both the sensor system and control system. The arrows indicate the directions of the main information stream. In fact all the information streams are bi-directional.

15.4. Application of Processing Maxims to Sensors

Introduction

The processing maxims as given in chapter 12 of this report are intended to be applied to the whole area of Computer Integrated Manufacturing. One of the most widespread components, that are extensively used in automated systems, are sensors. Therefore it is considered to be of importance to show the applicability of these maxims for sensors.

The aim for a manufacturer will be to have sensor systems implemented in such a way that an optimally reliable, flexible and economical manufacturing system results. The maxims are intended to be of help to obtain a system that approaches this goal. Due to the general nature of the maxims, one cannot expect that applying the maxims yields the complete set of answers for designing such an optimal system.

However, the aim of the maxims presented is that they are helpful in optimizing a system design. Given the needs as prescribed by the manufacturing process, a first design of a system can be made. The maxims now, will indicate points one has to pay attention to and also give guidance if the best solution among several has to be chosen.

In this way, the design can be optimized.

This now, will be illustrated by applying some of the processing maxims in the following sections.

Note however, that a further optimization might be possible with a set of more detailed sub-maxims, that are more specific to the field (in this case sensors).

The principle of intelligent local processing

Sensors are traditionally coupled to local processors. The reason for this is that sensors originate from stand-alone read-out instruments, that have to deliver a signal, in a format, ready to be perceived by the person (or system) receiving that signal. In order to be able to do this, most of these instruments, have to perform some processing tasks, such as noise reduction, filtering, gain controlling, averaging etc., all on their own.

Integrating sensor systems in the complex CIM system, is *not* a reason to shift these processes to a centralized processor. On the contrary, a very desirable feature in CIM, which improves the fail safeness and the flexibility of the CIM system, is the possibility to use subsystems that are able to run stand-alone. Local processing now, is the key to stand-alone sensor systems.

There are additional reasons to justify local processing for sensors.

1) It adheres to the fact that sensors have a connection with the exterior of the system.
2) The frequency of communication or the amount of data transmitted that would go with remotely processed sensors can result in high costs.
3) Quite some processes are specific to the sensor involved.

Based on these arguments, one can identify several processes, that lend themselves to be handled locally, such as common processes (e.g. sensor control and adaptation processes which are on the lowest level similar for most sensors), data reduction processes and even application dependent processes. Next we will illustrate some of these issues.

Several aspects that are related to sensors, are not related to the requirements that have to be fulfilled by these sensors. They typically are a result of the adaptation of a system to the outside world. For example, a vision system has to adapt itself to the illumination of the scene. If there are no special illumination facilities, the luminosity can vary, dependent on the time of day, the season, the weather conditions, an obstacle passing the scene etc.

Due to local processing, that part of the system that has to cope with this adaptation can be localized. For sensor systems there are transducer interface control and adaptation tasks, such as filtering, noise reduction, tuning, gain adjusting etc. These are more or less standard processes, that will have to be done for almost every type of sensor.

It is clear that these control and adaptation tasks will have to be performed as close to the sensor itself, not only to obtain the desired stand alone possibility, but also for reasons of simplicity, communication reduction, response time requirements etc.

By doing so, it might as well be advantageous to locally perform processes, that reduce the amount of data to be sent, such as event selection and/or data manipulations (e.g. averaging, estimating, selection, collection and conversion). This does not require a lot of extra investments in hardware, since often the same local processor that handles the standard processes, can handle these reducing processes as well.

Event selection and manipulation of the data locally not only has the important side effect of reduction of the communication requirements, but also simplifies interfacing with the rest of the system.

Local processing does not necessarily have to be restricted to these more or less application independent tasks. Also some application dependent tasks may be very well suited to be processed locally, provided these tasks do not have to interact very often with other processes that are not localized at the same place. This will result in an intelligent local system, that is easier to fit in with the overall CIM architecture. The more intelligent the local system, the more simple the interface with the rest of the system will be.
As an illustration we will give an example of a gripper system.

A gripper, being part of a robot, usually will be equipped with a force sensor, to be able to detect whether an object is held firmly enough. This force sensing can be done with an internal sensor, such as an air pressure measuring sensor for a pneumatic gripper, or by measuring the back-EMF of an electrically driven gripper. It also can be sensed with external sensors based on stress measurements, such as strain gages or even artificial skin.
Anyhow, the result of the force sensing will primarily have to be fed back to the controller of the gripper actuator. The standard processes, that are necessary in order to be able to measure the force, such as sampling and interpretation of the data, conversion to the correct magnitude etc., are not of interest to any other part of the system and thus clearly are processes that will have to be handled locally.
The decision of with what force an object has to be gripped, requiring knowledge of some parameters of that object such as the weight and surface roughness, is made elsewhere and also sent to the controller of the gripper actuator.

One can see that due to local processing, force sensing can be a completely local affair which is handled inside the gripper system, thereby exonerating the rest of the system. Force sensing however, is only one of the processes that have to be started, while grasping an object. Clearly, among the total package of processes that are needed to perform the grasping task, there are some application dependent processes, such as controlling processes. If these processes all are performed locally, it is possible to send simple instructions to the gripper system in order to control the gripper.
The gripper system is integrated in a more embracing local system, the robot system, which means one can distinguish several levels of locality. The robot will, due to the localization of the necessary processes, be able to handle the simple instruction to pick a part located at a specified place, with a specified gripping force. The robot controller will, completely locally, take adequate measures to locate the gripper and in turn can simply instruct the gripper system to grasp the object with the specified force.

It may be clear, that local processing is both sensor and application dependent. For instance, simple micro switches do not need noise reduction, gain adjusting etc. On the other hand, for certain applications, vision systems can have very high processing requirements.
In order to determine the processing requirements of the sensor at this local level, the classification of sensors, as given in section 15.2.1 and in appendix 4 section 1, can be a guide.

System design should aim for the distribution of processing

The maxim about distribution states that a system should be designed in such a way, that processes can be distributed among the system. Processing tasks employed by sensor systems, often are processes that can be made distributable.

Active-sensing devices, for instance, are sensors having their own source of radiation (see appendix 4), which will have to be controlled. Radiation control is a processing task that differs in nature and that runs rather independent from the processing required for handling and interpretation of the sensor data.
Another example is an adaptive sensor. Adaptive sensing (also described in appendix 4), is sensing combined with manipulation of the sensor or manipulation of the sensed object. Here also one can observe a separation between the handling and interpretation of the sensor data on one side and the control of action on the other side.

These examples of two processes running in parallel and rather independent of each other, are typical distributable processes. However, also pipeline-structured processes lend themselves to be made distributable. Pipeline-structured processes are processes that run sequentially and one process uses the output of its predecessor as input. A vision system, for instance, can be pipeline-structured. In such a system, a coordinate transformation may be followed by an edge detection process, a part position determination process and a part identification process.

Making these processes distributable, does not mean they also have to be distributed. The system specifications will have to justify a distribution. This can be based on timing constraints, the requirement to localize processes, the configuration of available processors, the reliability of the system etc.

System design should aim for deadlock-freedom

Sensor systems are input devices, which in order to provide sufficient flexibility to be used with arbitrary control systems, ought to be able to operate in three different modes. These three modes are:

1) Polling mode (i.e. sense once on request)
2) Sampling mode (i.e. sample continuously or at a given frequency)
3) Event mode (i.e. signal an alarm when an event is sensed).

The last mode, the event mode, now can be a cause of deadlock, if the system is poorly designed. In case of deadlock, the system persists waiting for an event that will never occur. The cause of this non-occurrence can be that the event is impossible, in which case the specification of the event as supplied by the control system is wrong. It can also be caused by a defective part of the system. Wrong specification of the event may be avoidable, but it will not be possible to design a system so that parts cannot be defective. Therefore, these situations will have to be avoided or side-stepped. One can, for instance, simply combine the event the system is waiting for, with another event, which is sure to happen (e.g. time-out, reset signal) and which on occurrence also unlocks the system.
This, however, may not be a sufficiently good solution for a real-time system, whereupon the decision can be made, not to allow event mode sensing in such an environment.

In a well designed system, a control system, ordering its subsystem (the sensor system) to sense in event mode, in no way can be deadlocked itself due to the event never occurring. This is conform the maxim about propagation of faults: the deadlock of the subsystem must never be able to propagate to other parts of the system. Because the control system ordered the subsystem to sense in event mode, it is the obvious part of the system to also take care of deadlock recovery (e.g. resetting) of the subsystem.

Now take for instance an automatic drilling machine. In some way or another, the depth of the drilled hole in the workpiece must be sensed.
The most simple mode of sensing may seem to be event mode sensing. In that case, the sensor system is programmed by its controller, with the depth to be drilled. On reaching this depth, the sensor system sends the alarm signal, upon which the automatic feeder system of the drilling device will have to stop.
This is all comfortably simple, but it is not deadlock free. For instance, consider this system having a defective sensor system. The depth to be drilled will, according to its sensor system, never be reached, and the automatic feeder system does not stop until the end of its range, damaging the workpiece and maybe also the drilling machine.
This deadlock situation can be side-stepped by also introducing a time-out mechanism. Knowing the feeder speed, the approximate time it takes to drill the hole is also specified. Now, whenever this approximate time is exceeded, the feeder system also has to be stopped, and the controller should do a diagnostic inquiry.
The depth of the hole will in spite of this time-out mechanism, (slightly) exceed the specified

depth, but this may be acceptable.
In this way, the combination of accurate depth sensing combined with an inaccurate time-out mechanism, may result in a sufficiently safe system.

Using the sampling mode for depth sensing, would even avoid deadlock situations caused by the sensor system. In this mode, the sensor system frequently senses the depth and sends the result to the controller. This controller compares the actual depth with the specified one and thereupon decides to continue (to slow down) or to stop the feeding.
This may require more sensing (and processing), but it has the advantage that with every sensed depth, the result can also be compared with the previously measured result, in order to check whether the value is acceptable (see also diagnostics). If the feeder system has not been stopped yet, this value must differ a known amount (the elapsed time multiplied with the feeder speed).
The controller will thus be immediately alarmed upon malfunctioning of the sensor, whereupon it can take protective measures. In this way the specified depth will not be exceeded. This can be a reason to decide for cumbersome sampling mode sensing instead of simple, but deadlock sensitive event mode sensing.

Every (sub)system for computer integrated manufacturing must provide diagnostics information

One of the most widespread tasks of sensors, is error detection, in which the only task of sensor systems is to supply information to a control system, that checks whether those parts of the system, that manifest themselves in the real world, behave (or did behave) as they should. In case of a misbehaviour, an error signal is generated. Upon this error signal, an advanced control system, that is a control system which is able to do diagnostics, will start inquiring diagnostic information, in order to ascertain the location, type, severity and cause of the fault. The sensor system, as available to this control system, may play an important role in this inquiry. Although diagnostic inquisition is possible with present-day technology, it certainly is not state-of-the-art in industry yet. It may occur in case the system is very complex, in order to speed up the recovery process. It may also occur when the attainability of the system is low due to hostility of the environment. Usually, however, the system will, upon error occurrence, simply be checked by a mechanic.

There are two basic modes of interaction of a system with the real world, to which sensor based error detection and diagnostics can be applied: manipulation and sensing.
The above mentioned error detection and diagnostics by means of sensors, is primarily intended for the first mode of interaction of the system with the real world, that is checking of transport systems, manipulators, machining systems etc. In this case, one can state that in order to be effective, the sensor system has to be more reliable than the system it checks.
This leads to the second mode of interaction of the system with the real world to which error detection and diagnostics can be applied to, namely the sensor system itself.
Since the maxim about diagnostics states that a system should provide its control system diagnostic information, we will have a look at what this means applied to sensor systems.

There are several ways in which a sensor can be checked for correct functioning.
Statistics can be very helpful for error detection. Quite often a reasonable indication of the acceptable number and the acceptable measure of deviations is known. Upon serious deviations of this acceptable number or measure, the system may start inquiring diagnostic information.

The system can also (as error detection method) compare a sensed state with a previously measured state and check whether the difference between them is acceptable, given the elapse time and relevant circumstances.
This can be a rather simple method, which can be performed by a local processor, making use of pre-fixed or variable limits, between which the differences of the states of two consecutive measurements have to find themselves in order to be acceptable. When variable limits are used, the method is slightly more complex and will require more processing and

communication.

As an example, think of a manipulator with a joint velocity sensor. Given the moment of inertia of the system and the maximum torque that can be supplied by the actuator, the maximum acceleration is fixed. Because of this, the maximum acceptable difference of velocity between two consecutive measurements will have to obey these prescribed limits.

Based on fixed limits, one has to take into account the minimum moment of inertia of the system and the maximum torque that can be supplied by the actuator. The minimum moment of inertia is the moment of inertia of the relevant part of the manipulator (i.e. from the joint of interest towards the gripper) without load and in a specific position (heavy parts as close to the joint as possible). Obviously, it is more exception than rule, that the manipulator is without load, with the relevant part in this specific position and with the joint of interest actuated with the maximum torque. As a result these fixed limits in practice can be set too wide. In fact they can even be set too wide to be an adequate error detection method at all.

The method can be improved if variable limits are used, by taking into account the current moment of inertia or, if this is unwieldy, an approximation (lower than the actual value!) of the current moment of inertia, and the current torque as supplied by the actuator. This, however, requires more computation and communication than needed with the method based on fixed limits.

A slightly different error detection method, is based on the comparison of the sensed state with the prediction of the state as evaluated in the world model of the system (see section 15.3.2). A slight deviation from the predicted state is acceptable and will result in an updating of the world model. A too large deviation, however, indicates an erroneous situation. It may be clear, that evaluating predictions of the state requires a lot more processing than the previously mentioned method, and generally will be justified only if these predictions also can be used elsewhere in the system. This is also a reason why this evaluation probably should not be done locally. On the other hand if prediction of the state is already prescribed by other processes, it is an excellent reason to use this prediction for error detection.

These methods, however, merely indicate a suspicious behaviour of the sensor. They do not *prove* it is the sensor that is defective, since it is *possible*, that there are other parts of the system that are defective, yielding the same error signal.

A well known method to directly check sensors upon correct working is making use of a calibration point. Calibration points are commonly used for calibration of slowly deviating sensor systems, but they also can be used for checking non-deviating sensors. Checking of the sensor by means of a calibration point can be done automatically every once in a while. In this case it is an autonomous checking method. It also can be done upon request, whenever the sensor shows a suspicious behaviour according to its control system.

Examples are a calibration rod for a length measuring sensor, or a calibration scene for a vision system.

In sensor complexes, the sensors can be used to check each other. Surprisingly at first sight, this does not necessarily mean that the sensors that check each other, have to be of the same converting energy class. It does mean, however, that they have to be able to measure the same quantity in an overlapping range and territory.

As stated in section 15.2.2, the sensor complex controller must be able to compare the overlapping information of the individual sensors. If the information of both sensors conflict, the controller may decide to accept the information of one of the two or a (weighted) average, or the controller may start inquiring diagnostic information, this all dependent on the severity of the conflict. Clearly, if one of the sensors is at the limit of its range, the information is more likely to be doubtful than information of that sensor in the middle of its range. So, if two overlapping sensors provide conflicting information, and one of the two is at the limit of its range, the severity of the conflict is not high, and the value produced by the other is probably acceptable. However, if both sensors are well within their range and the information differs considerably, at least one of them produces erroneous information.

As an example, consider a robot with a wide view, low resolution vision system and a short range, high resolution proximity sensor on its gripper. The vision system can be used to coarsely guide the the robot gripper towards an object. Whenever this object comes in the range of the proximity sensor, the two sensor systems should indicate the same distance between gripper and object. On the other hand, since the robot vision system does not have a very high resolution, its information is very likely to differ from that of the high resolution proximity sensor. The sensor complex controller, aware of this fact, concludes that the value the proximity sensor provides is probably correct, and it will deliver this distance to the robot controller. However, if the difference between the two systems outreaches the resolution of the vision system, the sensor complex controller should recognise this as an erroneous situation.

Basic components should be used where feasible

There are in CIM a number of equivalent sensing tasks, that show up in several parts of the system, such as proximity sensing. This kind of sensing shows up in robot grippers, along conveyer belts, with part and tool detection, in transport systems (e.g. automatic guided vehicles), for checking of doors, etc. Although these tasks may differ slightly, they all can make use of the same proximity sensor with interface (provided the range is of the right order). The difference in task, only shows up at the level of interpretation of the sensed distance.
From this, one can conclude that a proximity sensor with its interface, is an excellent example of a basic component, which can even be easily standardized. By doing so, all the advantages as mentioned in the narrative of maxim 9 chapter 12 become valid.
These are: no development costs, multi-usability, availability, known properties, etc. In spite of this, there still are sensors, that (although they allow to be treated as a basic component) do not allow standardization yet, because they still show a progressive development. It is a well known fact, that development and standardization are two conflicting issues.
It can, from one side, be rather tricky (and thus discouraging), to develop a truly standardized system any further, without coming in conflict with the accepted standard. From the other side, trying to standardize systems, that are still in development, is like crying for the moon.
Although aiming for standardized sensor systems, can be too far out of reach, it is, as stated above, worth while to use as much basic sensor systems as possible. Suppliers will tend to develop their products in a direction as prescribed by their customers and as compatible with older products as possible.

The principle of itemized processing

The principle of itemized processing states that processing for a certain task is only started when all information required is available. Applying this maxim on sensor systems has influence both on the design of the sensor system as part of a manufacturing system and on the processing specifications of sensor systems.

To illustrate the influence on the design of sensor systems, think of an automated assembly station. In order to be able to assemble the different parts that will form a product, this maxim states that all the parts will have to be present. Moreover, the parts will have to adhere certain specifications.
In order to assure this, there must be an adequate sensor system that is able to verify if these conditions are fulfilled.
This means that somewhere in the system, quality control will have to be performed on the parts. In-house produced parts, will be checked immediately after production in order to obtain directly information about the production process and to avoid senseless transportation of worthless parts. External produced parts best can be checked upon entering the system.
By doing so, it suffices that all the parts necessary for the assemblage are checked upon presence only, which is simple to do. An exception to this may be necessary for fragile parts.

The influence of the maxim on processing specifications results in the requirement that information as coming from a sensor system, no matter its complexity, must be complete before it is processed.

To do so is especially rewarding for complex sensor systems, in which information is supplied by several individual sensors. These individual sensors may operate with different cycle times and may deliver their information in a vastly different outline. If it is not made sure that information is itemized, the administration of the system may soon become unmanageable.

On the other hand, if the information *is* itemized, the administration of which sensor has not yet finished its delivery and what the outline of the information of that particular sensor should be like, can be completely left out. Moreover, the result of itemizing can be that (except for diagnostics) it is not necessary to be aware of where the information is coming from at all.

An individual sensor or sensor system, that supplies itemized information, typically is an example of a basic component (see processing strategy maxim 9).

15.5. Application of Communication Maxims to Sensors

Introduction

In the following sections the applicability of the communication maxims to CIM-applications involving sensors is illustrated. In the descriptions of the maxims their general applicability to communication in CIM was illustrated. In contrast to this approach, the purpose of the following sections is to show that the maxims often have specific value for some of the most important modules in Computer Integrated Manufacturing, as for example sensors.

Often, the communication maxims ensure that the information, gathered by the sensor system to resolve uncertainties about the physical state of a system indeed reaches its destination or, stated differently, they ensure that the sensor system is not counteracted by the communication system.

Almost every communication maxim has a specific value for sensor systems in CIM. A few maxims however, have, although they should be adhered to in Computer Integrated Manufacturing, no specific value for sensor systems. The approach followed in the next sections is that for almost every maxim one (or more) examples from sensor applications are given that clarify the need for the communication maxim.

Network modification should not lead to shut down

Sensor systems are used during the manufacturing process for control purposes. Individual sensors, independent of other sensors in the CIM system can be useful, but often complex CIM applications require various cooperating sensors to obtain the required information. A vital component, which is needed to realize this cooperation is the communication network.

In every factory once in a while the production process has to be changed. This is a consequence of the ever changing wishes of customers and the modernization and adaptation to the state of the art. Depending on the manufacturing process changes will be regularly or irregularly, but as long as total flexibility is not realizable, modifications are inevitable. These changes do not only include adjustment of machines and robots but sometimes other sensor complexes are needed to recognise the new products and/or the new environment.

At first sight it may seem unimportant whether the installation of new sensors will bring the network to shut down or not, because the production process has to be stopped anyway when the switch to a new product is made. In spite of that, if the installation of a new sensor system brings the network to a halt state, this will be undesirable. It is possible that various sensor complexes, related to several distinct production processes will have to share one (physical) network. From the point of view of efficiency this will not only be possible but even likely, especially in a hostile environment where expensive hardware might be used, and it would be unwise not to share it amongst several production processes.

Production losses during the adaptation of a production line will be unavoidable. If the introduction of new sensor systems implies a shut down of the network however, the costs of these changes can be much higher than absolutely necessary.

De facto standard protocols should be used in CIM as much as possible

Until recently, connection of sensors was often done in an ad hoc manner. Producers of sensors and sensor complexes were in many cases primarily interested in an improvement of the sensing capabilities of their product, and not that much in the communication to the outside world. Since it was unclear which protocols were going to be accepted by various standardization committees, they could not foresee via which protocol the communication had to take place. Therefore the sensors were often only usable in combination with other products of the same manufacturer.

Now that standardization evolves rapidly this situation should change as soon as possible. If standard protocols are used, it will be easy to replace sensors by sensors with different capabilities, as long as they adhere to the same communication standard. On the other hand, it will be possible to use sensors in different networks thus enabling the replacement of the network as well. The consequence is that sensors that are not required any more during a particular production process can be removed and used in another production line, even when this line is supported by a different communication network.

This is certainly not state of the art yet, but on the long run it will be cost saving.

The specification of a network should clarify which timing requirements can be handled by the network.

Before the installation of a new production line it will, in general, be known more or less precisely which sensors are used. Furthermore, it will be known how often these sensors have to produce information and how many bytes the information produced will be composed of. Furthermore, it will be known how fast information has to be transmitted. To give an example, if a robot is guarded by a sensor system, the information that is produced will be valid for only a short period, depending on the velocity of the robot etc. In this way a stable feed-back loop can be realised.

The consequence of this knowledge and the maxim is that it can be calculated whether a network can support the communication needed by the sensor system or not. By this, one important requirement of the communication network can be checked easily.

It is important to note the relation to processing maxim 1. This maxim states that local processing should be done where feasible. One of the results of local processing can be that the number of bits needed to carry information can be reduced, sometimes even enormously. A simple example will clarify this. If, for instance, the value of an unknown quantity can vary between 0 and 30,000 and it is also known that the difference between consecutive measurements can never be more than 100, a severe reduction of communication requirements can be realised by not sending the absolute value of the quantity of interest, but instead the difference with the 'last' value. Seven bits will suffice, instead of fifteen bits when the value itself is sent along the network (Note however that sometimes a check value is required too).

The result of this form of pre-processing is that the communication need is reduced which means that it will be easier to fulfil this need with the network.

A network should allow for a mixture of different hardware technologies

The application range of sensors is very broad. They are used by the manufacturing system, often attended with various sources of interference and electromagnetical pulses. The sensor system can also be used to gather information in non-hostile environments.

The information gathered in hostile environments and the information produced in non-hostile environments can be related logically. This means that both types of information combined give a complete view of the parameters of interest. It is also possible that information out of environments of different types of hostility has to be combined. Think, for example of a production line where the product has to find its way along electrical welders and paint sprayers.

Whenever possible, if a logical relation between data is present, it should be reflected in the physical structure of the network. In other words, the effort to make the software, which ensures that the information is combined properly, will be reduced if the information is available on one physical network.

Such a physical network will often be composed of different transmission media. Some transmission media can be well suited to a special environment but not at all to another environment or too expensive to use in all CIM networks. This means, that it is required to use different transmission media in one physical network to ensure that the information gathered by the sensors will be combined correctly.

Special precautions should take care of protection against unauthorized access and provide for data security

The second part of this communication maxim will be most important for sensor systems. Protection against unauthorized access is not considered to be very important for the application of sensors. However, if an 'intruder' manages to let information disappear the usefulness of the sensor system will disappear too.

A discussion of the most vital part of the above communication maxim is given now. The production process is, to a large extent, governed by the information produced by the sensor systems. Both local and global decisions are made according the information available, gathered by the sensors. If, due to unperceived errors of the network information is disrupted, other information will often become totally useless.

In this connection we might think of a moving crane that has to be slowed down in time. If the slow down message is lost, the influence on the production process can be disastrous.

A network should be designed in such a way that it can cope with environmental noise

This communication maxim seems to be stated especially because of the use of sensors in Computer Integrated Manufacturing. It seems almost exaggerated to explain its applicability any further. However, we will elucidate it further by raising some properties of sensor systems that are not mentioned earlier.

Because of their properties and their use in Computer Integrated Manufacturing sensors will be surrounded by various sources of interference. Special precautions should ensure that data gathered by the sensor system will be transmitted to a 'safer' place. According as the distance to the source of the noise is diminished it will be more difficult to ensure this.

Not only the sensor system itself can be influenced by the hostility of the environment, but also the communication network. Although local processing can be used to reduce the communication requirements of a sensor system, often a sensor and its related processor are part of a larger system, which means that there will still be a need for communication.

A trade off should be made between network performance and network reliability depending on the application

In the following section we will show how the trade off between network performance and network reliability can be made.

The size of the data produced by sensor systems can differ largely. On one side of the spectrum there are computer vision systems producing large bulks of data, on the other side sensors are available that do not produce more than a simple on/off indication. This has immediate consequences for the performance of the network. It may seem that the reliability of the network is not related to the required performance. This relation is however present, and we will illustrate this by an example from a control system.

A sensor (or a sensor complex) can be used to determine the distance between a robot and another object. The purpose of this measurement will be that the robot is enabled to grip the object. When information about the distance is produced regularly (say every millisecond) the influence of one erroneous distance will not be severe, since it can be compared with a prediction based on previously acquired information. If the prediction and the actual value differ too much, the actual value can be disguarded. Since the next distance will generally be correct, robot control can be done in an adequate way. On the other hand, if information is collected not that often, erroneous distances can, in the worst case, cause the destruction of robot and/or object.

To summarise, if the distance is measured often, the influence of one transmission error will be small, but if not, transmission errors should be avoided at almost any price.

Processing requirements will have a large influence on the number of times a parameter of interest is measured. Therefore, if processing power is the bottle neck of the system, it will be requisite to have a highly reliable network, and the performance of the network is of a second order of importance. On the other hand, if information is produced at high frequency (because the application needs it), the performance of the network is vital and the reliability constraints are of a second order of importance.

The network should allow the use of messages with different priorities

This is another example of the nice way the sensor system and the communication network can cooperate to perform the complex task of Computer Integrated Manufacturing. As stated earlier, high priority messages will be used mainly to recover from undesirable situations. Often errors will be detected by a sensor (e.g., in the production process). As soon as an error is detected, appropriate actions have to be carried out. It will depend on the nature of the errors whether these actions will be locally or globally. Decisions about this can often be made globally only.

The number of special application protocols needs to be as small as possible

The main reason for this communication maxim as related to sensor systems is that it enables an exchange of sensors. As long as the sensors adhere to the same application protocol, the replacement of a sensor causes only a small, local influence.

The maxim does not state any property of the way in which data is collected by the sensor. A new sensor can produce the data completely different from the way it was collected by the old sensor. (In fact, the more efficient way of data collection can be the reason for the replacement of the old sensor.) However, the maxim *does* state a property of the application protocol the sensors use, namely that it should be the same for both old and new sensor.

In this way sensor replacement will have positive effects only (i.e., the increase of efficiency). Negative effects, like the expensive replacement (or modification, which often tends to patching) of software can be avoided.

Chapter 16

Graphics Systems and Computer Integrated Manufacturing

Introduction

Computerized graphics systems are playing an increasingly important role in Computer Integrated Manufacturing (CIM).

Graphics systems are instrumental on computerizing the handling of technical information. Traditionally, technical information is kept in engineering drawings and special lists, describing materials, schedules, etc. The alpha-numeric components of technical data can be much more efficiently managed when using and storing them in agreed upon formats. For both the handling of data with a geometrical component (like most technical data), and the handling of precisely formatted data, graphics systems can give the support by providing the primitives to present the basic data items as well as the primitive functions for the manipulation of the data items or their lay-out.

One of the effects of integration in manufacturing is, that in all phases, from design through planning and programming to actual manufacturing decisions, decisions are taken more rapidly and at the same time based on more categories of data being available. In order for human beings to be able to effectively keep control over these complex data, efficient tools need to be developed. In all of these tools data presentation involving pictures and access methods to these pictures are essential. Moreover, these graphics functions need to be embedded in man-machine communication systems. The real-time capabilities of these systems and the quality of the user interface are greatly determined by the performance of the graphics support system.

The complexity of the manufacturing process and the economic necessity not to interrupt a production process have created the need for extensive simulation facilities. They support the engineer during decision making and problem solving by providing simplified and yet adequate models of production processes and the operation of products. In this way checking production without test runs or plant shut down is possible. Also, the need for costly product prototyping is reduced.

In almost any area in CIM, graphics systems are being used. The first applications of these systems in CIM have been in the area of design. Many applications in the design area are based on so-called drafting systems. The graphics system supports definition of drawings as the means of defining a design and producing engineering drawings for further processing.

In an integrated environment, the use of such systems shifts from drawing production to sophisticated data-entry of technical data. These technical data are used subsequently for engineering tasks such as analysis, production planning and actual manufacturing control. It goes without saying, that all man-machine interfaces for these numerous applications will be influenced by the initial data-entry method used.

Therefore, computer graphics stands for a very important class of processing facilities (hard- and software) and must be considered as one of the most suitable candidates to show, what effects the maxims, developed in previous chapters, may have on the development of complex and cooperating systems to be used in CIM environments.

Following an overview of existing graphics functions, applications, hardware and software systems, a number of the maxims is applied to frequently observed examples in computer graphics and its applications in CIM.

Finally, a sketch of an "ideal" basic graphics system is presented, in the sense, that it complies with all the maxims, and the idea is suggested, that such systems are technologically feasible now or in the near future.

It would be very economical indeed if the numerous graphics subsystems in CIM could be built using the same basic graphics package. The circumstance that most of these subsystems will be using the same technical data base, makes this feasible. Currently, there exists one paramount candidate for such a general basic graphics system. This is the first international standard in the field of computer graphics programming called GKS (Graphical Kernel System).

GKS is a primarily European based development which now has been successfully introduced in North America as well as in Japan. In all three areas it is rapidly replacing most other existing basic systems. GKS offers an excellent opportunity to close the gap in information technology development, because it is one of the few areas where European IT is leading.

The designers of GKS were particularly anxious that CAD/CAE applications could be well supported. This makes GKS a very good candidate to support CIM subsystems. The special properties of GKS which were a requirement for it being a standard system, will make it possible to have compatible graphics subsystems throughout CIM.

Therefore, the advanced state of the art in graphics in Europe could carry over to the area of CIM systems, especially where interaction is involved. As one might expect from a carefully designed system, it allows for adhering the processing and communication strategy maxims as illustrated below.

16.1. Basic Graphics Functions

Computer graphics deals with methods and functions for the creation, storage and manipulation of models of objects and their pictures by means of digital computers. Basic tasks in computer graphics are:

- visualization
- structuring
- interactive manipulation.

For the visualization of pictures, which may be defined using two or three dimensional coordinate systems, *basic functions* are needed to display points, lines, areas and character strings. Furthermore, in general *attributes* are associated with these functions, which determine at the moment the function is used

(i) the coordinate *transformations* that have to be applied
(ii) the *appearance* associated with the function (e.g. colour, linewidth, character font to be used)

The *structuring* of the picture deals with the methods to be used for the grouping of these basic functions, so that a number of them, which together may define an recognizable object, can be viewed and dealt with as a whole. These groups, called segments, are used for the creation of subpictures (possibly nested), that can be manipulated independent from other subpictures. Even more flexibility can be achieved by using a more sophisticated graphics data organization, known as "dynamic picture graphs". Here, segments may be linked freely with each other and by adding and deleting links and segments, one can make complex and rapidly changing pictures.

For interactive manipulation, hardware facilities are needed for pointing to specific locations of the picture, for example mouse and light pen. This implies the need for a set of basic input functions, so that an application program is able to access information, provided by these graphic input devices.

It has been found necessary to classify these input devices and provide basic input functions for each of these classes. A contemporary classification as used in GKS is:

1) locator (to point to a specific location of a picture)
2) pick (to point to a specific object of a picture)
3) choice (to select various options, e.g. using a function key box)
4) valuator (to obtain a value, e.g. the amount of rotation specified by the position of a thumb-wheel)
5) string (to obtain a line of text from a keyboard)
6) stroke (to obtain a sequence of points or connected lines).

Furthermore, it is necessary, that these functions can be used in various modes:

a) request (the application program waits until the desired input has been provided by the user)
b) sample (the application program fetches the momentary value of an input device without waiting for any user interaction)
c) event (the user determines the moment, that input is done, which is subsequently remembered by the graphics system and delivered to the application program, when the latter has reached a state, in which it is able to accept the input.

Today, graphics workstations have been designed, that provide for all or the most important of these basic functions.

At a higher level, a number of functions can be recognized, which are less generally used and which depend more on the particular application. Usually, these functions are implemented using general purposes computers. For example, if high-quality output is needed, then appearance transformations are necessary, such as transparency, shading and anti-aliasing.

To display objects, that are defined in three dimensions, a projection on a two dimensional plane is necessary and in many cases also hidden line/surface removal.

In solid modelling, facilities are needed to specify a representation technique, for example constructive solid geometry, boundary representation or wire-frame representation. This is discussed in more detail in Appendix 3.

This hierarchy of functions also implies different datastructures at each level. These have to be chosen with utmost care, since they determine many important properties of the resulting graphics application system, such as:

- ease of use
- suitability for manipulation
- unambiguity of resulting pictures

For example, in solid modelling wire-frame representations do not have the property of unambiguity, in contrast to unambiguous representation schemes like constructive solid modelling and boundary representations (which are computationally more expensive).

16.2. Computer hardware

The hardware environment for computer graphics currently shows a tendency towards more uniformity. On one hand, low cost graphics devices all interface to applications through the same basic packages such as GKS. On the other hand, high performance graphics devices exhibit a striking similarity in functional interfacing, although the architectures may be quite different. Due to microprocessor technology, most of these devices can be made to achieve the same functionality. Then the hardware properties such as resolution, drawing speed, picture buffer size, dynamic support for geometric transformations, colour range and dynamic feedback for interaction, that determine which graphics workstation to put where. The cost for a highly powered workstation is rapidly decreasing. Typically a low cost medium-cost workstation includes a Megabyte of memory and a dedicated one Megaflop microprocessor.

The applications attempting to exploit these possibilities benefit from the opportunity to download many graphics functions to the workstation. However, the sophisticated support needed for new applications (especially in 3D picture creation) or to increase the picture quality (high resolution) leads to high demands on computing power from the application host, in spite of the downloading of basic graphics functions.

For optimal (i.e. cost-effective) use, it is necessary to provide for sufficient resources to allow, that about 80% of all CAD tasks are handled locally, especially the interactive tasks.

Since the computer power, needed to solve the increasingly complex CAD problems, is growing fast, mainframe-like machines will continue to exist in the future.

16.3. Graphics Applications

A rapidly increasing use of computer graphics, at any level of sophistication, can be observed in a number of different areas in industry at present.

The application of computer graphics in CIM can be divided in four major application areas. For each of these areas one can find examples throughout a CIM-system; a particular application category is not restricted to a specific CIM area. The four major application areas are:

- Design, technical data handling:
  The definition and manipulation of geometrically structured data, visualized through pictures.
- Business graphics:
  The presentation and handling of primarily administrative data trough pictures. In this case the efficiency of presentation and the understandability for naive users are the major characteristics.
- Simulation:
  Realistic models of three dimensional real world situations are presented for analysis and programming (e.g. production scheduling, robot programming)
- Process control:
  Online real time monitoring and control of manufacturing processes (e.g. transportation, assembly)

For the first two categories of applications the same basic graphics system (like GKS) can be used. The same holds for the majority in the other two categories. Only when the quality of simulation or the severity of the real-time demands in process control require extremely powerful and expensive graphics workstations, it is necessary to use a dedicated basic graphics system, though the standardized graphics packages of the next generation will also accommodate those.

16.4. Graphics Systems in Computer Integrated Manufacturing

In this section the use of graphics systems in CIM will be discussed in more detail by showing how a number of the maxims developed in previous chapters is applied.

An important property of all graphics systems is, that the quality of the man-machine interface is essential for the usefulness and efficiency of these systems. Most of the examples given below are chosen to show, how the maxims, when applied properly, can be an aid in achieving a man-machine interface of acceptable quality.

The principle of intelligent local processing

The scope of the local processes should incorporate as far as possible all tasks, that are to be done interactively. For example, if transformation of a picture on the display of a workstation is necessary as part of a user interaction, computations needed to visualize the transformed picture should be done locally in that workstation rather than on a time-shared host, where the response time tends to be unpredictable and long. As a result, users would tend to avoid using transformations even if it is most natural to use them.

Furthermore, the set of local tasks should comply with a functional interface. For example, if an unintelligent graphics terminal is being used, the graphics software system should be designed in such a way, that, when the terminal is replaced by an intelligent workstation, the logical functions, such as transformations, previously provided for by the graphics software system, can be easily moved to the intelligent workstation, and vice versa.

Systems design should aim for the distribution of processing

The usefulness and expected life-time of any graphics software system (and naturally also applications built on top of that) can be increased considerably, if distribution of tasks is incorporated from the first day of its design. This is a stronger statement than that in the previous section, since it also comprises the ability of parallel processing on distributed computer systems, whenever it is necessary or advantageous to do so.

Also, dynamic reconfigurability will be very desirable to allow for (temporarily) increasing processor capacity by distributing heavy computing tasks, which occur from time to time, over a pool of processors shared in a network.

As an example, GKS has been designed in such a way, that implementations can choose the most suitable distribution point for division of functionality between host and workstation. In this way, all available hardware facilities inside the workstation can be exploited. All missing functions can be emulated in software external to the workstation. Examples of such functions are those to define and manipulate segments. Introduction of new workstations with the corresponding redistribution of functions is always possible, provided the implementation has been designed for redistribution.

Systems design should aim for deadlock-freedom

This maxim is especially important in distributed environments, where because of parallel operation, processes have to synchronize (e.g. by exchanging messages). If one process is waiting for a message of another, which never arrives, deadlock may easily occur. An example of deadlock prevention in GKS is the possibility to restrict workstations in their synchronization of input and output. If a user wants to stop a program from sending further output, an adequate signal must arrive at the workstation at all times. However, it is possible to postpone input until all output buffers have been processed. In the case of relatively slow output devices, the requirement of "before next interaction" is dropped from the options.

Another example in graphics systems, where deadlock situations might occur, is at the man-machine interface. If an application program prompts for input, and if all output is buffered by a low-level software layer, all buffers should be processed, so that the prompt is actually displayed. Otherwise situations might occur, where the user is not aware of the fact, that the application program is waiting for input.

Finally, when user input is mandatory, for example in process control, deadlocks in user interaction can be bypassed using time-out mechanisms.

Every (sub)system for Computer Integrated Manufacturing must provide diagnostics information

Any subsystem in a CIM environment, like anywhere else, may occasionally exhibit abnormal or unexpected behaviour. Since these subsystems are mostly very complex, the cause of this behaviour may not be immediately apparent.

In general, three categories of causes can be distinguished:
1) hardware breakdown
2) software design and implementation faults
3) wrong user interaction.

In order to find these causes, the following strategies can be applied.

To ensure correct operation of hardware facilities, almost any modern graphics workstation is equipped with extensive stand-alone diagnostics software, which in general consists of a carefully selected and often varying mix of simple tests, that execute and check step-by-step basic hardware features.

Often during the development stage of a software subsystem software design errors and even more implementation faults occur. This happens so frequently, that a number of these faults will not have been detected when the subsystem has been completed, which leads to all kinds of unexpected behaviour of these subsystems when used in production. It is therefore currently a major task in Computer Science to develop methodologies to prevent and detect these kinds of faults.

As an example, in GKS for every function diagnostic information is unambiguously defined in the specification of that function. Furthermore, an extensive set of inquiry functions is available (e.g. to allow the application program to read the GKS state variables), which can never return errors. Thus, fault analysis is possible without introducing new errors.

Finally, a common source of errors is incomplete or inconsistent user input. Any good graphics system therefore must be designed in such a way to

(i) minimize the chance, that a user makes errors by consequently prompting and echoing input data (e.g. if a locator is moved beyond an allowed area on the screen, the echo of the locator should change to reflect this situation)
(ii) validate input data to the extent possible.

Faults should not propagate

An example, which clearly illustrates the dangers, resulting from propagation of faults is given in Appendix 3 and will briefly discussed here.

A very important property of representations schemes, that are used in geometric modelling systems to represent models of solid objects, is the *validity* of the resulting representations. If a geometric modelling system would allow to produce nonsense objects, that is, apparently possible objects, that are in fact meaningless, because they cannot be realized, all kinds of unwanted and even dangerous things might happen in an unpredictable way.

If a geometric database contains such an object, then, for example, the logic of other programs used in the design process for the analysis of these might be unable to recognize this situation, resulting in a program crash or the production of strange and inconsistent output.

Even worse is the situation, where the fault is so subtle and complete validity checking mechanisms are not available, that the fault is not recognized in the design stage and penetrates into the production engineering stage, so that wrong decisions about possible expensive investments easily can be made.

This elucidates the importance of data-integrity at each interface between modules anywhere in CIM: each module should produce valid data only, the receiving module should check for this and any fault should be resolved locally to the extent possible.

Real-time processing tasks should be well-specified

For a good and productive man-machine interface, it is very desirable, that all interactive actions of a graphics system are handled instantaneously. Otherwise, if the time-behaviour is unpredictable, users will get tired soon, increasing the chance of errors in the objects they are designing. One way to achieve this is to use real-time programming techniques, that are commonly used to implement actions, that should be completed within a known short time (< 1 sec.).

Which task should be considered real-time and which not, depends on the application; generally actions, that occur "frequently" as perceived by the user, like line-drawing, transformations, manipulations with segments etc., should be done in real-time.

One obvious way to achieve this, is to exploit local processing power for real-time tasks. But also the timing characteristics can be affected by unexpected events, like garbage-collection (e.g. because of the increasing complexity of an object, that is being designed, recovery from transmission errors etc.) or because of heavy access to overloaded time-shared resources like databases.

While in a design environment the occurrence of such events is undesirable and certainly should not occur often, when graphics systems are used for process control, everything possible must be considered to prevent them.

To achieve simplicity, extendibility, portability, flexibility and reliability high-level programming languages should be used
and
Basic components should be used where feasible

Besides the well-known general advantages of high-level languages (orientation towards application problems, facilities for modular construction of processing tasks and data organization, machine independence), it is also a desirable to use high-level languages to achieve distribution of processing tasks (moving tasks from host to workstation).

Also, this enables the use of basic components, such as standard graphics packages like GKS, which is a truly functional standard and can only be used when a high-level language is available, for which a language binding exists.

As a result, application programs become better portable and a higher degree of programmer independency can be obtained.

The principle of lazy evaluation

This principle, which states that processing should be done only, when it is certain, that the output of that processing is needed, can often be applied in graphics environments on a microscopic scale. This means, that the principle is not to be applied to overall system design, but for the purpose of smoothing incremental changes. These are typical for interactive sessions, where both user and system exchange information by adding little pieces at the time, which somehow may have, but only from time to time, large consequences. This implies that during these little increments, the system's response must be fast but may be simple. Whenever certain key elements have been communicated, more has to happen. But the user, expecting this, is then prepared to wait. The principle of lazy evaluation now implies, that on such occasions all postponed updating will take place.

The GKS graphics system can be implemented in such a way, that this (important) mechanism is supported. The way the system is specified with an explicit state vector allowing for very precise semantics with respect to incremental evaluation constitutes a firm basis for such an implementation.

The principle of itemized processing

The principle of itemized processing plays a very important role in inter process communication.

This can be illustrated using the GKS-Metafile as an example. In a GKS-Metafile picture descriptions are stored for the purpose of exchanging and filing of computer generated pictures (e.g. engineering drawings).

Each graphics system capable of reading a GKS-Metafile must take actions, which are equivalent to itemization of the input stream. It must build up a picture state description (choose colours, display layout, text fonts etc.) before it can actually produce output. This implies, that a large part of the metafile may have to be processed and converted into state descriptions, before any output actually takes place.

Then each output primitive will pick the relevant part of the state and this combined information constitutes one complete item, which can be passed to the drawing device.

It would make the graphics system extremely complex, if this itemization were to be done internal to the system. Imagine, for instance, that the system is merging two picture files. Then it would have to maintain several states simultaneously. This means, that non-itemization will add a parameter to every internal routine, explaining according to which environment the internal itemization must take place.

Therefore, either the itemization is guaranteed in the communication protocol, which means, in this example, that it is the responsibility of the metafile generating subsystem, or the free metafile format is to be combined with a preprocessor, which, using local buffer space, for each metafile creates complete items before giving them for further processing to the graphics system.

16.5. Basic graphics support.

At present, commercially available software packages exist (e.g. DISSPLA, DI/3000), covering a wide range of existing graphics hardware. However, software packages like this, although very useful in the way they are, have the disadvantage, that they are strongly oriented towards configurations consisting of mainframes to which graphics terminals are connected with a rather limited set of input facilities.

A recently developed and standardized system like GKS is oriented towards more sophisticated graphics workstations with a richer set of input facilities. Furthermore, in order for GKS to be the basic system for a wider range of applications varying from very simple to very complex ones, GKS has a level structure. This level structure separates the GKS functionality in a number of upwards compatible basic systems. Each application can be provided with an adequate basic system by selecting the most appropriate level, but can avoid the overhead associated with unneeded extra functionality. Considering this, there is almost no graphics area in CIM, where GKS could not be the basic graphics system.

Implementations of GKS, that satisfy the maxims mentioned, exist or are in a mature state of development.

Systems like this have a broad application scope, are relatively easy to maintain and to use and thereby constitute a solid foundation on which reliable and sophisticated complexes can be built.

Appendix 1

Flowcharting conventions

In order to define the basic processes that are involved in the manufacturing area as the first step towards the Design Rules project, the complete CIM area has been divided into five major activity areas, namely:

Computer Aided Design - CAD
Computer Aided Production Engineering - CAPE
Computer Aided Production Planning - CAPP
Computer Aided Storage and Transportation - CAST
Computer Aided Manufacturing - CAM

Having divided CIM into these five areas the basic processing involved in each area was investigated and documented in the form of a flowchart.

Many of the symbols normally used by Data Processing staff were used in the preparation of flowcharts. The conventions that were used are described below:

Symbols Used

The following symbols have been employed in the preparation of the flowcharts:

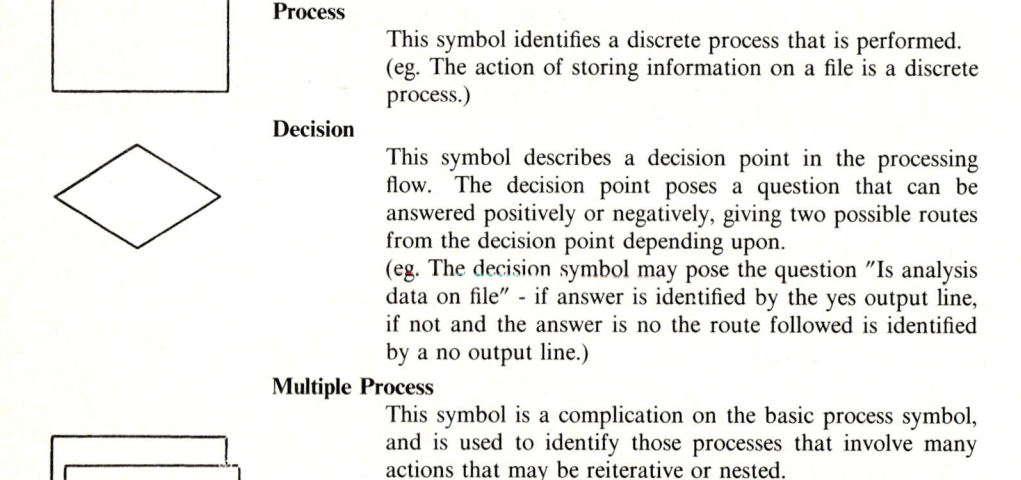

Process
This symbol identifies a discrete process that is performed.
(eg. The action of storing information on a file is a discrete process.)

Decision
This symbol describes a decision point in the processing flow. The decision point poses a question that can be answered positively or negatively, giving two possible routes from the decision point depending upon.
(eg. The decision symbol may pose the question "Is analysis data on file" - if answer is identified by the yes output line, if not and the answer is no the route followed is identified by a no output line.)

Multiple Process
This symbol is a complication on the basic process symbol, and is used to identify those processes that involve many actions that may be reiterative or nested.
eg. The process "redesign" can be a simple change to a specification or a major set of changes that can involve other processes on the same process many times.)

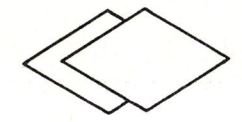

Flowcharting conventions

Multiple Decision
This symbol is used to describe a decision support activity rather than a simple yes/no question. Typically this symbol would describe a reiterative decision process asking what-if? questions.

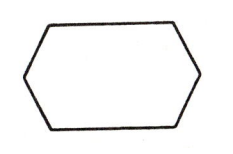

Initiating Process
This symbol is used to describe a process that initiates (sets-off) other processes, eg. a manual start-up or computer load command.

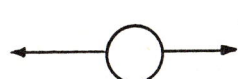

Connecting Node
This symbol is used to connect one area of the flowchart with another flowchart node that may be on another page or in another major CIM area.

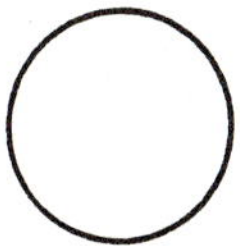

Other Activity Connecting Node
This symbol is used to indicate those areas where it is necessary within the process flow to exit or enter from another company area. The areas of the company used within these flowcharts are as follows:

> MARKT - Marketing
> PAMS - Purchasing and Material Supply
> MAINT - Maintenance
> PERS - Personnel
> QUAL - Quality
> SALES - Sales

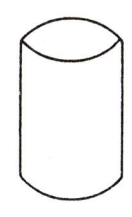

Information
This symbol is used to identify information that is used by, or created by a processing node.

Node numbering

Each processing node, decision node, multiple processing node, and multiple decision node has been numbered in the following standard way:

>   E   10
>   Node id   numeric id

Where the node id identifies the major area of CIM.

eg: Computer Aided Design is D

Computer Aided Production Engineering is E

Computer Aided Production Planning is P

Computer Aided Storage and Transportation *and* Computer Aided Manufacturing are M, Y, X and T

The numeric id is a relative node number. Initially nodes have been numbered in increments of 10, eg. 10, 20, 30, etc.

eg: D 10 is the node number 10 within the CAD area.

The connecting node symbol contains the node number where flow is going to (destination node) or the node number where flow is coming from (origin node).

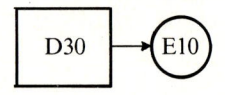

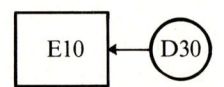

$$\boxed{D30} \rightarrow (E10)$$

$$\boxed{E10} \leftarrow (D30)$$

The processing flow is from D30 to E10

There are instances where the information identified has not been created by a node within the scope of this project, but has been considered necessary to include.

Two types of information fall within this category, that of standard information (eg. company practices), and that of temporary work information. These two types have been indicated on the flowchart as:

      Area STD - standard information (eg. D STD)

      Area TEMP - temporary work file (eg. D TEMP)

Information numbering

Each information node identified the information passed, and the node that created the information (eg. D10, DSTD, DTEMP, etc.).

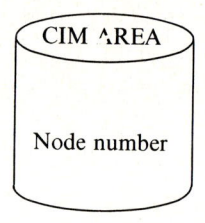

$$\begin{array}{c} \text{CIM AREA} \\ \text{Node number} \end{array}$$

Information description of items.

The node number follows the general conventions and identifies the node creating the information (eg. D10, DSTD, DTEMP, etc.).

The CIM area identifies the major CIM area (eg. CAD, CAM, etc.) that created the information.

415

Appendix 2

Selection processes

Generalised selections - narrative

1   Narrative

1.1   Summary Statement

A generalised selection process will **extract**, from a complete set of options, that sub-set of **eligible** options which possess **required properties or characteristics**, it will sort them into **order of preference**, and it will present the process-user with the **best choice or choices** in the **required format**.

1.2   Eligibility - flowchart nodes 1-3 & 9

It may be desirable to proscribe the selection of particular options, even though they might posses the properties and characteristics demanded by the selection criteria. Proscribed options are notified to the process by unique individual identifiers, any attempt to proscribe options according to common characteristics is merely a refinement to the selection rules - q.v.

Although the flowchart shows selection constrained to a set of eligible options (i.e. removal of ineligible options occurs as the first operation), it is equally valid to apply eligibility tests simultaneously with selection tests, or to remove ineligible options from the resulting sub-set after unconstrained selection has occurred.

1.3   Selection Rules - flowchart nodes 4, 5 & 7

Each time a selection process is executed the user has to state the properties and characteristics which will determine the selection or non-selection of each option. Obviously different selections would be expected to result from different statements, and different selections could result from two executions using identical statements if changes to the options set had occurred between the executions.

A property may only be included in a statement if a datum representing that property-type appears in the recorded data for some (but not necessarily all) of the options.

   eg: "Yellow" may only be included in the statement of required properties if at least some of the options posses a "colour" datum.

The required properties and characteristics are referred to as the "Selection Rules".

1.4   Extraction - flowchart nodes 7-12

The selection process is essentially a dichotomisation of the eligible options into selected options and unselected options. This may be effected in several ways such as:

- Physical Separation of unit records (such as punched cards or edge-punched cards)
- Copying selected records into a new extract file (with or without reformatting)
- Chaining together records on a directly accessed file

The process may need to scan all the eligible options, or only a portion of them, depending on the method of structuring the options file, and the manner in which the selection rules are specified.

1.5   Ranking (Preference) - flowchart nodes 4, 6, 8, 10, 11, 13 & 16

Each selected option will be assigned a rank (or preference rating). An option's rank is determined when it is selected. Rank may depend on one property of an option, or on a combination of several properties. The same property (or properties) may be used in the determination of both an option's selection and its rank, or distinct properties may be used.

- eg: Selected options could be ranked purely in ascending unit-cost order, or in ascending cost within descending horsepower. With either alternative a maximum cost could have been specified as a selection rule.
- NB: Although the flowchart shows a physical sort to put selections in preference order, it is obviously equally valid to retrieve the records directly from the options file by some method which ensures their proper sequencing such as an alternative index based on rank, or a direct-address chain specific to this selection.

1.6 Cut-Off Criterion - flowchart nodes 14, 15 & 20

The sub-set of selections may be larger than the user wishes to consider; a "cut-off criterion" enables the user to reduce output to a convenient maximum number of entries. The sorting of selections into preference order prior to curtailment ensures that the user is always presented with the "best" options. If, for any reason, the user is not satisfied with the best options, he may redefine the cut-off criterion to obtain the next best choices.

A cut-off criterion is either a simple number (the top 3 selections say), or is expressed as a value for the most significant component of rank.

- eg: If the selections are ordered as ascending cost within descending horsepower, a cut-off criterion could be stated as "200hp".
- NB: If the user then decided to examine selections down to 150hp, he need only redefine the cut-off criterion; whereas if he had included a selection rule "horsepower = 200", he would need to re-initiate the entire selection process.

1.7 Output Presentation - flowchart nodes 15-19

The output from an SP may need to include only a sub-set of each option's properties.

- eg: "Last Advice Note Date" may be a property-type which is recorded against each option but which is of no significance to an SP user.

The output from an SP may include properties other than those which determined selection or preference.

- eg: "Cost" may be of interest to the user although it was not taken into account during the extraction operation.

The output may need to include properties which are not held as such on the options file, but are derivable from options which are.

- eg: If "Unit Cost" and "Stock" are property-types recorded against each option, it is obviously possible to include a property-type "Value of Stock" in the output from the process.

The property-types required in an SP's output may vary from execution to execution.

An "Output Format Specification" enables a user to redefine, for each execution, his precise requirements for the presentation of the output.

- NB: Although the flowchart shows the formatting of the output occurring after the declaration of the cut-off criterion, it is obviously equally valid to reformat option records during an extraction or sort operation.

Selection processes   417

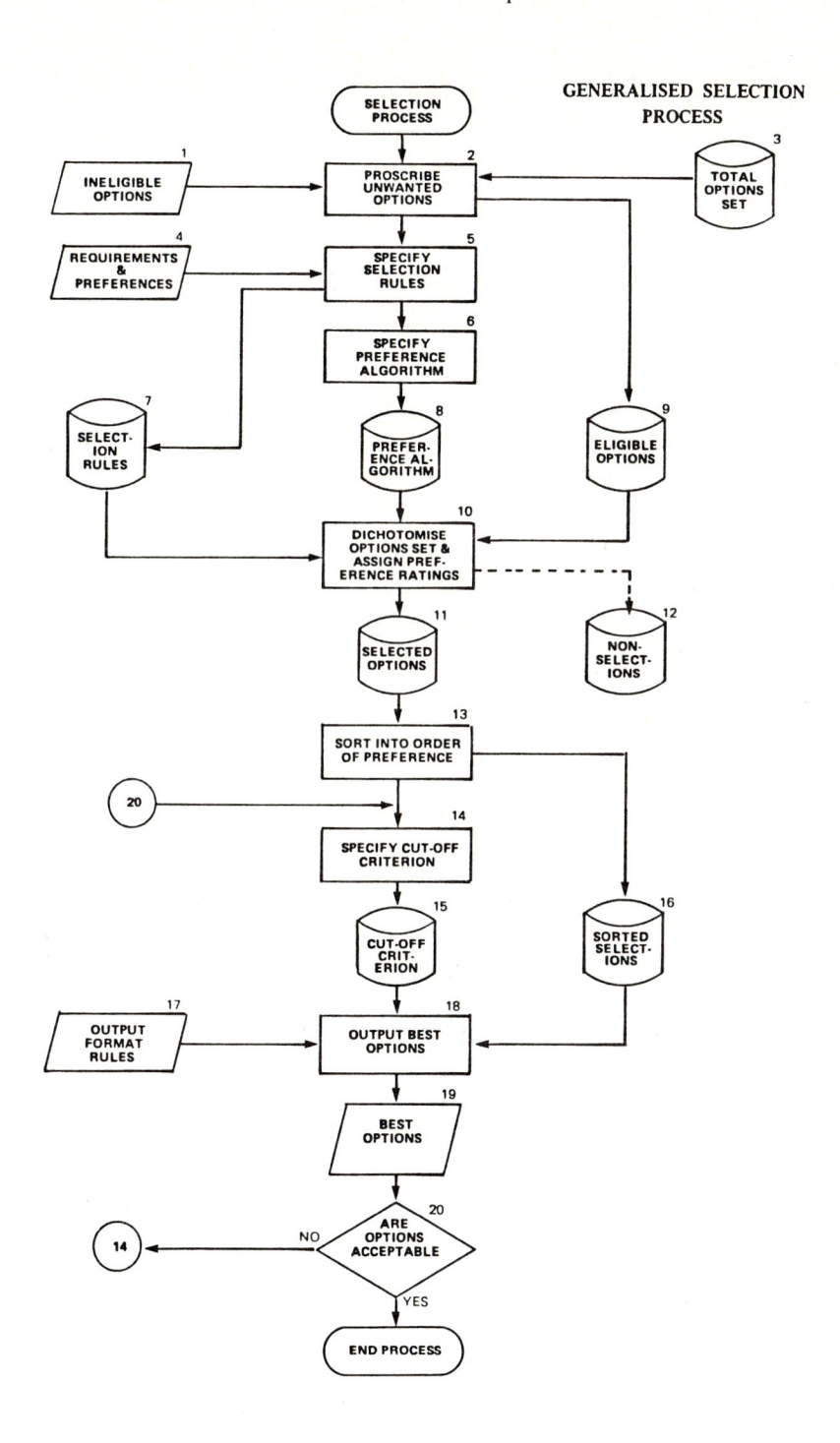

GENERALISED SELECTION PROCESS

- 1. INELIGIBLE OPTIONS
- 2. PROSCRIBE UNWANTED OPTIONS
- 3. TOTAL OPTIONS SET
- 4. REQUIREMENTS & PREFERENCES
- 5. SPECIFY SELECTION RULES
- 6. SPECIFY PREFERENCE ALGORITHM
- 7. SELECTION RULES
- 8. PREFERENCE ALGORITHM
- 9. ELIGIBLE OPTIONS
- 10. DICHOTOMISE OPTIONS SET & ASSIGN PREFERENCE RATINGS
- 11. SELECTED OPTIONS
- 12. NON-SELECTIONS
- 13. SORT INTO ORDER OF PREFERENCE
- 14. SPECIFY CUT-OFF CRITERION
- 15. CUT-OFF CRITERION
- 16. SORTED SELECTIONS
- 17. OUTPUT FORMAT RULES
- 18. OUTPUT BEST OPTIONS
- 19. BEST OPTIONS
- 20. ARE OPTIONS ACCEPTABLE — NO → 14; YES → END PROCESS

Generalised selections - rules

2   Design rules

2.1   Selection Constraints

A Selection Process (SP) must allow a process-user to proscribe the selection of individually identified options, it must not allow the user to proscribe the selection of generically defined options.

   ie: The user must only quote key values which uniquely identify the ineligible options, and not a single value for a property-type which may be common to several ineligible options.

2.2   Selection Rules (Property-Types)

An SP must allow a user to define his criteria for selection by nominating a set of property-types, and quoting a value, or range of values, for each property-type.

An SP must allow a user to specify selection rules for any property-type for which values are recorded against any of the options.

   eg: If "age" is a property-type owned by some of the options, the SP must accept a selection rule meaning "20 years or younger".

Ideally an SP will allow a user to specify selection rules for generated property-types.

   eg: If "Usage Rate" and "Unit Cost" are property-types appearing on the options file, a user could create a selection rule for (UsageRate) x (Unit Cost) and nominate this generated property-type "Usage Value" say.

2.3   Selection Rules (Occurrence)

An SP must allow a user to create as many or as few rules as he wishes whenever he initiates a selection.

A user must be able to create more than one rule for a single property-type if he wishes.

   eg: Rule 1: 10mm = Length = 50mm
         Rule 2: Length 40mm

A user must be able to create rules for as many property-types as he wishes.

2.4   Selection Rules (Value Ranges)

An SP must allow a user to define a range of values by quoting a series of discrete values.

An SP must allow a user to define a continuous range of values by quoting range limit values.

Ideally an SP will also allow a user to define ranges of values by quoting maximum values, minimum values, proscribed values and partial values.

eg:

Property-type	Value Range	Value Type
Diameter	4mm, 5mm, 6mm	discrete
Length	20-30mm or 40-45mm	continuous
Weight	10Kg or under	maximum
Power	100Hp or above	minimum
Colour	not "BLACK"	proscribed
Description	contains "DRILL"	partial

2.5 Output Sequence

An SP must allow the user to define the order in which the selections are to be presented.

A user must be able to define any concatenation of property-types owned by the selections as the sorting sequence, and each of the numeric property-types included may be requested as ascending or descending.

> eg: Possible sorting sequences for selections whose property-types include "Weight", "Power", "Cost" and "Country of Origin" are:

- ascending weight
- descending power
- ascending cost within country-of-origin
- ascending cost within ascending weight within descending power, etc.

Ideally a user may define a generated property-type as the sorting sequence, or as a component of the sorting sequence.

> eg: A user may wish to define a property-type "Value" (which is not held on an option's record) as the product of "Cost" and "Stock" (which are held), and sort his selections by ascending value.

2.6 Output Presentation

An SP must allow the nomination of property-types which are to appear in the output.

An SP must accept the nomination of any property-type which may be found on the options file, whether or not that property-type is used in the specification of selection or sequencing requirements.

Ideally an SP will allow the nomination of user-generated property-types - see 2.2 and 2.5 above.

Ideally an SP will allow run-time definitions of editing requirements, such as justification, zero-supression, rounding etc.

Ideally an SP will allow run-time definition of layout requirements, such as headings, spacings, pagination etc.

2.7 Output Volumes

An SP must allow a user to set a maximum limit on the number of options output.

Ideally a user may specify a value for the senior component of the sort sequence as a cut-off criterion.

Appendix 3

Computer Aided Design of Solid Objects

1. Solid Modelling Systems

In recent years, a number of powerful systems have been developed specifically suited for the design of three dimensional solid objects known as "Solid Modellers". [Requicha1982, Requicha1983a] These systems originate from some academic sites, for example Build-2 [Hillyard1982] and PADL-1/2 [Brown1982] and from airplane and automobile industries, for example GMSolid [Boyse1982] The latter system is a direct offspring of the PADL-system developed at the University of Rochester.

The term *Solid Modelling* denotes a wealth of theories, techniques and systems aimed at constructing *informationally complete* representations of solids. Such representations enable, at least in principle, that any well-defined geometrical property of any represented object can be calculated automatically.

This is important, since the geometrical properties of a designed part, especially its shape, play an important part in any design phase. Usually a designer starts with a shape definition. The identification of sub-components can be done easiest by referring to the geometry. Also, non-geometrical data (such as materials, strength, tolerances, etc.) need the geometrical model for the identification of the component they apply to. The production designer refers to the same geometry for specifying machining and assembly operations.

Therefore, it can be argued, that the *same* geometric model is the base for design support in almost every design. Hence, the capabilities of the geometrical modelling methods determine to a large extent the degree of integration possible for the components of the Computer Integrated Manufacturing (CIM) system.

Increasing the degree of integration may result in a far higher level of flexibility in industrial automation, so that in course of time the "A" in CAD/CAM might stand for "automated" rather than "aided".

2. Representations of solid objects

Since a geometric model of an object is the coupling between CAD (where it is created), CAE (where it is analyzed) and CAM (where it determines the final result of the manufacturing process), geometric modelling must be regarded as a vital module in terms of Information Technology with respect to Computer Integrated Manufacturing. Within geometric modelling, the representation schemes available for representing solid objects can be regarded as a key issue, since their properties determine the feasibility to automate activities needed in CAD, CAE and CAM and the associated processing characteristics.

To justify this, the common representation schemes presently used in geometric modelling will be explained. Moreover, it will be shown, that particular representation schemes are advantageous or even necessary in particular application areas. Hence, during the course of the whole CAD/CAE/CAM process, representation conversions are needed at carefully selected points in that process.

When discussing the properties of representation schemes it is advantageous to distinguish those properties, which can naturally be *formalized* from those, for which that cannot easily be done. When the former are adequately and consistently defined, it will be possible to make unambiguous and verifiable statements. To show, how this can be achieved, some basic notions and definitions will be discussed.

Since the material discussed in the following paragraphs is complicated and based on several extensive mathematical theories, the treatment as given here is necessarily rather superficial. For detailed information, the reader is referred to textbooks, for example [Encarnacao1980, Nowacki1982, Encarnacao1983] and reviews, for example [Requicha1980, Requicha1982, Requicha1983a] The discussion, presented here, closely follows the treatment by Requicha. [Requicha1980]

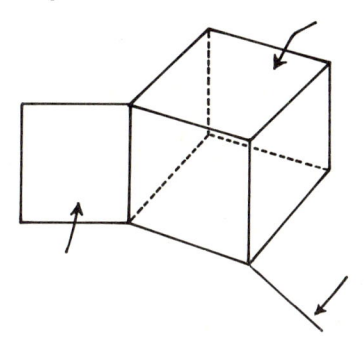

Figure A3.1 : "Dangling" face and edge.

3. Properties of abstract solids

An *abstract solid* can be described as a pure mathematical notion, which should have the following properties in order to be of practical use:

(1) *Rigidity.* An abstract solid should have an invariant shape, which is independent of its position and orientation in space.

(2) *Homogeneous three dimensionality.* An abstract solid must have a well-defined interior and a solid's boundary cannot have isolated or "dangling" portions (figure A3.1).

(3) *Finiteness.* A solid must occupy a finite portion of space.

(4) *Closure* under rigid motions and a number of Boolean operations. Translation and/or rotations or operations that add or remove material should, when applied to solids, produce other solids.

(5) *Finite describability.* Solids must have distinct finite aspects (e.g. a finite number of faces) in order to be representable in a form, suitable for computer processing.

(6) *Boundary determinism.* The boundary of a solid must determine unambiguously what is "inside", and hence comprises the solid.

4. Formal definition of representation schemes

In order to obtain mathematical models having the properties listed in the preceding section, it has been shown, [Requicha1977] that suitable models are (congruence classes of) subsets of three dimensional Euclidean space which must be bounded, closed, regular and semianalytic. Requicha calls these sets *r-sets* (regularized sets). These mathematical notions are well-defined, [Kuratowski1976] and may be explained intuitively. The set in figure A3.1 is closed because it contains its boundary, but it is not a regular set because of the existence of "dangling" portions, which do not define an interior. Such "dangling" portions could easily result from Boolean operations. Figure A3.2b contains a two dimensional example. This example also shows, that r-sets are not closed under conventional set operations, since the set depicted in figure A3.2b is not an r-set, because a line segment does not determine an area.

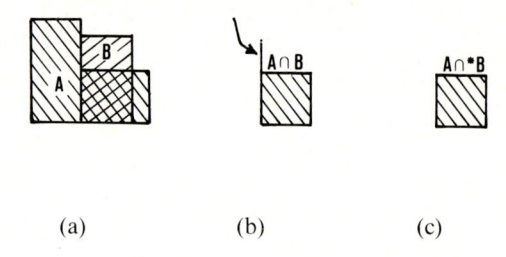

Figure A3.2 : Regularization after a set operation

However, Tilove and Requicha have shown, that, without loss of mathematical rigour, one may define a *regularization operator*, which, when applied after each of the Boolean operations taking union, intersection, difference or complement, result in a regular set. Informally, regularization can be described as: take everything inside the set, excluding its boundary, and cover this with a "skin" to form a closed set. [Tilove1980] More formally, one may define the *regularization* of a set X as $r(X) = ki(X)$ where i and k denote the interior and the closure of a point set according to the conventional definitions. [Kuratowski1976] In this way one may remove effectively point sets such as the "dangling" line segment in figure A3.2b to obtain the regular set in figure A3.2c.

One may now postulate a *mathematical modelling space M* whose elements are the r-sets, which can be seen as mathematical models of abstract solids.

Further, one may define syntactically correct representations as finite symbol structures constructed with symbols from an alphabet according to syntactical rules. The collection of all syntactically correct representations is called a *representation space R*.

Now, a representation scheme is defined formally as a relation $s: M \to R$ with domain $D \subset M$ and value range $V \subset R$ (figure A3.3).

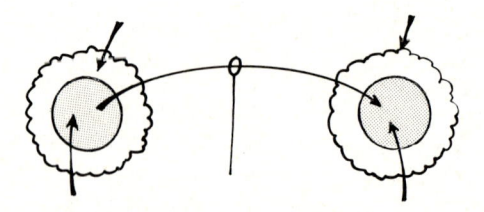

Figure A3.3 : Domain and range of a representation scheme

Any representation r in V is said to be *valid* since it is both syntactically (it belongs to R) and semantically correct (it has one or more elements in domain D). A representation r in V is said to be *unambiguous* (or *complete*) if it corresponds to a single-element subset of D. It is called *unique* if its corresponding objects in D do not admit representations other than r in V. A *representation scheme s* is unambiguous if all of its valid representations are unambiguous and, similarly, it is unique if all of its valid representations are unique. Note, that unambiguity of a representation neither implies nor is implied by its uniqueness. Informally, this means that a valid representation is ambiguous if it corresponds to several solids and a solid has non-unique representations if it can be represented in several ways by means of a particular representation scheme.

5. Formal properties of representation schemes

The definitions of the preceding section have some important practical implications for the representation schemes, which are being used in actual Geometric Modelling Systems (GMS).

5.1. Domain

The domain of a representation scheme is the set of entities, that can be represented by the scheme and therefore characterizes its *descriptive power*.

5.2. Validity

The range of a representation scheme is the set of representations which are valid. It is obvious, that *representational validity* is very important to ensure the integrity of the databases on which an actual GMS is operating. Those databases should not contain nonsense objects, otherwise such invalid information could result in a system crash, obvious suspect results, or, in the worst case, apparently credible results in fact being meaningless and worthless.

In the latter case, it may take considerable time before the invalidity of the result is determined so that the fault might be propagated in an unpredictable way through the whole CAD/CAM process. This must be considered very dangerous and should be precluded at all cost.

Since ensuring validity of representations by humans, for example by inspecting a drawing on a graphics display, is error-prone or hardly possible , it is essential to check validity by automatic means. The most straightforward way to achieve this is using representation schemes in which all syntactically correct representations are valid (e.g. Constructive Solid Geometry). This reduces validity checking to a parsing problem.

5.3. Unambiguity

Engineering drawings, the traditional means for specifying solids, can be viewed as informal means of communication among humans. Because engineers and technicians possess a vast amount of "world" knowledge, they usually interpret these drawings correctly but sometimes they make errors or, inversely, they interpret incorrect drawings as if they were correctly, perhaps without even noticing the error. Automated machines, for example programs that interpret drawings, unfortunately do not have these capabilities in general and therefore, unambiguous representation schemes are mandatory.

Until recently, commercially available geometric modellers were mostly based on ambiguous so-called "wire-frame" representations. [Voelcker1977] For example, a polyhedron, when represented by a list of vertices, is not uniquely defined in such a scheme, as can be seen from figure A3.4. Even a list of edges does not represent an unique polyhedron (figure A3.5).

Therefore, a necessary condition for the automatic integration of modellers and other automated systems, is, that object information should be represented according to an unambiguous representation scheme. As shall be shown later, such representation schemes exist and are successfully implemented in recently developed practical modelling systems known as *"solid modellers"*. [Requicha1982] Some authors, for example on page 240 of [Encarnacao1983] point out, that simple but ambiguous representation schemes like wire-frame models have useful applications. For example, the relative simplicity of the algorithms make such schemes suitable for hardware implementation, so that the model or the viewing point can be changed in

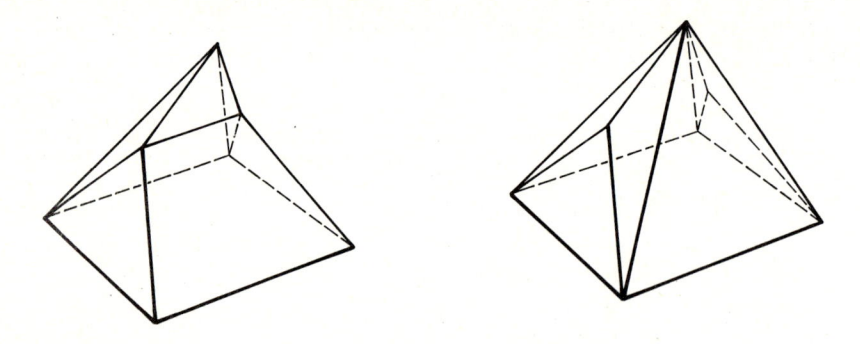

Figure A3.4 : A polyhedron is not uniquely defined by a list of vertices.

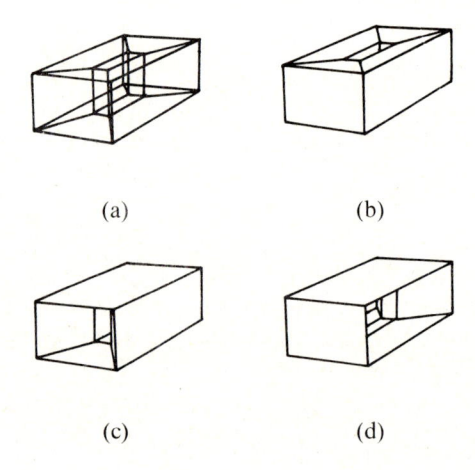

(a)

(b)

(c)

(d)

Figure A3.5 : Ambiguity of a wireframe representation.

real-time, as perceived by the designer, enabling him to investigate many design variants or aspects of the design in a short time.

However, since it is quite possible to construct a wire-frame representation from an exact representation automatically, [Requicha1983a] while the reverse is not true, it seems more appropriate to store the complete model using an exact representation. Such a general, as complete as possible representation should be the starting point for extracting additional representations which, although no longer unambiguous, are better suited for particular kinds of processing (e.g. real-time display).

5.4. Uniqueness

The uniqueness of a representation scheme is an important property for establishing the equality of objects. For example, programs for Numerically Controlled machining must eventually produce objects equal to those specified by product designers.

As another example, planning programs, that search through a space of alternatives, may loop indefinitely if previous situations cannot be recognized. In practice, most representation schemes are not unique for at least the following reasons:

(1)   substructures in a representation may be permuted
(2)   different representations may correspond to differently positioned but congruent copies of a single geometric entity

While both cases are conceptually easy to distinguish, determining whether two structures contain the same elements may be computationally expensive (especially when these structures are large) and the design of algorithms, that decide whether two geometric entities are congruent is a complicated task.

From the above discussion, it is apparent that the primary representation of a model should be both valid, unambiguous and unique.
When a justifiable need exists in a particular application, representation conversions can be used to obtain a representation, that is suitable to perform a particular function on that model.

6. Informal properties of representation schemes

Besides the above properties, which can be formalized, as shown, there are also a number of important properties, which cannot be easily formalized in an adequate way.

6.1. Compactness

Representation schemes, which naturally result in compact data structures, have obvious advantages, since they can easily be stored or transmitted over data links. However, compactness should not be aimed at to the extreme. Carefully chosen redundancy may have important advantages such as:

- wider applicability
- informationally completeness
- possibility to easily detect and correct syntactic errors in representations
- often dramatical improvement of computational speed by storing rather than computing repeatedly needed data.

It is often difficult to predict, whether or not data storing is to be preferred over repeated computing. This may depend heavily on a particular application or on a particular situation created by the user. A better strategy would be to determine dynamically (i.e. during run-time) which approach is most desirable, for example by measuring the real-time spent in a particular computation over the same dataset.

More generally, a good general purpose CAD system has to decide dynamically about a great number of optimum configurational parameters and the data management must, more or less independent of the representation scheme used, allow for dynamic extensibility whenever it is evidently advantageous to use less concise information schemes (dynamic memory allocation).

6.2. Ease of use

Since CAD systems should be characterized as *interactive systems,* where a software system interacts heavily with a human being to accomplish a complex task, the interface with the user is of crucial importance. Simple and concise representations are generally easier to create and to manipulate than extended ones, but the former may not convey enough adequate information. Using an extensive representation scheme will be a tedious and error-prone job unless the designer is assisted by elaborate input mechanisms.

Moreover, such modelling systems should also contain mechanisms to ensure the validity and consistency at any significant stage in the design process. It is quite feasible, that knowledge-based analyzing systems may become of great help for the designer in the near future.

7. Important representation schemes used in geometric modelling

7.1. Wireframes

The first useful representation schemes are the wireframe representations. They first appeared as simple, interactive 2-D systems for the design of printed circuit boards and for 2-D mechanical drawings. The internal representations used were mostly simple lists of lines and arcs. Later, in the 1970's, they were extended to lists of segments of 3-D curves. These can be computed so that it is possible to obtain orthographic, isometric and perspective views. Although 3-D wireframe systems are clearly useful, they have some serious deficits. In figure A3.5 is illustrated, that the wireframe (a) is ambiguous, since it may represent any of the three objects (b), (c) or (d). Other examples are given by Markovsky and Wesley. [Wesley1980] Furthermore, a wireframe systems may tolerate nonsense objects and there is no possibility to check the validity of the generated representations. Finally, they are not concise and users have to supply a lot of low-level data to describe something as simple as a cube.

7.2. Boundary representations

Boundary representations are those schemes, that represent a solid in terms of its boundary or enclosing surface. These boundaries are usually represented as the union of *faces*, while each face is defined in terms of its boundary (normally a union of edges) and additional information defining the surface in which the face lies (figure A3.6).

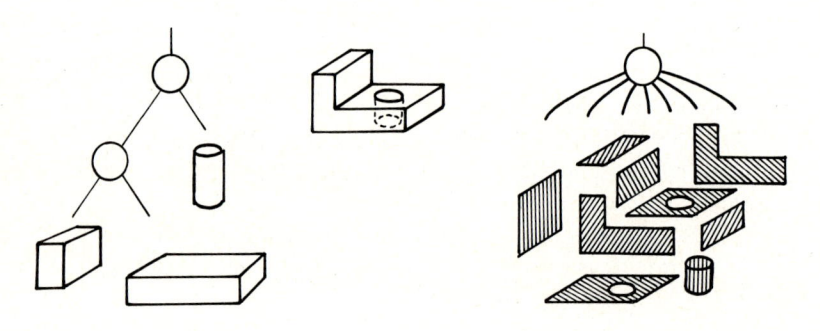

Figure A3.6 : A CSG and a boundary representation.

If these faces are represented unambiguously, boundary representations are, in contrast to wireframes, also unambiguous. This follows from mathematical theorems, proving that regularized sets are unambiguously defined by their boundaries. [Requicha1977] Furthermore, under an extensive number of conditions, which are not easy to verify in practice, they can be guaranteed to be valid. However, in general, they are not unique.

7.3. Constructive solid geometry (CSG)

Constructive Solid Geometry (CSG) denotes a family of representation schemes, where complicated solid objects may be represented by ordered additions, subtractions or regularized Boolean operations of simpler solids. The simplest objects of such a scheme (e.g. blocks and cylinders) are usually bounded primitive solids.

These schemes are unambiguous and, moreover, the algebric properties of regularized sets guarantee that any CSG-tree is valid if the primitive leaves (i.e. the primitive solids) are valid.

However, in general, CSG representations are not unique. The domain of a CSG scheme is largely determined by the primitive objects and the operations supported. Generally, they are an order of magnitude more concise than boundary representations and validity checking is almost trivial compared with the latter.

Boundary representations, on the other hand, have advantages for generating line drawings and graphic displays, since the important data for these purposes are face, edges and the relations between them.

It may be quite laborious to construct a boundary representation because of the vast amount of data to be specified by the user. In a number of actual geometric modellers, CSG representations are used as an input technique and the data thus generated are subsequently converted to a boundary representation because of its advantages with respect to the processing necessary for interactive display.

8. Research directions and future trends.

Although the solid modelling approach is to be regarded as a significant step forwards when compared to the classic wireframe modellers, neither of the representation schemes discussed are completely satisfactory. Moreover, there are a large number of open issues, which are studied intensively at a number of research institutions. Some of them are discussed briefly in the following sections. This material is based on recent papers by Requicha and Voelcker. [Requicha1982, Requicha1983a]

8.1. Sculptured surfaces

Since the solid modelling representation schemes discussed only support (possibly complex) combinations of primitive rigid objects, which have a simple and regular structure, they are not well suited for the representation of objects with sculptured surfaces. Such objects are often adequately described using Bezier curves or B-spline techniques. It is desirable to incorporate these in solid modellers and to obtain good algorithms to support Boolean operations, mass property calculation, interference analysis and other capabilities, normally found in solid modellers, also for sculptured solid objects.

8.2. Representation of tolerancing information.

For many design and production activities it is essential to have facilities for the representation of tolerancing information. Contemporary solid modellers do not possess such facilities, unfortunately, but studies on this subject are being undertaken at various places. [Requicha1983b, Requicha1983c] Also, representation of stressed objects is beyond the scope of solid modelling, but highly desirable from an engineering point of view. [Myers1982]

8.3. Improving computational speed.

Although presently available solid modellers support high-level and very powerful operations, they are often perceived by users as being relatively slow. Both improving algorithms (e.g. conversion algorithms between various representation schemes) and the development of special purpose hardware are obvious methods to attack this problem. An example of the latter are octree representations , which can be used to approximate solids for purposes of display and the generation of meshes for analysis by Finite Element Methods. Because of the regular structure of such representations, they are suitable candidates for hardware implementations. [Meagher1982]

8.4. Applications

Since solid models are unambiguous it is possible, in principle, to support fully automatic algorithms for any geometric application. However, for most application areas, these algorithms are still under development. A few examples will be given.

Interference analysis.

Static interference checking can be defined as follows: if A is a collection (e.g. an assembly) of solids $S1, S2, \ldots, Sn$ then take all pairwise regularized intersections of the solids in A. If all these intersections are empty (i.e. they all generate the null-object), they do not interfere. Thus, a system is able to perform static interference checking if

(a) it is able to represent objects as regularized Boolean compositions

(b) it is able to compare represented objects with the null-object. [Boyse1979]

Dynamic interference checking, where components are swept through spatial trajectories, is a much harder computational problem and requires the ability to detect a null-object in four dimensional space-time. [Esterling1983]

Graphics display.

While all modelling systems have features for displaying objects being designed, techniques are improving, for example to generate realistic illuminated and/or shaded displays using ray-casting techniques, for example, [Cook1982] or to generate pictures with real-time motions.

Finite Element Mesh generation.

The Finite Element Method (FEM) is one of the most important engineering tools for the analysis of mechanical structures, that exist today. Normally, FEM-programs operate on collections of geometrically simple elements, like cubes or tetrahedra. These collections are commonly called "meshes", and are different from the representations used by Geometric Modelling Systems. Since it is a tedious and error-prone job to generate these meshes manually, it is very desirable automate this task and various algorithms to achieve this have been reported, for example [Wordenweber1981]

Numerical Control.

The automatic generation of Numerical Control programs from geometric models is a complicated problem, especially when multiprocessing programs for machining centers have to be generated. Several investigations in this field have been undertaken, for example [Armstrong1983]

Simulation.

Usually, solid modellers are restricted to representations of rigid objects and assemblies, but it appears, that they can be extended quite easily to represent moving mechanical objects such as industrial robots, as reported by. [Tilove1983]

Robotics.

It is evident, that the application of robots in modern industries is vitally important. Today, usually they are being programmed off-line, by teach-in methods. In the near future, however, on-line programming will become essential, and it is being studied in the context of solid modelling. [Pickett1983] A difficult problem is the automatic generation of robot action plans from high-level task specifications, which involves issues such as path planning for collision avoidance. [Lozano-Perez1981]

References

Armstrong1983. Armstrong, G., "A Study of Automatic Generation of Non-Invasive NC Machine Paths from Geometric Models," PhD dissertation, Dept. of Mechanical Engineering, Univ. of Leeds (Apr. 1983).

Boyse1979. Boyse, J.W., "Interference Detection Among Solids and Surfaces," *Comm. ACM* **22**(1) pp. 3-9 (Jan. 1979).

Boyse1982. Boyse, J.W. and Gilchrist, J.E., "GMSolid: Interactive Modeling for Design and Analysis of Solids," *IEEE Computer Graphics and Applications* **2**(3) pp. 27-40 (March 1982).

Brown1982. Brown, C.M., "PADL-2: A Technical Summary," *IEEE Computer Graphics and Applications* **2**(2) pp. 69-84 (Mar. 1982).

Cook1982. Cook, R.L. and Torrance, K.E., "A Reflectance Model for Computer Graphics," *ACM Trans. Graphics* **1**(1) pp. 7-24 (Jan. 1982).

Encarnacao1980. (ed.), J. Encarnacao, *Computer Aided Design: Modelling, Systems Engineering, CAD-Systems,* Springer-Verlag, Berlin (1980).

Encarnacao1983. Encarnacao, J. and Schlechtendahl, E.G., *Computer Aided Design (Fundamentals and System Architectures),* Springer-Verlag, Berlin (1983).

Esterling1983. Esterling, D.M. and Rosendale, J. Van, "An Intersection Algorithm for Moving Parts," *Proc. NASA Symp. Computer-Aided Geometry Modeling,* pp. 129-133 (Apr. 20-22, 1983).

Hillyard1982. Hillyard, R.H., "The Build Group of Solid Modelers," *IEEE Computer Graphics and Applications* **2**(3) pp. 43-52 (March 1982).

Kuratowski1976. Kuratowski, K. and Mostowski, A., *Set Theory,* North-Holland Pub. Cy., Amsterdam (1976).

Lozano-Perez1981. Lozano-Perez, T., "Automatic Planning of Manipulator Transfer Movements," *IEEE Transactions on Systems, Man & Cybernetics* **SMC-11** pp. 681-698 (Oct. 1981).

Meagher1982. Meagher, D., "Geometric Modeling Using Octree Encoding," *Computer Graphics & Image Processing* **14** pp. 249-270 (June 1982).

Myers1982. Myers, W., "An Industrial Perspective on Solid Modeling," *IEEE Computer Graphics & Applications* **2**(2) pp. 86-97 (March 1982).

Nowacki1982. Nowacki, H. and Gnatz, R., *Geometrisches Modellieren,* Springer-Verlag, Berlin (1982).

Pickett1983. Pickett, M. and Tilove, R.B., "Robo Teach: An Off-Line Robot Programming System Based in GMSolid," *Proc. General Motors Symp. Solid Modeling,* (Sept. 26-27, 1983).

Requicha1977. Requicha, A.A.G., "Mathematical models of rigid solid objects," Techn. Memo. 28, Production Automation Project, Univ. Rochester, Rochester, N.Y. (Nov. 1977).

Requicha1980. Requicha, A.A.G., "Representations for Rigid Solids: Theory, Methods and systems," *Computing Surveys* **12** pp. 437-464 (Dec. 1980).

Requicha1982. Requicha, A.A.G. and Voelcker, H.B., "Solid Modeling: A Historical Summary and Contemporary Assessment," *IEEE Computer Graphics and Applications* **2**(3) pp. 9-24 (March 1982).

Requicha1983a. Requicha, A.A.G. and Voelcker, H.B., "Solid Modeling: Current Status and Research Directions," *IEEE Computer Graphics and Applications* **3**(7) pp. 25-37 (Oct. 1983).

Requicha1983b. Requicha, A.A.G., "Toward a Theory of Geometric Tolerancing," Tech. Memo. 40, Production Automation Project, Univ. of Rochester (Mar. 1983).

Requicha1983c. Requicha, A.A.G., "Representations of Tolerances in Solid Modelling: Issues and Alternative Approaches," in *Proc. General Motors Symp. Solid Modeling*, , Detroit, Mich. (Sept. 26-27, 1983).

Tilove1980. Tilove, R.B. and Requicha, A.A.G., "Closure of Boolean operations on geometric entities," *Computer Aided Design* **12**(5) pp. 219-220 (Sept. 1980).

Tilove1983. Tilove, R.B., "Extending Solis Modeling Systems for Mechanism Design and Kinematic Simulation," *IEEE Computer Graphics and Applications* **3**(7) pp. 9-19 (May/June 1983).

Voelcker1977. Voelcker, H.B. and Requicha, A.A.G., "Geometric Modeling of Mechanical Parts and processes," *Computer*, pp. 48-57 (Dec. 1977).

Wesley1980. Wesley, M.A. and Markowski, G., "Fleshing out Wire-frames," *IBM J. Research and Development* **24**(5) pp. 582-597 (Sept. 1980).

Wordenweber1981. Wordenweber, B., "Automatic Mesh Generation of 2 and 3 Dimension Curvilinear Manifolds," PhD dissertation, Computer Laboratory, Univ. of Cambridge (Nov. 1981).

Appendix 4

Sensor Applications

Introduction

The intention of this appendix is to supply some background information about sensor applications.

The first section of this appendix handles about classification of sensors. The classification of sensors already has been presented in chapter 14 of this report. In this appendix a short description of the characteristics of each class is given.

In the next two sections, the state of the art of sensing in chemical and in mechanical industry is discussed. It is considered useful, not only to discuss the state of the art of sensing restricted to CIM applications, but also to have a look at sensing in the chemical industry. The reason to do this is that the chemical industry has a much longer experience with sensors than the mechanical industry. As a result of this, more advanced sensors are used in chemical industry, for more complex tasks, such as process control. The mechanical industry only recently makes use of sensors for higher level tasks.

The contrary is true when it comes to computer-vision. No doubt mechanical industry triggered the development of this most advanced form of sensing. The chemical industry hardly has applications for and experience with computer-vision.

The difference between the two industries is that chemical processes usually are continuous, whereas mechanical processes usually are discrete. This is a result of the fact that chemical industry primarily works with fluids, whereas mechanical industry primarily works with solids. This of course has influence on the types and morphological aspects of sensors used in both industries. Nevertheless, mechanical industry can take advantage of the experience obtained by the chemical industry.

A part of sensing in the mechanical industry is the special field of robot sensing. This field of sensing is discussed in a separate section due to the important role robots may have in Computer Integrated Manufacturing.

Section 5 handles about computer-vision, a subject that has an outstanding importance in automation and an outstanding measure of complexity.

Finally section 6 handles about future trends in sensing.

1. Sensor Classification

The following eight aspects of sensors are frequently chosen as the basis of a classification (see fig.A4.1): converted energy; transduction principle; influence on the environment; adaptability; connection with object; relation of signal and parameter; relation to manipulator; property of output signal.

Aspect	Classification
Converted Energy	Radiant Mechanical Thermal Electrical Magnetical Chemical
Transduction Principle	Self-generating Modulating
Influence on Environment	Passive Active
Adaptability	Non-adaptive Adaptive
Connection with Object	Contact Non-contact
Relation of Signal/Parameter	Absolute Relative Linear Logarithmic Inverse .....
Relation to Manipulator	Internal External
Output Signal	Analogue Digital

Figure A4.1 : The different classifications of sensors based on eight different aspects.

Converted energy
The energy forms that can be converted, can be grouped in six domains, leading to the six signal domains [Lion1969] : *radiant* signals (light intensity or frequency), *mechanical* signals (pressure, level), *thermal* signals (temperature, heat flow), *electrical* signals (voltage), *magnetical* signals (magnetic field) and *chemical* signals (PH, composition).

Transduction principle
It appears that based on the transduction principle there are two different kinds of sensors. Some of them, called *self-generating sensors*, are sensors which generate an electrical output without an auxiliary source of energy. They directly convert one of the above-mentioned forms of energy into electrical energy. Examples of these are conventional tachometers, thermocouples and solar cells.
The other sort, called *modulating sensors*, are sensors which need an auxiliary (electrical)

energy to be able to transform the physical quantity of interest. This transformation is based on the fact that these sensors cause a modulation of the auxiliary electrical energy related to the modulation of the form of energy that is converted. Examples of these are microphones, temperature dependent resistors etc.

It can be noticed that the electrical energy produced by the self-generating sensors is always smaller than the amount of energy put into it. This is different for modulating sensors, where the energy of the output signal can be higher than the energy of the input signal.

Influence on the environment
Based on their influence on the environment a distinction can be made between *passive* and *active* sensors. The first, as the name implies, are sensors which work with the given environmental condition (e.g. illumination), whereas the latter are sensors consisting of a source and a detector to provide their own source of radiation (for instance sonar systems or structured lighting vision systems).

This feature makes active sensors more versatile than passive sensors, since they are less dependent on environmental conditions. However, one has to take notice of the fact, that the radiating source of the active sensor can unintendedly change the physical condition of the environment.

Adaptability
Another classification is based on the interpretation of the clean sensor signal. This can appear as a *non-adaptive* or as an *adaptive* mechanism. Non-adaptive mechanisms of interpretation use fixed sensors and the interpretation is based on information obtained from a relatively small number of observations.

Adaptive sensing mechanisms utilize the ability to alter either the object, or the sensor in order to enhance the acquired representation and supplement the information provided. Obviously, to obtain this flexibility, real-time analysis of the sensor signals is necessary.

For sensor control, the distinction between adaptive and non-adaptive sensors is very important.

Connection with object
Especially in the wide assortment of devices and systems that exist for robot sensing, it is useful to make a distinction between *contact* and *non-contact* sensing.

Contact sensing is the domain of tactile sensing, defined [Harmon1980] as a continuously variable touch sensing over an area within which there is a spatial resolution, and of force or torque sensing, which is usually a simple vector resultant measurement at a single point.

Non-contact sensing is the domain of optical, sonic and magnetic based sensors. The most complex and no doubt very powerful optical sensor is pure vision. Other examples of non-contact sensors are beam interrupters, range finders etc.

Relation of signal and parameter
The previous classifications are based on the method of sensing. In general the signal does not indicate by what method it was obtained. On the level of signal interpretation it is even desirable to be shielded off from the details of the sensor hardware.

It is very important to decide which aspects of the sensing signal must be preserved and which aspects should disappear above a certain level of interpretation.

For instance, at the lowest level of the interpretation it will be important to know the relation between signal and measured parameter. In other words, whether sensors provide *absolute* or *relative* values, and whether there is a *linear, logarithmic, inverse* or other relation between the signal and parameter.

At higher levels, these differences in relations will, as a result of preprocessing in the lower levels, usually have disappeared.

Relation to manipulator
Sensors serving a manipulator control system can be classified according to their relation to that manipulator. This leads to the distinction between *internal* and *external* sensors.
Internal sensors are used to monitor the state of the manipulator. Examples are: joint-position or joint-velocity sensors and gripper-force sensors. Internal sensors usually are part of a closed control loop, which implies a high sampling and processing rate.
On the other hand, external sensors provide information about the relevant environment of the manipulator. The information is fed to a higher control level, which usually can do with a lower sampling speed. This is in agreement with the fact that these external sensors, such as vision, usually have a high processing requirement. Other examples of external sensors are distance and special purpose sensors (such as temperature sensors).

Property of output signal
Since sensor signals in CIM eventually will have to be coupled to a digital read-out system, a useful distinction that can be made is the distinction between *analogue* or *digital* sensors.
Digital sensors (for example a digital angular encoder) do not require a sometimes expensive analogue/digital conversion circuit. Reasons to use analogue sensors can be: available technological experience, the non-existence of digital equivalents and specific demands on their properties such as sensibility, resolution and dynamic range.

2. Sensing in Chemical Industry

In the chemical industry, sensors nowadays play a vital role. Their main task is to measure and control the chemical processes. The interesting parameters for this industry are pressure, temperature, flow, level and composition.
The following is a discussion of some methods used to measure these parameters.

2.1. Pressure

There are several sensor devices to measure pressure. These are in order of their frequency of occurrence: Bourdon, diaphragm, bellow and silicon pressure transducers. All these types of sensors are based on the principle that pressure is translated to displacement of a flexible element. This displacement, which is related with the pressure applied, is in all four types sensed with some sort of displacement transducer. The distinction between the four types is based on their difference in translating the pressure into displacement.

Bourdon. The bourdon pressure transducer is based on the principle that a bended tube will try to straighten itself when an internal pressure is applied. One end of the tube is fixed and the other end is connected to a displacement transducer. The outside of the tube is exposed to atmospheric pressure, which makes this transducer a relative pressure sensor.

Diaphragm. A diaphragm pressure transducer consists of two chambers, separated by a flexible partition, which is connected to a displacement transducer. The first chamber is exposed to the unknown pressure, the second chamber usually is evacuated. In that case the transducer measures the absolute pressure.
The second chamber also can be exposed to a second pressure, in which case this type of transducer can be used for differential pressure measurements.
If the first chamber is omitted, the diaphragm pressure transducer is a dedicated device for measurement of the atmospheric pressure.

Bellow. A bellow pressure transducer consists of one chamber, the inside of which is exposed to the pressure to be measured. The outside of the chamber is exposed to the atmospheric pressure. The chamber can expand itself in the same sort of way a bellow can. This expansion, measured by a displacement transducer, is made proportional to the applied pressure due to added elastical forces.

Silicon. Silicon pressure transducers usually are of the diaphragm type. The reason to mention them separately is that the above-mentioned transducers are mechanical devices, constructed of several materials. In silicon pressure transducers, the flexible element is integrated with the displacement transducer on one chip, usually combined with a signal processing and temperature compensating circuit.

It is interesting to notice that the frequency of occurrence shows a remarkable relation with the "age" of the transducer type. The older transducer types, are mechanical devices and thus complex, sensitive to wear etc. The remarkable fact is that one would expect that modern silicon transducers by now would have taken their place in a lot of cases, since they are available at low cost, for different applications and able to measure pressures of up to $\sim$35 kPa.
It seems on the contrary, that they still are in the minority, probably due to conservatism and the fact that there is a lot more experience with the older types of transducers.

2.2. Temperature

Nearly every electrical property of a material or device varies as a function of temperature and could in principle be employed as a temperature sensor. Unfortunately, no material or device has a temperature dependence that over a wide temperature range is a combination of maximal linearity, maximal sensibility and maximal reproducibility. Furthermore properties as size, cost, speed and unassailability can be of importance. For this reason a sensor with the optimal combination of properties has to be selected for each particular application.
Temperature can be measured with temperature dependent resistors, temperature dependent capacitors, piezo-electric crystals, thermocouples, silicon sensors and optical techniques.

Temperature dependent resistors. Thermistors are semi-conductor resistors with a high negative temperature coefficient. Obviously they belong to the so-called modulating sensors. To measure the temperature, these thermistors can be connected to a constant voltage or a constant current supply. Measuring the current flow through the thermistor in the former case or measuring the voltage across the thermistor in the latter case, will give an indication of the resistance of the thermistor and thus of the temperature. In order to measure the correct temperature, care will have to be taken to limit the power dissipation in the thermistor. A high power dissipation could be the result of choosing a too high constant voltage or current, to what one is tempted in order to increase the sensitivity of the sensor.
Thermistors are cheap and available in a wide variety of resistances, shapes and sizes (down to microscopic), which explains for their popularity. A disadvantage is that their temperature dependence is not linear but exponential. This non-linear response can be compensated by several methods, such as employing a logarithmic amplifier, which is a costly method, or the cheap way by simply adding a fixed resistor in parallel, a linearization method with the disadvantage that the in this way created linear temperature range is limited.
Other popular resistance thermometers are platinum, nickel and tungsten. They have a wide temperature range (platinum from -250 to 1100°C, nickel from -100 to 300°C), have an almost constant positive temperature coefficient and are chemical resistant.
Germanium and carbon resistors are useful at temperatures below 50°K where they exhibit an exponential temperature dependence like thermistors.

Temperature dependent capacitors. The temperature dependence of capacitors can be used to measure the temperature. A common read out method is to let the capacitor be the frequency determining part of an oscillator circuit. In this way the frequency shift of this oscillator becomes a function of the temperature.
Advantages of this method are the simplicity and the fact that frequencies can be easily converted to digital numbers by means of a frequency counter. Disadvantages are that capacitors as thermosensors are not very sensitive, their reproducibility is low (excluding them from exact applications) and further their response-time is long.

Piezo-electric crystals. Another method based on a temperature dependent frequency is making use of specially for this purpose fabricated piezo-electric crystals. These crystals as a part of an oscillator circuit, resonate at a frequency which shifts proportional with temperature.

Advantages of this method are linearity, reproducibility and simple converting to digital numbers. Disadvantages are the large size and cost of the crystal and the cost of the frequency counter, which must, due to the small frequency shift per degree ($\sim 0.01\%/°C$), be very stable.

Thermocouples. A junction between two dissimilar conductors produces a small voltage difference, due to the difference in the effective concentration of electrons in the two conductors, which is almost linear with the junction temperature. Thermocouples are based on this property. They usually consist of two junctions, one is kept at a reference temperature (usually 0°C), the other is the temperature measuring sensor.
Thermocouples have a small size, a rapid response, a wide temperature range and are very simple. On the other hand they are not linear (although better than thermistors), the reference junction has to be kept at a constant temperature and they are self generating sensors, which means that their output energy is very low.

Silicon sensors. A forward biased diode also has a temperature dependent junction voltage. If the current through the diode is kept constant, this temperature dependence is exponential. Because the temperature characteristics for normal diodes are not highly reproducible, these diode sensors are not suitable to be used as high-accuracy thermometers. In applications with less demands, such as overheating detectors, they can serve very well.
A more advanced silicon sensor makes use of this temperature dependence of the junction voltage. The temperature response of the silicon diode is linearized by a signal processing circuit added on the same chip. The in this way accomplished linearization can be in the order of 0.1°C in the temperature range -55 to +125°C.

Optical techniques. The thermal radiation of objects can be used to measure the temperature of these objects with infra-red sensors. This optical way of temperature measuring can be used in a very wide temperature range with a very high accuracy (e.g. one sensor can cover the range -25 to +1400°C with a accuracy of 0.2°C).
This non-contact method clearly has the advantage that it is non-invasive and even can be used at large distances. It led to the development of infra-red vision. With infra-red vision an image of an object can be produced, revealing the temperature distribution of that entire object.
Obviously, infra-red camera's give powerful abilities to industry, medical science etc.
The disadvantage of the infra-red method is the complexity of the signal processing circuit, especially for infra-red camera's. This makes the method less useful for control applications.

2.3. Flow

Existing flow measurement methods are based on measurement of: differential pressure, heat exchange, mechanical effects, Doppler effect or cross correlation.

Differential pressure. If the flow is measured with the differential pressure method, the pressure difference between two points along the stream, if necessary separated by a constriction or other aerodynamic mechanism, combined with the flow-through area and some physical properties of the fluid or gas (such as viscosity) is used to calculate the flow. Difficulties that can occur with this method are: calibrating problems and variations in viscosity due to variations in material density, composition and temperature.

Heat exchange. A thermistor with a deliberately high applied voltage or current and thus high internal power dissipation, is cooled by a flowing fluid. The cooling rate is proportional to the square root of the velocity of the fluid. By keeping the resistance (and thus the temperature) of the thermistor constant by means of a feedback circuit, the applied voltage or current can be used for determining the flow.
Usually, a second thermistor is needed for compensation of temperature changes of the fluid.
Several silicon sensors, with advanced compensating and signal processing circuits, based on this principle are developed. If a heating device is centralized in a circle of thermal sensors, not only the velocity but also the direction of the flow can be determined.

Disadvantages of this method are the need for a compensating thermistor, the method is sensitive for variations in material density, it is invasive and it is not possible to combine a fast response time, which is satisfied by a small fragile thermistor, with rigidity demands, which in turn will lead to solid thermistors which have a long response time.

Mechanical. The mechanical method is based on measuring the revolving speed of a turbine that is localized in the flow. This method is very invasive and problems such as sticking, choking-up, corrosion wear etc. can occur.
The advantage of this method is its simplicity.

Doppler effect. The Doppler flow-measuring method is based on the well known Doppler effect. This method can utilize for ultra-sonic and electro-magnetic waves.
This modern advanced method has the advantage that it is non-invasive and has an order of magnitude less dependence of viscosity, density, temperature and composition.
The disadvantage of this method is the higher complexity compared to the above-mentioned methods.

Cross-correlation. The cross-correlation flow-meters, [Beck1983] measure the transit time of a tagging signal in the flow between two separated sensors along the stream. Dependent of the sort of tagging signal, these can be optical, sonical, thermal or electrical sensors. A tagging signal can be already available in the flow due to inhomogeneities or it has to be inserted (such as a thermal fluctuation).
The advantage of this method is that the flow of very inhomogeneous liquids or gasses can be measured. There is no need for calibrating the sensors since only the transit time is the relevant parameter. The optically detectable tagging signal is also non-invasive.
The disadvantage of this method is the higher complexity, it is even more complex than the Doppler method. Due to the reduction of the cost of VLSI-circuits and microprocessors, this method has become economically realistic.

It is a remarkable fact that the differential pressure method, in spite of its problems, still is the most widespread flow measuring method. The simple mechanical method is second in order of occurrence. The Doppler method is beginning to find its way in industry, but the most advanced cross-correlation method still seems to be in development.

2.4. Level

There are several methods to measure the level of liquids. This can be done in a *discrete* or a *continuous* way. In the first case, the sensor only indicates if the level of the liquid is above or below its position. In the second case the sensor gives an indication of the exact level of the liquid.
Among the wide variety of level sensors, one has to choose the proper sensor for a certain application dependent on the property of the liquid and its container.

Thermistors and piezo-electric crystals can be used for discrete level sensors. The *thermistor* as a discrete level sensor is based on the difference in cooling effect between liquid and air, which can be detected like the cooling effect in the heat exchange based flow measuring method. The *piezo-electric crystal* as a part of an oscillator circuit can be used as a discrete level sensor by detecting the resonance shift of the frequency of the crystal or the lowering of the amplitude of the oscillation, caused by the surrounding liquid.

Continuous level measuring methods are: differential pressure, displacement of a float, resistance measurement, capacity measurement and optical techniques.

Differential pressure. Among these measurement methods, the differential pressure method, based on the pressure caused by the fluid, is the most widespread. This method has the same sort of disadvantages as its analogy in flow measurements, such as problems caused by calibration, variations in material density, composition and temperature.

Displacement of a float. The mechanical method, based on displacement of a float connected to a potentiometer or other sort of displacement transducer, also is a very common level sensing method. The disadvantages of this method are that it is an invasive method and problems such as corrosion, wear and sticking can occur. The advantage of this method and the reason why it is so frequently used is its simplicity.

Resistance measurement. The resistance measurement method is a simple method which can be used with conducting liquids. The resistance measured between a dip-stick and the liquid is inversely proportional with the level.
Variations in conductivity of the liquid can be compensated by adding a reference pair of electrodes in the liquid.

Capacity measurement. The capacity measurement method is a method that is based on the dielectrical properties of the fluid.
Varying properties of the liquid, can be compensated by making use of a reference capacitor in the liquid.
This method cannot be applied with conducting liquids. It is a rather simple method, but due to demands on the electrical properties of the liquid less frequently used than the above-mentioned methods.

Optical techniques. Level measurements with optical techniques have the advantage of being non-invasive, but problems such as sticking and obscuration and (if the level method is not a simple "reach level" measurement) also calibration problems can occur. In spite of this the optical method seems to be attractive enough to explain its increasing application.

2.5. Composition

In the chemical industry a lot of effort is put in the development of special dedicated sensors used for measuring compositions of solutions (e.g. PH-measurement). Usually these sensors are based on measuring the voltage difference of two different electrodes, sometimes shielded from the solution by a membrane, permeable to specific ions to be measured.
This sort of sensing has aspects specific for chemical industry and is considered to be of less interest in this context.

3. Sensing in Mechanical Industry

In the mechanical industry sensors are used to measure parameters which are in general different compared to those measured in the chemical industry. These parameters typically are temperature, distance, presence, force, torque, friction, length, area, angle, displacement, surface roughness and texture.

The major tasks of sensors in mechanical industry nowadays are detection tasks, such as detection of abnormal conditions, detecting the presence of objects, detection of tool wear etc. The result of these tasks usually is a right/wrong message.

Sensors are also used for measuring tasks, in order to check on the work being undertaken. In this case the decision about right and wrong can be taken at a higher level. This does imply, however that more information has to be gathered and send to this higher level than in the case of simple detection. This is the reason why this is considered a more complex task than simple detection.

Recently, even more complex tasks such as localizing and recognition of workpieces and quality control have become possible. It may be clear that these rather complex tasks require collecting of even more information.

All of these tasks are accomplished by means of measuring one or a combination of the above-mentioned parameters with sensors suited for the application.

So far, the mechanical industry has shown little interest in using sensors for the more complex tasks. Apparently the current attitude is: as long as limit switches can do the job there is no need to replace them for more complex systems.

Using more advanced sensors in a system can result in a more flexible system. For instance if the decision about right and wrong is taken at a higher level, this usually is a software level and thus easy and cheap reprogrammable.

No doubt the cost of a system with more advanced sensors initially will be higher than a system with simple sensors, but the increase of flexibility obtained and also cost reduction due to less constraints on part position, can in the long term result in economical benefits.

Next a discussion is given about some methods used to measure some of the above-mentioned parameters.

3.1. Temperature

Temperature measurements in mechanical industry require, due to the different aspects of this industry compared to those of the chemical industry, a different approach. Nevertheless the basical principles of temperature measurement are the same for both industries and the same sort of sensors can be used, although probably with a different design. Sensors that can be used for temperature measurement are described in section 2.2 of this appendix.

3.2. Distance

Distance of objects can be measured with inductive and capacitive proximity detectors, ultra-sonic methods and optical methods.

Inductive proximity detectors. The inductive proximity method is restricted to objects with certain magnetic properties.

This method can be based on measuring the reluctance of a probe coil, that indicates the distance of ferromagnetic objects.

It can also operate by subjecting electrical conducting objects to an alternating magnetic field. This field generates circulating currents (eddy-currents) in these objects. These eddy-currents in turn create a magnetic field, that interacts with the subjected field, thereby affecting the impedance of the coil that generates the subjecting field. Measuring this change in impedance can result in an indication of the distance of the object.

This sort of distance measurement of ferromagnetic or electrical conducting objects is only applicable for rather short distances (up to 10 cm) and in a magnetic distortion-free environment.

Capacitive proximity detection. Capacitive proximity detection is based on measuring the capacitance formed by electrically conducting objects and a probe. This capacitance is inversely proportional with the distance between them.

It may be clear that this method puts heavy electrical constraints on the environment. For this reason it is not a widespread method.

Ultra-sonic method. The ultra-sonic method is based on measurement of the elapsed time between the sending and receiving of a (modulated) ultra-sonic wave. This ultra-sonic wave is send out by a source and is reflected by the object back to a detector. The elapse-time together with the knowledge of the traveling speed of the sonic waves, defines the distance of the object.

This ultra-sonic method has proven to be a reliable system for range-finding applications in the range 0.01-10 meter. The method hardly puts constraints on the environment and material of the object.

Although the method is commercially viable and already used in numerous industrial applications, quite a lot of effort is still put into research and development. The main development effort is put into the direction of making the method applicable for robots.

An interesting direction of research is the field of acoustic imaging. Primitive shape recognition is possible using a matrix of ultra-sonic transducers. It is also possible to do texture determination using ultra-sonic methods.

Optical Method. Optical methods to measure distances are based on the triangulation principle. A beam of light is cast on the object and the angle between the direction of the light-source and detector and the distance between them is used to calculate the distance of the object. Range finding based on this principle can also be one of the tasks of a complex general-purpose vision system.

A number of dedicated range finders are developed using this method. These range finders in most cases make use of laser beams, in order to reduce necessary constraints put on the environment.

3.3. Presence

The probably most widespread type of sensors in the mechanical industry are sensors used for presence/absence detection. These type of detectors are often very crude, in which case they merely consist of *contact switches* or *optical beam interruption switches.*

More elaborate systems can distinguish metallic and non-metallic objects. This can be done with simple *metal contact detectors,* making use of the electrical conducting properties of metals or with inductive and capacitive proximity switches, making use of the electrical and/or magnetical properties of metals like distance sensing.

It is surprising to see how much is achieved using this simplest form of sensing.

3.4. Displacement

The change of position of an object is an important parameter, which can be measured with displacement transducers. There exists a wide variety of these transducers, based on different principles. They can be subdivided in linear, which sense the displacement along a line, and angular displacement transducers, which sense rotation about an axis.

A very common application for this sort of transducers, is to be combined with some flexible element, that changes its position under influence of some external parameter. In that case, not the displacement itself, but the external parameter that causes the displacement is the subject of interest. Examples of this are discussed in section 2.1 (pressure sensing) and in section 2.4 (level sensing) of this appendix.

There are some displacement transducers, which do not require a mechanical contact with the object, but most of them do. In that case one has to be careful to select a transducer that does not noticeably influence the measurement due to friction or elastical forces.

Some of the displacement transducers are: potentiometers, inductive and capacitive displacement transducers, strain gages and digital displacement transducers.

Potentiometer. A potentiometer is one of the simplest displacement transducer. A constant voltage is applied across the potentiometer and it delivers an output voltage proportional to the displacement. Obviously, due to its construction, it can easily be used as an angular displacement transducer. By means of a cable system however, a potentiometer can quite simple be transformed in a linear displacement transducer, with a range in the order of 10cm. Disadvantages of potentiometers are friction and a rather poor resolution ($\sim 0.1\%$). The advantage of using a potentiometer as a displacement transducer is that a simple readout circuit suffices and the method is linear in its full range.

Inductive. Inductive displacement transducers are, like the inductive proximity detectors (section 3.2) based on the change of inductance of a coil. In this case however, the ferromagnetic material that changes the inductance by changing its position is part of the transducer itself. The object, of which the displacement has to be measured, is mechanically in contact with this ferromagnetic material. Due to this, there are no requirements put on the material of which the object is made.
Inductive displacement transducers usually are linear displacement transducers. These transducers need a more complex read-out circuit than the potentiometer. They have a limited range in which the transducer has a linear behaviour.

Capacitive. Just as the inductive displacement transducers show a resemblance with inductive proximity transducers, capacitive dislacement transducers show a resemblance with capacitive proximity detectors (section 3.2). In this case the moving electrode that changes the capacitance is also a part of the transducer and is mechanically connected with the object. These displacement transducers can, dependent on their construction, be used for angular or linear displacement measurement.
An advantage is their high sensitivity, disadvantages are the rather complex read-out circuit required and non-linearity.

Strain Gages. If an object is subjected to stress, this object is deformed. This deformation can result in a change of the length of a part of its surface. This change of length can be measured with a strain gage, cemented on that surface.
A strain gage is an electrical resistor, which changes due to strain. This is the result of the change of cross-section of the conducting strip which forms the resistance. If the gage is made of semi-conducting material, the strain influences the resistivity of the material itself, usually resulting in a higher sensitivity.
Strain gages are very sensitive and do not require a complex read-out circuit. On the other hand, they are also very sensitive to temperature, which may cause requirement of compensating measures. [Wobschall1979]

Digital. No doubt a transducer, that provides a digital output signal, is desirable in this world of increasing use of digital systems. There are both angular and linear displacement transducers that provide such a digital output signal. The angular digital encoder is a displacement transducer. It consists of a disk, which has a number of tracks on it, and a read-out system. Each track contains a bit of a digital coded signal, which represents the absolute angular position. This coded signal can be binary or some sort of cyclic code and is usually read-out optically. Joint position sensors as used in robots, usually are of this type.
The linear encoder is based on the same principle.
The resolution of both types of transducers is determined by the number of bits (and thus tracks) used (this may be up to 14). Using an additional track which produces a sine wave output, can increase the resolution with 5 bits. [Woolvet1983]
Another digital transducer, based on the same principle, but with one single track, is the incremental type. In this type the number of pulses that the disk produces as it turns round are counted. In this way, the number of revolutions or the relative angular displacement can be detected. The resolution of this type can be very high. Linear incremental displacement

transducers are similar.

3.5. Computer Vision

There are a number of dedicated computer-vision systems commercially available. They operate successfully on low level problems of verification, inspection, recognition and determination of object location and/or orientation. Low level problems are problems that are based on one or two dimensional images of objects in less structured environments.

A very attractive application of computer-vision in the mechanical industry is non-contact inspection of objects. It has the following advantages compared to the more commonly used contact inspection methods.

- It can be faster because the surface or contour of the object does not have to be scanned mechanically.
- The object cannot be damaged by the optical method.
- There is no wear of the probe, which would require recalibration.
- The object usually does not have to be positioned.
- Non-contact inspection can be a safe method, especially in the case of hazardous objects.

Even the most sophisticated computer-vision system cannot handle problems such as recognition in complex structured environments in real time. Another restriction is that they are not equipped to handle three-dimensional analysis for recognising objects from arbitrary viewpoints.

There is no general purpose vision system commercially available today.

More on this subject in section 5 of this appendix.

4. Robot Sensing

The first generation of robots (around 1965) were used in limited area's of industry. This was mainly due to the fact that they were equipped with internal sensors and the only interaction with their environment was for synchronization. Internal sensors usually are rather simple sensors such as joint position sensors. This usually results in rather simple and thus inexpensive control systems. On the other hand, using just internal sensors requires a high positioning accuracy of the objects to be handled, which results in costly object fixing systems.
Current robots [Gevarter1982] may have interface provisions for external sensors, to be used for feedback control. The feedback facility of these second generation robots enables them to respond to changing operating conditions, which allows expansion of their task-range. External sensor systems can be rather complex, such as force, proximity or vision systems. Clearly, using these sensor systems will imply higher cost due to the real-time demands on the processing of the sometimes rather complex signals they provide. On the other hand, these external sensors yield a much higher flexibility in the tasks the robot can perform and a reduction on part positioning accuracy. This can be the reason that using these external sensor systems may turn out to be economically beneficial.
This is the first step towards robots of the third generation. These are robots which are able to determine their own actions based on their perception and planning abilities.

When it comes to robot sensing quite often the distinction between contact and non-contact sensing is made. Non-contact sensing provides important information for manipulation control. The most powerful non-contact sensor system is vision. Some specific information needed for manipulation however, cannot be obtained with vision systems at all, or only with extreme effort (and cost). This is typically the kind of information that can be provided by means of contact or near-contact sensing.

4.1. Contact Sensing

The simplest contact sensors used for robots are switches that stop arm motion and open and close grippers. More sophisticated contact sensors can measure slip, force, torque, displacement, temperature and surface roughness. They can even be used to recognize objects.
Contact sensing for industrial robots is still at a rather primitive level. If force is measured at all, it is measured with simple mechanical sense-probes based on measuring displacement caused by stress, or by measuring internal signals such as air pressure in pneumatically operated gripper systems. It is surprising that contact sensing does not always take place in that part of the robot, where the robot mechanically makes contact with the outside world, that is its gripper fingers. This is especially the case for the less sophisticated force sensing systems. They usually are build in the wrist of the robot and merely indicate whether the gripper is touching an object. Usually they can tell in what direction, with what force the gripper is pressing against this object.

In various research laboratories however, rather sophisticated touch sensing systems are being developed. [Stanton1983] Most of these touch sensors are based on measuring resistance variations between array's of electrodes, induced by deflection of for instance a pad of conducting foam. [Christ1982] Other recently developed touch sensors are based on arrays of sensors made of semi-conducting material.
A big advantage of these sensors is that they can be mounted on the gripper-fingers. With these systems it is possible to detect shape, texture, slipping of objects lifted by the robot and compliance of objects by touch alone.

Perhaps one of the most vital problems for robots nowadays is the problem of soft grasping of objects. If grippers are made compliant to insure that objects are not damaged, this unfortunately results in undesirable effects such as a reduced frequency response and hysteresis of the force sensor and uncertainty about the position of the sensed object. If the compliance is imitated by the software to eliminate these effects, this requires costly high bandwidth sensors, hardware and software.

In industry this problem is evaded by using grippers that are dedicated to the particular application. It might be clear, that this need for dedicated grippers is a limiting factor for the universal applicability of robots for manipulation.

4.2. Non-Contact Sensing

Beside robot-vision, there are a number of non-contact sensing methods, that have shown to be very useful for robot applications. Most of these non-contact sensing methods however, are restricted to short distance sensing. For this reason they are often referred to as near-contact sensors or even as remote touch. Examples of how to accomplish this near-contact sensing is described in section 3.2.

This sort of sensing is specially useful for avoiding obstacles and for positioning the gripper prior to grasping.

5. Computer Vision

A rigorous definition of computer-vision does not exist. There exists a statement of its goal in anthropomorphic form, namely, computer-vision is the art and science of developing systems to imitate human visual perception. For robotic-vision (a subset of computer-vision) this can be further narrowed to the definition of a black box whose input is a 'scene' and the output is a set of labels assigned to sub-images of the scene. Generally these labels belong to a knowledge base (or a pool of a priori knowledge) supplied by the user. These labels can have varying degrees of depth for example a simple label (an object), or a deeper label (a pump housing with four mounting holes lying on the conveyor belt with its axis 57° off the belt axis). This could help in performing such common industrial tasks as:

- Recognition of workpiece and/or tool,
- Determination of their pose,
- Salient feature extraction for spatial reference,
- In-process inspection, quality control.

To accomplish these tasks, robotic-vision can be partitioned into two distinct modules: *image acquisition* and *image understanding*.

In image acquisition a 3D-scene is projected onto a photo-electronic device as a 2D-photon distribution which is converted into electrical signals, usually in the form of a matrix of digitized grey values.

In image understanding this matrix is mathematically processed, often in conjunction with a resident knowledge base, to arrive at the labels.

Historically, image acquisition has been the domain of photo-electronic hardware, whereas software (and the associated computer hardware) reigned in image understanding. As a result, these two fields developed rather independent of each other.

After a brief review of these fields, we shall attempt to show that the recent activities toward smart-sensors is merging these two. There are, for instance, photo-electronic devices that can produce from a scene not only a matrix, but also labels, albeit shallow ones.

This is a major development that promises to play a significant role in robotic-vision.

5.1. Image Acquisition

In robotic-vision, image acquisition by definition, is a real-time activity and thus the traditional means of imaging, for example photography, will not be considered. There is a large number of devices available for image acquisition. They differ widely in their characteristics and can meet diverse demands. There is no unique classification scheme to group them into well-defined categories. We shall use device physics and the following general properties to classify them.

- *Spectral sensitivity.*
  This is the ability to respond to radiations of different wavelengths. It can be measured as the signal output per input photon of a given wavelength. The wavelength can vary from macroscopic length (i.e. about a meter) for a side looking radar, to 10^{-17}m for nuclear radiations. If robotic-vision applications are envisaged from space explorations of the earth to nuclear plant operation, then this entire range has to be kept in mind.
  In intensity, the incident photon-flux could range from single photons (photon-counting) to solar levels (10^{+18} photons). Obviously, no single device has the dynamic range to deliver such performance.

- *Resolution.*
  This can be taken as the number of spatially resolved elements (pixels) into which a scene can be divided by the sensor. A high grade, off-the shelf TV-sensor can now deliver up to one million pixels per scene (photographic film gives about 1000 times more pixels), [Blouke1983] This is sufficient for almost all current applications of robotic-

vision, where usually the bottleneck is the lack of computer power to digest so many pixels in real-time. In current robotic-vision practice, an image of more than 64k pixels is seldom considered.
- *Speed.*
  This is the sensor's ability to follow scene motion. Since the present-day robots are still very anthropomorphic, a fraction of a second can be taken as a benchmark time interval. An ordinary TV-sensor with a frame time of 25 ms is fast enough for robotic-vision. Again, owing to the limited real-time computer capacity available presently, a quantum jump in the speed is not expected and the sensor speed is not likely to be taxed soon.
- *Signal mode.*
  This is a catch-all phrase to describe how and in what form is the video signal generated. A TV-sensor gives its signal in a sequential (raster) mode, however, with some modifications, it can address the scene as a random access video memory. With further modifications the video signal can be read with 'logic', that is, some low level image processing can be done at the sensor.
  Such developments have strong implications for robotic-vision and one should be alert to advances in device-physics.

Based on the physical principles of operation, the robotic-vision sensors can be divided into two broad categories, *electron-beam addressed sensors* and *charge transfer devices*.

Electron-beam sensors.

Electron-beam sensors are the old war horses of the vacuum tube era. [Lubszynski1978] In spite of the latters demise, electron-beam sensors are still unchallenged in many applications. The common principle of operation shared by these devices is that the incident light from the scene, is converted into photoelectrons and onto an integrating solid state target. This charge distribution is 'read' out as an analogue signal by an electron-beam that scans the target. In this sense these devices can be likened to an electron-beam addressed analogue video memory.

Whereas the 'read' mode for these devices is always the same that is electron-beam, there is a great diversity in the 'write' mode. From this stems the unchallenged versatility of the electron-beam sensors. Using the 'write' mode as a distinguishing feature, the electron-beam sensors can be placed into two large classes as discussed below.
- Vidicons.
  In these devices the light is incident on one side of a thin target and converted into a charge distribution. An electron-beam scans the target from the opposite side and 'reads' the charge distribution (in a destructive manner), to produce the analogue video signal. These are all-round robust devices of simple construction and with good characteristics for general purpose applications. These include such devices as antimony-vidicons, plumbicons, saticons, newvicons etc. [Goto1974]
  With moderate light sensitivity, they offer the highest resolution. For robotic-vision it is the first choice as an economic and good performance sensor.
- Orthicons.
  This is a large family of rather complex, but highly specialized sensors. The 'write' needs, generally, three components:
  1. A photoelectric target that converts the incident light into free electrons.
  2. A relay of electron-optics that focuses, accelerates, pans and zooms the photoelectron beam.
  3. A solid state target integrates (or differentiates) the incident photoelectron beam into a charge distribution which is stored until it is read out by the electron-beam.

By selecting various combinations of these steps a wide variety of operating characteristics can be achieved to meet the most critical applications. Notable examples of this family are, silicon intensified target, image dissectors, forward looking infra red, faint object camera of space lab, secondary electron conduction, intensified charge coupled device. For robotic-vision, the most noteworthy feature is that in spite of (or because of) its complexity, it offers great possibilities for on-site logical operations at the sensor.

Charge Transfer Devices

The development of a self-scanned solid-state image sensor, to equal the performance of an electron-beam sensor has been an elusive goal for the last 20 years. Even the most advanced chip technology has not been able to produce a cost effective sensor whose picture quality is comparable to a modest electron-beam sensor. [Weimer1983]

Although the sensitivity of a charge transfer device is superior, its resolution does not reach the level of electron-beam sensors. Charge transfer devices have, thus, not found wide applications in closed circuit or broadcast TV. However, a large number of robotic-vision applications do not place such exacting demands on resolution and can thus benefit from the unique features of charge transfer devices.

A solid-state charge transfer device sensor, along with the photon conversion and charge storage functions of an electron-beam sensor, also has to have a mechanism for scanning the charge distribution. To accomplish this, the (quasi)continuous silicon wafer is graticulated with composite elements that act as both photoelectric converters and charge storage elements. In one form or the other, these elements are reverse-biased photo-diodes. These individual elements, the potential pixels, are arranged in either a lattice (to furnish a 2D-detector) or a linear array (facsimile scanners, linear-scanners). The incident photon flux from the scene is sampled at these sites and converted into photoelectrons. These photoelectrons are trapped in the diodes. There are two basically different modes to handle these photoelectrons, either by digital multiplexing or by charge transfer. Correspondingly there are two generic types of charge transfer devices, namely, *charge coupled devices* and *charge injection devices*.

- Charge Coupled Devices
  In these devices along with the diode-complex, a buffer memory in the form of a one-line memory buffer register, or a whole frame buffer memory is provided. The charge stored in the reverse-biased photodiodes is transferred into the buffer memory which in turn is 'read' separately.

- Charge Injection Devices
  In these devices there is no buffer memory on the chip but instead there is an x-y address bus. The reverse-biased photodiodes act as memory cells and can integrate low-level light. For all practical purposes a charge injection device can be considered as a random access analogue video memory.

For all such silicon-based sensors the resolution is limited by the number of diodes. Presently there is an economic limit of 64k pixels, though chips with $\sim$0.5 Mpixels, (800x800) are available for critical applications. A further limitation is that the spectral response is limited to that of the silicon-substrate which may be too limiting for some applications.

On the other hand, efforts are underway to make the optical elements (diodes) form a part of a dedicated image processing chip. In this manner one can calculate fourier transforms or correlations at unprecedented rates. In the context of robotic-vision the sensor hardware is in an advanced stage of development but much effort is needed to find the appropriate software for image understanding and then integrate it into the sensor chip.

5.2. Image Understanding

The human vision system follows a bottom-up approach to image understanding. We have, at the lowest level the physical perception of the image by the retina (sensor) and at the highest level cognition of the image. The robotic-vision activity can also be organized into a loosely ordered range of representations of higher and lower levels.

One can categorise [Kanade1983] the representations into the following four levels:*iconics, segmentation, geometry* and *relations*.

Iconics

In this step the sensor signal is processed to produce iconic (image-like) structures. This is the first step to which the sensor data is subjected. Since different types of sensors may be used, this step is sensor-sensitive. For instance, laser-triangulation data has to be treated in a different way than CCD data. In general, these are domain independent processes such as smoothing, contrast stretching and compression, histogram modification, entropy maximization, thresholding, erosion, dilation, skeletonization, chain coding, various types of gradient enhancements etc. This is a very numeric-intensive step. The number of steps is proportional to the number of pixels raised to a power, commonly ranging between 2 and 4. Obviously, the sensor resolution is important here. Often a higher sensor resolution is desirable and physically possible to deliver, however, it may require prohibitive amount of number-crunching steps and is hence not advisable. On the other hand, it is possible to carry out these operations quite efficiently if parallel architecture is used. Therefor, there is a major research and development effort to develop dedicated parallel machines to carry out complex iconic tasks. On the whole, whatever else is developed in robotic-vision for higher level tasks, if no efficient solution for this step is found, all the other efforts will be largely futile.

Segmentation

In this step, using some a priori knowledge, some interesting features of the image are sought and enhanced. This includes such image processing operations as, edge detection, curve (and straight line) detection by hough transform, graph search, contour following, region growing, split and merge, texture analysis etc. The result of this step is to replace the original image-matrix by a set of geometrical primitives such as, lines, curves, boundaries. This leads to a very significant data compression, and can considerably ease the subsequent analysis. The operations required here are again number-crunching combined with knowledge base operations. Many of the comments, offered for iconics, are also valid here, in particular, a non-Von Neumann, parallel machine is quite desirable.

Geometry

After the interesting features, that is the geometrical primitives, have been isolated in segmentation, the next step is to combine these primitives into geometrically identifiable surfaces and objects. This calls for 3D-reconstruction, boundary representations, polyhedra occlusion, strip trees, region representation, run length coding, spline-surfaces, shadows, structured light, computer generated Moiré, line drawings etc. There is a very strong overlap with computer graphics at this stage.

Much attention is devoted to this step in such robotic-vision problems as, bin-picking, pose and general assembly tasks. In a slightly different manner this step is also very important for quality control. Using special sensors and/or special software, features relevant to quality control are enhanced and isolated and then the quality of the product can be assessed. Commonly, this task is not considered to be a part of robotic-vision but relegated to in-process quality control. However, with some modifications this could be integrated into robot operations, for instance a paint spraying robot that performs quality control of the finished surface while painting it.

Relations

Once the interesting objects have been isolated and reconstructed, one tries to 'understand' the scene. This is generally done with the help of a knowledge base. This is, at the moment, the most difficult task. There is hardly a commercial robot which can 'understand' a non-trivial scene. The operations used here are the domain of AI and include such items as analogic and prepositional representations, semantic nets and inference matching, graph-theoretic matching, backtrack, decision tree etc. In general these operations seem to possess an inherent serial control hierarchy and need large knowledge bases. Current industrial robotic-vision practice does not yet extend much this far. The trend there is to come up to the geometrical part and then devise some other configurations such as strip lighting, stereo-cameras, and engineered environment etc., to avoid this AI bottleneck.

Commercial robotic-vision systems, available today, are almost entirely based on binary image analysis of silhouetted objects (engineered environment). The principle mode of operation is to start with a 'shallow' grey image (2-4 bits) and threshold it to obtain binary silhouettes of the objects. For this binary image, simple geometrical properties such as perimeter, area, curvature, low order moments etc. are calculated and a look-up-table type of feature-match is made.

6. Future Trends

There is no doubt that there is still a lot to be done to improve all types of sensors. Higher reliability, higher durability, non-invasiveness and a low cost are some of the properties, the industry thinks most desirable. A higher speed, a better resolution and a larger dynamic range are the main properties desired by the development laboratories. These are demands on the sensor hardware itself.

Even more desirable however, is improving the sophistication of the sensor signal processing.

For CIM, it is important that modular systems with standardized interfaces between them will be developed. This would make integrated systems of a variety of robots, tools, sensors and control systems possible. For sensor systems this would mean the standardization of sensor interfaces. Clearly due to the large difference of complexity of sensor systems, it does not seem advisable to design one standard interface covering the whole area.

It might be better first to design standards for a simple sensor interface, which is commonly thought to be ready for standardization.

The general point of view about the next step, the design of a complex sensor interface standard, is that this would at this moment slow down the development of complex sensor systems.

In laboratory environment, tactile sensor systems already have proven to be able to offer the necessary versatility and flexibility for robot grippers. This will allow robots to be used for more complex assembly applications.

One can expect that due to the amount of research in progress, it probably will not be long before tactile sensor systems are ready to be used in industry.

References

Beck1983. Beck, M.S., "Correlation in Instruments: Cross Correlation Flow Meters," in *Instruments Science and Technology*, ed. B.E. Jones, Adam Hilger Ltd, Bristol (1983).

Blouke1983. Blouke, M.M., "800*800 Charge Coupled Device Image Sensor," *Optical Engineering* **22** p. 607 (1983).

Christ1982. Christ, J.P. and Sanderson, A., "A prototype Tactile Sensor Array," The Robotics Institute TR 82/14, Carnegie Mellon University, Pittsburg (1982).

Gevarter1982. Gevarter, W.B., "An Overview of Artificial Intelligence and Robotics," NBSIR 82-2479, National Bureau of Standards, Washington, DC (1982).

Goto1974. Goto, N., "High resolution vidicon," *IEEE Transactions on Electron Devices* **ED-21** p. 662 (1974).

Harmon1980. Harmon, L.D., "Touch Sensing Technology: A Review," Society of Manufacturing Engineers (1980).

Kanade1983. Kanade, T., "Geometrical aspects of interpreting 3D-scenes," *Proceedings IEEE* **71** p. 88 (1983).

Lion1969. Lion, K.S., "Transducers: Problems and Prospects," *IEEE Trans. Industr. Electron. Contr. Instrum.* **IECI-16** pp. 2-5 (1969).

Lubszynski1978. Lubszynski, A., "Review of TV camera tubes and electron optics," *Advances in Electronics and Electron Physics* **64** p. 278 (1978).

Stanton1983. Stanton, M.D., Hill, H.S., and Crisp, N.D., "Grippers-Sensors and their Control Aspects," Proceedings of the 6th British Robot Association, Birmingham (May 1983).

Weimer1983. Weimer, P.K., "Image sensors for TV and related applications," *Adv. in Image Pickup and Display* **6** p. 177 (1983).

Wobschall1979. Wobschall, D., *Circuit Design for Electronic Instrumentation*, McGraw-Hill Book Company, New York (1979).

Woolvet1983. Woolvet, G.A., "Digital Transducers," in *Instruments Science and Technology*, ed. B.E. Jones, Adam Hilger Ltd, Bristol (1983).

Appendix 5

CIM in the small firm

European manufacturing industry includes a very large number of small companies, employing less than 250 people in total. Companies within this sizable and important sector of manufacturing industry typically employ few professionally qualified specialists in particular disciplines and rarely operate Data Processing or Information Technology departments. It was recognised at the very beginning that small companies had special needs which might conceivably impact the design rules for CIM systems used in such companies. In recognition of this need the Department of Industrial Management of the University of Dublin were invited to participate in the project for the express purpose of addressing the special CIM Systems Design Rule needs of small companies. As the team developed the flow charts depicting the basic procedures and activities which need to take place within any manufacturing industry, these were passed to the University of Dublin for study. These charts were then assessed against the equivalent procedures and activities which are necessary within small companies - being validated where appropriate, against a sample of several companies from different sectors and backgrounds.

Particular emphasis was placed on the consideration of those procedures and activities which were catagorised under CAPP and CAPE since these were identified at an early stage as being of particular importance and interest to small companies. As a result of this work the University of Dublin were able to confirm that the design rules, as identified and described by ISTEL, were, without exception, entirely appropriate to small companies. These studies did however indicate that whilst the identified system design rules, which are concerned entirely with sub-system functionality, were commonly applicable to companies of all sizes, small companies required computer systems to posses other attributes which were not essential to larger and more sophisticated organisations.

In short, the study concluded that CIM sub-system functionality was universal but that CIM sub-system products for small and larger companies needed to be differentiated. Whilst this document is concerned exclusively with the functionality of the CIM sub-systems, some of the more significant system design attributes have been identified and briefly mentioned below as these may be of some interest to those involved in the technical design of CIM system products for small companies.

1   Systems for small companies should be capable of being understood with an absolute minimum of tuition. Ideally, such systems should be designed to be self-teaching. and therefore not require that extensive literature or other training material be used or understood.

2   Systems for small companies should not presume advanced levels of professional expertise in respect to the business activities addressed by the system. Nor should such systems require high levels of education or knowledge of particular academic subjects. Many small companies are created and managed by very well educated and professionally qualified individuals - many more are not!

3   Systems for small companies must include simple-to-use provisions for the creation of user defined reports and listings. These would ideally include facilities to transfer files to other computers such as personal computers, for ad hoc interrogation, report generation, analysis etc.

It must however be emphasised that these attributes do not form part of the main study and that much more research would be required before any complete or definitive list of the attributes essential for the products destined for a small company could be identified and described.

Index

Addressing scheme   381
Attribute Data   33
Bill of Materials   13, 24, 113
CAD Administration   13, 24
Capacity Planning   3, 91
Coding and Classification   24
Communication Strategy   7, 295, 360
Computer Vision   387, 401, 431
Computer Aided Design   3, 304, 324, 360, 412
Computer Aided Storage and Transportation   106, 147, 187, 257
Concept Design   12
Constructive Solid Geometry   406, 423, 426
Control systems   12, 149, 317, 388, 443, 450
Cutting Tool Selection   65, 96
Data Effectivity   302
Data Levels   306
Data Strategy   7, 295, 326, 371
Delivery Management   147, 247
Design Analysis   12, 23, 36
Design for Manufacturing   65 - 69
Design Modification   12, 24, 296
Design Proposal   12, 23, 34, 59, 69
Detailed Design   12, 24, 66, 297
Diagnostics   314, 329, 378
Digitised Geometry   33, 40
Distributed Capacity   348
Distribution of Data   310, 326
Engineering Change   12, 24
Engineering Design   12, 22, 59, 324
Engineering Test   12, 23, 36
Environmental noise   372, 382, 401
Equal resource utilization   346
Fixture Design   64, 150
Flexible Manufacturing   3, 150, 308, 333
GKS   340, 404
Geometric Data   13, 30
Geometric Modelling   319, 409
Graphics systems   12, 30, 320, 402
Intelligent local processing   323, 392, 407
Interface rules   290
Itemized processing   349, 397, 410
Lazy evaluation   344, 410
Local Area Networks   355

Long Term Forecasting   105
Machine and Routing Selection   65
Machine Control Systems   149
Machine Selection   60, 82, 293
Management Information Data   304
Material Requirement Allocation   187
Material Handling   5, 50, 150, 271
Material Management   153, 179, 206, 249
Medium Term Planning   116
Numerical Control   3, 27, 49, 149, 308, 428
Numerically Controlled Machining   22, 424
Order Assimilation   122
Part-Programme   149, 209
Part-Programming   50, 209
Parts Classification   3
Plant Layout   50
Prime Authorship   10, 300, 371
Process Planning   3, 50, 347
Process Selection   63
Processing Strategy   6, 295
Product Functional Specification   12
Production Management   153, 234
Production Planning   3, 59, 105, 122, 184, 270, 307, 362, 403
Production Sequencing   105
Programming Systems   341, 388
Replication of Data   308
Resource Scheduling   147, 251
Robotics   2, 428
Sensor Classification   385, 432
Sensor Systems   383
Short Term Planning   120, 141
Solid Modelling   22, 406
Text Data   33
Tool Library   61, 96, 160
Tool Management   6, 149, 251
View Data   33
Wide Area Networks   359
Wireframes   426
Workcentre Management   253
World model   389